MULTIPURPOSE PLANTS

MULTIPURPOSE PLANTS

DR. P.K. SRIVASTAV

Deputy Director,
Regional Sericultural Research Station
Sahaspur, Dehradun

SCIENTIFIC PUBLISHERS (INDIA)

P.O. BOX 91 JODHPUR (RAJ.)

Print: 2020

Published by:
Pawan Kumar
Scientific Publishers (India)
5-A New Pali Road,
P. O. Box 91,
Jodhpur-342001 (Raj.)

E-mail: info@scientificpub.com
www.scientificpub.com

ISBN: 978-81-7233-553-3

Cover Design: Tanay Sharma

Lasertype set: Rajesh Ojha
Printed in India

Dedicated to

1. Prof. Krishna Mohan Srivastava,

Ph.D., D.Sc., F.N.A.Sc., F.R.A.S.: *Brother*

2. Prof. Krishna Gopal Srivastava, Ph.D.:

Brother

3. Mrs. Kamla Srivastava :

Sister

4. Late Kailash Kumar Srivastava :

Brother-in-law

Who helped me during my education.

— **Dr. P.K. Srivastav**

सत्यमेव जयते

गोपाल कृष्ण प्रसाद भा.व.से.
महानिदेशक

G.K. Prasad IFS
Director General

भारतीय वानिकी अनुसंधान एवं शिक्षा परिषद्
(पर्यावरण एवं वन मंत्रालय – भारत सरकार की एक स्वायत्त संस्था)
पो.ऑ. न्यू फॉरेस्ट, देहरादून – 248 006

Indian Council of Forestry Research & Education
(An authonomous body of Ministry of Environment and Forests, Govt. of India)
P.O. New Forest, Dehra Dun - 248 006

FOREWORD - I

Biodiversity refers to all types of variability among living organisms and the ecosystem complexes in which they occur. It has great intrinsic value to human existence, industries and commerce. Out of estimated 12.5 million living forms on earth, only less than 2 million have been identified and studied. In angiosperms (flowering plants), 2.50 lakh species have been identified so far and of these only 17,500 species have been identified in India. Besides supplying food, plants protect environment, beautify our surroundings and provide fodder, fiber, timber, tannins, resins, medicines, dyes etc. Ornamental plants are also important commodity in international trade. Of late, the importance of silk producing, medicinal and aromatic plants has also been realized.

Plant diversity as a global resource has immense potential for development of natural products for sustainable development of human beings. However, excessive exploitation combined with heavy habitat destruction has led to the decline of major commercial species. We must conserve life support systems or ecological processes that keep our plants healthy through scientific understanding of species and ecosystems. Even general public should follow principles of conservation and sustainable development and commit themselves to **"save the plants that save us"**. Trees like *Ficus religiosa* (Peepal), Ficus *bengalensis* (Vat), *Terminalia arjuna* (Arjuna), *Anthocephallus cadamba* (Kadamba) and *Ficus krishnae (*Krishna's butter cup) were worshiped in ancient India and *Tulsi, Bilva, Garike, Shreegandha,* Lotus, *Javakusum* and *Gulanchi* became indispensable components for all the temples and religious places and have also been commonly used for treatment of various human ailments.

We will have to re-establish our friendship with nature if we want to survive and improve our economy. A mass awareness programme is utmost necessary for achieving this target and the first step in this endeavor is to educate the people about our plant wealth. It is in this context that Dr. P. K. Srivastav has written this Book on detailed description of "**Multipurpose Plants**", particularly non-agricultural plants.

This Book contains scientific names, popular names in different international / national / regional languages; systematic position according to three systems of plant classification viz., Bentham & Hooker, Engler & Prantl and Hutchinson; geographical distribution, morphological features, floral formulae, flowering, fruiting, methods of propagation / cultivation and various uses of 159 multipurpose plants of Indian subcontinent, used in sericulture, forestry, pharmaceutical, timber, horticulture, match, tannin, leather and petroleum industries.

It is hoped that the Book would foster attention of all concerned to conserve and facilitate sustainable utilization of natural resources through conservation measures and integration of various industries besides helping students, teachers, scientists, foresters, naturalists, industrialists and farmers.

(G.K. Prasad)

दूरभाष/Phones: (0135) - 2759382 (O), 2754748 (R)
2759646, 2754301 - 02 Extn. 555 (O)
2757021 - 26 Extn. 4509 (R)

ई मेल / E-mail : prasadgk@icfre.org
फैक्स / Fax : (0135)2750298
मोबाइल / Mobile : (0) 98370 54435

HEMWATI NANDAN BAHUGUNA GARHWAL UNIVERSITY
SRINAGAR GARHWAL, UTTARANCHAL - 246 174 U.A.

Prof. S.P. Singh, FNA
Vice Chancellor

Ph. 01346 252167
Fax 01346 252174
Ref.No. VC/2006/
Dated - 16-01-2006

FOREWORD - II

Biodiversity means the variety and variability of plant and animal species. Biodiversity has ethical, social, cultural and economic values, which have long been recognized in religion, art and literature. The concept of biodiversity has recently gained momentum in GATT era. India ranks amongst one of the 12-megabio-diversity countries of the world due to immense biodiversity. Out of several hot global biodiversity centers, two fall in Western Ghats and North-eastern region of India. India is endowed by nature with 45,500 species of flowering plants, out of which 17,500 species have been described. These plants provide oxygen, shelter, food, medicine, fodder, fuel and clothing for our daily needs and raw materials for industries. Human predominance, interferences and population explosion have led to the depletion of valuable genetic resources and shrinking of vegetation blanket throughout the world. Although extinction is a slow natural process, species once lost is lost for ever with its several million years of evolution. Once a species becomes extinct, it is lost to the society together with its potential contribution to sustainable development and human welfare. With extinction of species, simultaneous extinction of the sub-populations and varieties results in irretrievable loss of genetic diversity. Hence, biological diversity is essential for the survival of human race on this earth and needs to be conserved on war footing.

Utilization of biological resources is a prerequisite for human sustenance, yet it is equally important to conserve these resources for posterity through their sustainable utilization. India has the distinction of having close linkages with nature over the centuries and Emperor Ashoka the great, possibly enacted the first recorded conservation measures as back as in third centuries B.C.

Convention on biological diversity held at Rio-de Janeiro in June, 1992 has three main objectives:

i. Conservation of biological diversity.
ii. Sustainable utilization of the components of biological diversity and
iii. Fair and equitable sharing of benefits arising from utilization of genetic resources.

For achieving these goals, generation of mass awareness is highly required. The **"Multipurpose Plants"** prepared by Dr. P.K. Srivastav, which is comprising of 159 multi-purpose plants, is indeed an utmost need of the hour for generating awareness among the public about these plants.

I congratulate the author for his timely unique venture, which will not only help in conservation but also facilitate sustainable utilization of these plants. I hope this book will be immensely useful to the students, academicians, scientists, sericulturists, research scholars, environmentalists, planners, policy makers, industrialists and managers engaged in development of bioresources industries.

S.P. Singh

भारत सरकार
पर्यावरण एवं वन मंत्रालय
वन शिक्षा निदेशालय
डा0घ0. न्यू फॉरेस्ट, देहरादून

GOVERNMENT OF INDIA
Ministry of Environment & Forests
Directorate of Forest Education
P.O. New Forest, Dehra Dun Pin : 248 006

PREFACE

The importance of plants for environmental protection, ornamentation and as a source of cereals, vegetables, fruits, spices, condiments, oils, fuel, fodder, fiber, paper, timber, tannins, resin, dyes, other raw materials for various industries and medicines has been realized since the dawn of various civilizations. History of association of plants with human beings has lost in antiquity. The indigenous societies of various regions across the globe discovered various uses of plant resources around them. Such traditional knowledge was based on their necessities, instincts, observations, trial and error and above all on long experiences. During Egyptian and Mesopotamian era, science of medicine evolved for the first time and showed a distinct sign of breaking free from complete dependence on Gods. Simultaneously, China developed Acupuncture, the remedial procedure to regulate the flow of life forces called "Chi" through the twelve meridians in the body for achieving balance of the two elements "Yin and Yang". Similarly, the agricultural village based societies of the Indo-Aryan civilizations gave birth to "Ayurveda", the natural herb based remedial measures in India. The importance of plants as a source of medicines has been realized since the time of the "Rig Veda" which is supposed to be 5000 years old. The use of plants as a source of medicines has developed extensively in different systems of medicines such as Ayurveda, Unani, Allopathy, Homeopathy and others. "Ayurveda" the science of life is still considered as a safe system of medicines as most of their preparations are herbal based and harmless even if used for longer periods in proper dose. The supply base of 90% herbal raw drugs used in the manufacture of medicines in these systems comes largely from the wild species of higher plants. Indian sub-continent has more than 3000 species of medicinal plants, out of which approximately 600 find major use in herbal drugs and medicines. The demand for Ayurvedic medicines is increasing. The international market of medicinal plants related trade of US $ 60 billion/year is growing at the rate of 70% annually. Although China and India represent major sources of medicinal herbs, India presently holds only 2.3% share of global market, which shows tremendous prospects and potentialities of medicinal and aromatic plants for building India a developed nation.

The diversity of wild plant resources promises numerous discoveries of useful products either already identified or yet to be identified. They serve as host plants of

Tel.: 0135 - 2757326, 2750127 (O), 2755163 (R) Fax : 0135-2750125, 27571658 e-mail: thedfedun@yahoo.com useful wild insects viz., tropical tasar, temperate tasar, muga and eri silkworms and lac insects also. They are source of recreation and aesthetic and religious pleasures to many people. They are deeply embedded in folklore and our shared heritage. They inspire works of art and expressions in language and figure prominently in religion and legends. This great plant wealth has intrinsic value to our existence and commerce. Yet, it is finite and hence needs judicious sustainable exploitation as well as conservation for posterity. Today we are facing the greatest challenge from gene erosion as only 30 plants provide 90% of global nutritional requirements as against 1500 wild species used by ancients. Over exploitation of timber and fruit trees, medicinal plants, flowering / foliage ornamental plants, orchids, cacti and other succulent plants has posed a severe threat to plant diversity. As pointed out by International Union for Conservation of Nature and Natural Resources (IUCN) as many as 60,000 plant species may be in danger of extinction or serious gene erosion during the next 30-40 years. Hence, it is every body's concern to conserve plant diversity through sustainable utilization. There should be a call to all people to save the plants that save us and for achieving this Herculean task a mass awareness programme is utmost need of the hour. The first step in this endeavour would be to educate our people about our plant resources.

Several books have been published separately on vegetables, spices and condiments, timber and fodder and flowering trees, cereals, pulses, fruits and medicinal and aromatic plants. It is very difficult as well as time consuming for an ordinary man to go through all these books for acquiring knowledge in a nutshell. Further, it is also observed frequently that various organs of the same plant have been used as food, fuel, fodder, fiber and medicine for treatment of various human ailments. Therefore, an attempt has been made in this treatise to deal with various aspects viz., names in different languages, classification, morphological features, flowering, fruiting, methods of propagation and various uses of 159 multipurpose plants of Indian subcontinent.

It is hoped that such attempt of Dr. P.K. Srivastav will facilitate generation of awareness among the general public, industrialists, policy makers, administrators, scientists, educationists, students and media personnel about these natural resources and help their judicious sustainable utilization in our country.

Anil Kumar

Anil Kumar
Director Forest Education
Dehradun

ACKNOWLEDGEMENT

I express my deep sense of gratitude to the following persons whose kind cooperation, discussion, literature, encouragement and keen involvement in one or the other way enabled me to prepare and publish this treatise entitled **"Multipurpose plants"**:

1. Shri, G.K. Prasad, ex Director General (Forests), Govt. of India, C.G.O. complex, New Delhi and the ex D.G., Indian Council of Forestry Research & Education.
2. Dr. S.P. Singh, ex Vice Chancellor, Hemwati Nandan Bahuguna University, Srinagar, Uttarakhand.
3. Shri Anil Kumar, ex Director, Forest Education, Forest Research Institute, Dehradun, Uttarakhand.
4. Dr. S.N. Raina, ex Head, Department of Botany, University of Delhi, Delhi.
5. Late Prof. H.C. Arya, ex Head, Department of Botany, Jodhpur University, Jodhpur, Rajasthan.
6. Late Dr. K. Sengupta, ex Senior Director, Central Silk Board, Bangalore, Karnataka.
7. Dr. Inder Deo Arya, Scientist F & Incharge, Tissue Culture Laboratory, Forest Research Institute, Dehradun, Uttarakhand.
8. Dr. Sarita Arya, Scientist E, Tissue Culture Laboratory, Forest Research Institute, Dehradun, Uttarakhand.
9. Dr. B. Saratchandra, Director (Technical), Central Silk Board, B.T.M. Layout, Madiwala, Bangalore, Karnataka.
10. Dr. S.B. Dandin, Director, Central Sericultural Research & Training Institute, Mysore, Karnataka.
11. Dr. Raje Urs, Director, Seri-biotech Research Laboratory, Bangalore
12. Dr. M.A. Khan, Scientist E, Central Sericultural Research & Training Institute, Pampore, Srinagar, Jammu & Kashmir.
13. Dr. Jagdish Singh, Director of Sericulture, Government of Uttarakhand, Dehradun, Uttarakhand.
14. Shri M.M. Bhat, Scientist D, Regional Sericultural Research Station, Sahaspur, Dehradun.

15. Dr. M.K. Tayal, Scientist C, Regional Sericultural Research Station, Sahaspur, Dehradun, Uttarakahnd.

16. Mr. C.A. Manjunath, Stenographer, Central Sericultural Germplasm Resources Centre, Hosur, Tamilnadu.

I am also grateful to the members of my family viz. **Mrs. Kanchan, Ms. Shruti** and **Ms. Srishti** who patiently co-operated though missed my help in either family or their educational affairs during preparation of this treatise for a very long span of time.

Last but not least, I am also grateful to **M/s Scientific Publishers (India), Jodhpur-342 001** for publishing and adding this book in the form of a drop to the ocean of knowledge.

Dr. P. K. Srivastav

LIST OF ILLUSTRATIONS

INTRODUCTION

The variability among living organisms and the ecosystem complexes in which they occur has great value for the existence of human beings, industries and commerce. Out of estimated 12.5 million living forms on our earth only less than 2 million have been identified and studied. In angiosperms (flowering plants), out of estimated 2,40,000-7,50,000 species only 2,50,000 species have been identified so far. Of these, only 3000 have been regarded as a food source and around 200 plant species have been domesticated for food purposes. India has identified 17,500 species of flowering plants.

Plants protect environment, beautify our surroundings and serve fuel, fodder, fiber, paper, timber, tannins, resin, clothes, medicines, dyes and silks. In international trade, ornamental plants are recognized as an important commodity and their cultivation surpasses the variety of plants commonly grown for food around the world. The importance of medicinal, aromatic and silk producing sericegenous plants has also been realized now. Plant diversity as a global resource has immense potentials for development of natural products for sustainable development of human beings. But, heavy exploitation combined with massive habitat destruction has led to the decline of major commercial species. The "Earth Summit" at Rio-de-Janeiro suggested for 20% cut in green house gas emissions, restriction on felling of forest trees in biodiversitically rich tropical countries, measures and steps to control population and remove poverty for reducing pressure on forests, technology transfer to clean up pollutant and improve energy efficiency and conservation based sustainable development. Therefore, we must conserve life support systems or ecological processes that keep our plant fit for life through scientific understanding of species and ecosystems.

Our association with nature, particularly with trees, dates back to time immemorial since they quenched our thirst, provided us food, covered our body and provided shelter to us even when we were ignorant of agricultural crops and cultivation techniques. However, changing scenario of human civilization destroyed our relationship with nature and trees were cut indiscriminately. We had not realized that destruction of forests, trees and other plants will annul us since there would not be any agriculture for feeding, oxygen for breathing and water for drinking if there will be no forests. Hence, we will have to retrieve our past mistakes and reestablish our friendship with nature again, if we want to survive and inherit a balanced ecosystem to our descendents besides improving our economy by sustainable development through balanced exploitation of our natural resources.

The general public should also follow the principles of conservation and sustainable development through judicious use of biological resources. A mass awareness programme is utmost need of the hour for achieving this Herculean task and the first step in this endeavour should be to educate our people about our plant wealth. It is in this context I have written this book on **"Multipurpose Plants",** particularly non-agricultural plants.

The book contains scientific names, popular names in different international / national/ regional languages, legendary/historical significance, systematic position according to three systems of plant classification *viz.,* Bentham & Hooker, Engler & Prantl and Hutchinson; geographical distribution, morphological features, floral formulae, flowering, fruiting, methods of propagation/cultivation and various uses of 159 multipurpose plants of Indian subcontinent in sericulture, forestry, pharmaceutical, timber, horticulture, match, tannin, leather and petroleum industries, if any. While all attempts were made to compile maximum information on these multipurpose plants hitherto scattered in various journals, magazines and books, lapses might have crept in unknowingly. Author has taken liberty to ascertain the primary / secondary / tertiary status of food plants on the basis of his vast experience in sericulture and due recognition has also been given to earlier reports in lac culture.

It is hoped that the book will create interest even among the common man and invite attention of all those concerned to conserve and facilitate sustainable utilization of these natural resources through conservation measures and integration of various industries besides helping students, teachers, scientists, foresters, naturalists and industrialists in formulating various projects and farmers for increasing their income through plantation and adoption of various schemes/ avenues related to these plants.

P.K. Srivastav

CONTENTS

2. *Accacia auriculiformis*

3. *Aegle marmelos*

1. *Accacia arabica*

4. *Areca catechu*

5. *Artocarpus heterophyllous*

6. *Azadirachta indica*

7. *Bambusa arundinacea*

8. *Bauhinia variegata*

9. *Bixa orelana*

11. *Borassus flabellifer*

10. *Bombax ceiba*

12. *Butea monosperma* syn. *B. frondosa*

13. *Calotropis gigantea*

15. *Carica papaya*

18. *Cedrella toona*

14. *Careya arborea*

16. *Cassia alata*

17. *Cassia fistula*

20. *Cinnamomum zeylanicum*

19. *Cinnamomum tamala*

21. *Cocos nucifera*

22. *Dalbergia sissoo*

23. *Datura metel*

24. *Dendrocalamus strictus*

25. *Embellica officinalis* syn. *Phyllanthus embellica*

26. *Eucalyptus globulus*

27. *Ficus benjamina*

28. *Ficus elastica*

29. *Ficus glomerata*

30. *Ficus religiosa*

31. *Hardwickia binata*

32. *Heteropanax fragrans*

33. *Lagerstroemia indica*

34. *Lagerstroemia speciosa*
syn. *L. flox-regjnae*

35. *Lithocarpus dealbata*

36. *Litsaea polyantha*

38. *Machilus bombycina*

37. *Litsaea salicifolia*

39. *Malus pumila*
syn. *Pyrus malus*

40. *Mangifera indica*

41. *Manihot utilitissima*

42. *Melia azedarach*

43. *Michelia champaka*

44. *Moringa oleifera*

45. *Morus alba*

46. *Murraya koegnii*

47. *Parkia javanika*

48. *Pinus insularis* syn. *P. khasya*

49. *Plumeria acutifolia*

50. *Quercus acutissima*

51. *Quercus grifithii*

52. *Quercus. leucotrichophora*

54. *Saraca indica*

53. *Ricinus communis*

55. *Shorea robusta*

56. *Syzizium cumini*

57. *Tamarindus indicus*

58. *Tectona grandis*

59. *Terminalia alata*

60. *Terminalia bellerica*

61. *Terminalia chebula*

63. *Terminalia glabra*

62. *Terminalia crenulata*

64. *Zizyphus mauritiana*

1. BABOOL

Botanical: ***Acacia arabica*** Wild; Syn. *A. nilotica* (Linn) Delie; Syn. *Mimosa orabica* Lam.

Hindi: Babool, Kikar	Gujarati : Baval	Malayalam: Karivelam
Bengali: Babool	Kannada: Jali, Gobbli	Manipuri: Chinggonglei
Marathi: Kikar/Babul	Punjabi: Kikar	Sindhi: Babar
Tamil: Karu Velam	Telugu: Nellatuma	

SYSTEMATIC POSITION

Bentham & Hooker (1883)	**Engler & Prantl (1899)**	**Hutchinson (1959)**
Phanerogams	Phanerogams	Angiospermae
Dicotyledones	Dicotyledoneae	Dicotyledones
Polyptalae	Archichlamydeae	Lignosae
Calyciflorae	Rosales	Rosales
Leguminosae	Leguminosae	Leguminales
Mimosidae	Mimosoideae	Mimosaceae

Geographical distribution: Indigenous to Sind (Pakistan), Afghanistan and Northern India. Wild or cultivated and naturalized throughout India but not in the most humid regions on the Western coast and in the extreme North-west where frost is too severe in winter. Also found in tropical Africa and Arabian countries.

Habit: A moderate sized, thorny, almost evergreen, wild or cultivated xerophytic tree with a short trunk and a spreading crown of feathery foliage (Fig. 1).

Habitat: Tropical deciduous – subtropical deciduous forests along the rivers and their tributaries in the drier but not in the too cold tracts.

Root: Tap, much branched, deep.

Stem: Erect, aerial, woody, solid, cylindrical, branched, bark dark brown, rough with deep narrow longitudinal fissures, heartwood pale red; when fresh cut nearly colour less, turning reddish brown on exposure.

Leaves: Stipulate, cauline and ramal, stipules modified into spines that are straight, 1.5-5.0 cm long, turning white in the older branches, petiolate, alternate, 2.5-8.0 cm long, bipinnately compound, pinnae 3-6 pairs, subsessile, opposite, cup shaped glands at the base of lowest and uppermost pair, leaflets small, oblong, opposite, subsessile, entire, obtuse 1.5-4.0 cm long, 10-20 pairs, venation unicostate reticulate.

Inflorescence: Axillary cymose heads globose, on short or medium sized peduncles, heads about 1.5 cm in diameter, yellow.

Flower: Ebracteate, subsessile, small, complete, hermaphrodite, regular, actinomorphic, hypogynous, cyclic, yellow, fragrant.

Calyx: Sepals 5, gamosepalous, campanulate, 5-toothed, slightly petaloid, inferior.

Corolla: Petals 5, gamopetalous, tubular, tube twice longer than calyx, yellow, inferior.

Androecium: Stamens indefinite, polyandrous with long filaments and anthers, minute, yellow, dithecous, versatile.

Gynoecium: Carpel 1, ovary superior, unilocular, sessile, green, hairy, oval with small white glands, style long, filiform, stigma minute, placentation marginal, ovules numerous.

Fruit: A lomantaceous pod, much constricted between seeds at both sutures, whitish, tomentose, 8-15 cm long, stalked, fleshy, grey, seeds non-endospermic, 8-12 in a single row.

Flowering: Rainy season and cold season (June-December)

Sprouting: Never leafless but young foliage comes out in Feb.- April.

Floral formula: $\oplus$ ⚥ $K_{(5)}\ C_{(5)}\ A_{\alpha}\ G_{\underline{1}}$

Varieties: Var.1- **Kaulia** or **Kowri** babul found in Afghanistan. Also found in Poona, Sholapur and Satara and known as Vedi Babul. Spines white, stout, up to 7.0 cm long, flat pod on short stalks, 1.5-2.0 cm broad, very little constricted between the seeds, wood considered fit only for firewood, bark deeply cracked and exfoliating, crown formed by twisted interlacing branches. Var. 2 – **Telia** or **godi** babul found in Maharashtra. It is ordinary babul with moniliform pods. Bark less deeply cracked and exfoliating, wood highly prized.

Uses: It is one of the most important nitrogen fixing trees (Chundawat and Gautam 1993). The pods are excellent cattle fodder. Gum exudes largely from wounds in the bark. The bark is used as a tan. The natural colourant obtained from dried bark has recently been used for dyeing silk successfully since synthetic dyes are carcinogenic, mutagenic and allergic (Patel *et al.* 2003). The gum obtained from the bark is of good quality and used in confectionery. Baskets are made of young, green twigs. Gum is used in diarrhoea and dysentery. Bark is astringent and demulcent. (Pandey 1981, Brandis 1906). The tender leaves are used as blood purifier. The Cassie perfume is obtained from flowers (Shukla and Mishra 1979). The twigs are used as tooth green brushes (Datun). Ayurveda based toothpaste like Babool are made from its ingradients. The decoction of bark is used as a gorgle and that of pods in urogenital diseases (Sinha 1996).

Thorns are used for fencing. Wood (52 lb/cft) is an excellent fuel and gives good charcoal. It is also used for making cart hubs and agricultural implements. It is commonly planted on fields bunds as it has a deep taproot and its crown casts only a light shade. 30 years old plantations with girth of 150 cm yield about 50 tonnes of wood/ha. (Sagreiya 1967). The mean annual yield of biomass after three years has been found to be 22.84 t/ha in the plantations of *A. nilotica* raised at 1.4 x 1.4 meter spacing. The specific gravity and calorific value of fuel wood of *A. nilotica* has been found to be 0.60 – 0.75 and 4880-4950 K.cal./kg. respectively (Chaundawat and Gautam 1993). It may be extensively planted as excellent wind-break and soil stabilizer in sub-humid Sutlej, alluvial plains and humid and semi-arid western ghats of Karnataka plateau (Chundawat and Gautam 1993).

Cultivation: Through seeds and seedlings. The seeds are collected from more than 8 years old trees during May (Suresh *et al.* 2001) and sown during May-June for raising nursery. The seedlings may be transplanted during July. Alternatively, the seeds may also be directly sown during July. Recently, Indole Butyric Acid (IBA) at 4000 ppm mixed with talcum powder successfully induced 20.56% rooting in air layers after 70 days (Sharma *et al.* 2002). Hence, *A. nilotica* may also be propagated through air layerings.

2. AUSTRALIAN PHYLLODE

Botanical: ***Acacia auriculiformis*** A. Cunn. ex Benth.

English: Australian Phyllode Acacia		Hindi: Australian Babool

SYSTEMATIC POSITION

Bentham & Hooker (1883)	**Engler & Prantl (1899)**	**Hutchinson (1959)**
Phanerogams	Phanerogams	Angiospermae
Dicotyledones	Dicotyledoneae	Dicotyledones
Polypetalae	Archichlamydeae	Lignosae
Calyciflorae	Rosales	Leguminales
Rosales	Leguminosae	Mimosaceae
Leguminosae	Mimosoideae	
Mimosoideae		

Geographical distribution: Indigenous to Australia, largely cultivated in Bihar, Orissa, Uttar Pradesh and other Northern and North-eastern states.

Habit: A very handsome evergreen tree (Fig. 2).

Habitat: Xerophytic, dry and rocky localities.

Root: Tap, branched, deep.

Stem: Erect, aerial, woody, solid, cylinderical, branched, branches pendulous, bark dark grey to black cracked.

Leaves: True leaves lacking, leaf like glabrous phyllodes present which are slightly curved and bear several parallel veins.

Inflorescence: Terminal panicled racemes.

Flowers: Complete, hermaphrodite, actinomorphic, hypogynous, yellow, and pentamerous.

Calyx: Sepals 5, gamosepalous.

Corolla: Petals 5, gamopetalous, yellow.

Androecium: Stamens indefinite, polyandrous, anthers dithecous, filaments large.

Gynoecium: Monocarpellary, ovary superior, unilocular, placentation marginal.

Fruit: Lomentous pod, curved, turgid, twisted, 5.0-7.0 cm long.

Flowering and fruiting: Three times during the summer, rainy and winter seasons (Oct.-Nov.).

Floral formula: $\oplus$ ⚥ $K_{(5)}\ C_{(5)}\ A_{\alpha}\ G_{\underline{1}}$

Uses: It is very beautiful, nitrogen fixing, drought resistant, evergreen tree, most suitable for dry denuded localities and even rocky areas. It is cultivated along roadsides and excellent avenue tree. Suitable for parks, open spaces and small home gardens also. Wood serves as firewood. It yields 12.4 tons/ha. of biomass in 10-12 years and the calorific value of its wood as 4800-4900 K.cal/kg. It yields tannin and its ash is used as washing soda for clothes (Randhawa 1983, Chundawat and Gautam 1993).

Cultivation: Through seeds and seedlings. The seeds are collected during March-April, sown during April-May and seedlings are transplanted during July (Chundawat and Gautam 1993).

3. KHAIR

Botanical: *Acacia catechu* (*L.f.*) Wild. Syn. *A. sandra,* Bedd., *Mimosa catechu,* Linn, *M. sundra* Roxb.

Bengali : Khair

English: Catechu

Hindi: Khair, Kattha, Khadira

Burmese: Iha

Kannada: Kugli, Kachu

Marathi: Khair

Tamil: Korangalli

Telugu: Tella Tumma, Sundra

SYSTEMATIC POSITION

Bentham & Hooker (1883)	**Engler & Prantl (1899)**	**Hutchinson (1959)**
Phanerogams	Phanerogams	Angiospermae
Dicotyledones	Dicotyledoneae	Dicotyledones
Polypetalae	Archichlamydeae	Lignosae
Calyciflorae	Rosales	Leguminales
Rosales	Leguminosae	Mimosaceae
Leguminosae	Mimosoideae	
Mimosoideae		

Geographical distribution: Sub-Himalayan tract from Indus eastwards ascending in the valleys up to 3000 ft. ASL. Common in the Aravali hills (Rajasthan), Punjab, Madhya Pradesh, Uttar Pradeh, Bihar and Western Peninsula. Also found in Myanmnar.

Habit: A moderate sized thorny deciduous tree.

Habitat: Gregarious on islands and on the banks of rivers at their entrance into the plains, not found in very moist regions.

Root: Tap, much branched, deep.

Stem: Aerial, erect, woody, solid, cyclinderical, armed with spines, branched, bark dark coloured, peeling off in long strips, which remain hanging; branchlets slender, glabrous, brown or purple, shining. Prickles twin, infra-stipular, compressed, brown, shining.

Leaves: Stipulate, stipules, not spinescent, compound, common petiole 7.5-10.0 cm long armed with prickles, prickles/spines infra-stipular, bipinnate, pinnae 10-20 pairs, leaflets 30-50 pairs, linear – oblong, entire, glabrous or pubescent, less than 0.6 cm long, green turning brown on drying.

Inflorescence: Cylinderical spikes, yellow-pale yellow.

Flower: Complete, hermaphrodite, actinomorphic, hypogynous, regular, bracteate, small, cyclic, pale yellow.

Calyx: Sepals 5, gamosepalous, companulate, 5 toothed.

Corolla: Petals 5, gamopetalous, tubular tube three times longer than the calyx, pale yellow-yellow.

Androecium: Stamens indefinite, polyandrous, filaments long, anthers small, dithecous.

Gynoecium: Monocarpellary, ovary superior, unilocular, placentation marginal with numberous ovules.

Fruit: Pod lomentaceous, thin, brown, shining, dehiscent.

Flowering: May - October

Floral formula: $\oplus$ ⚥ $K_{(5)}\ C_{(5)}\ A_{\alpha}\ G_{\underline{1}}$

Varieties: Prain (1898) recognized three varieties in *A. catechu* (see Brandis 1906)

(1) ***Catechu* proper:** Calyx, petals and rachis covered with spreading hairs. Found in Punjab, Garhwal, Kumaon (Uttarakhand), Bihar and Ganjam (Orissa). Also found in Pegu and upper Myamnar.

(2) *Catechuoides*: Calyx and petals glabrous, but rachis puberulous. Found in Sikkim Terai, Assam, Karnataka (Mysore) and Nilgiris. Also found in upper Myanmar.

(3) *Sundra*: Entirely glabrous calyx, petals and rachis. Found in Western Peninsula and upper Myanmar.

Uses: Wood is used for house posts, cartwheels, tool handles, agricultural implements etc. It is good firewood and yields excellent charcoal. The heartwood is exploited for obtaining Kattha and cutch; the former is well known ingredient of paan and the latter is used as a dye and as a boiler compound. (Sagreiya 1967, 5th edition 1994). Kattha is generally used as a masticatory, dyeing and preserving agent. Kattha industry is mostly found in Bareilly, Bombay and Gwalior. Kattha content is found maximum (6.31%) in 20-21 years old tree and minimum (2.22%) in 10-11 years old tree. The cutch yield does not vary significantly between 15-16 and 20-21 years age classes of Khair trees (Vasistha and Kaushal 2002). Catechu is also used as a raw material for plywood adhesives. It is extensively used for dyeing canvas, fishing nets and ropes (Pandey 1981). It also serves as the host plant for the lac insects, *Kerria lacca* Kerr. (Shukla and Mishra 1979). Besides it is fodder yielding, nitrogen fixing and soil ameliorating/protecting tree (Chundawat and Gautam 1993).

Cultivation: Through seeds and seedlings. The seeds are collected during Jan-Feb., and sown for raising nursery during May-June. The seeds are either directly sown or seedlings are transplanted during June-July (Chandawat and Gautam 1993).

4. SHIKAKAI

Botanical: *Acacia concinna* DC.

English: Soap Pod | Hindi: Shikakai, Banritha
Burmese: Subok | Bengali: Ban-ritha | Gujarati : Chikakai
Kannada : Siege | Malayalam: Chikaka | Marathi: Shikakai
Tamil: Shikai | Telugu: Chikayi

SYSTEMATIC POSITION

Bentham & Hooker (1883)	Engler & Prantl (1899)	Hutchinson (1959)
Phanerogams	Phanerogams	Angiospermae
Dicotyledones	Dicotyledoneae	Dicotyledones
Polypetalae	Archichlamydeae	Lignosae
Calyciflorae	Rosales	Leguminales
Rosales	Leguminosae	Leguminosae
Leguminosae	Mimosoideae	Mimosaceae
Mimosoideae		

Geographical distribution: Sub-Himalayan tract from Oudh eastwards, Assam, Bihar, Andhra Pradesh and West Coast. Also found in Myanmar, Java and Malaya Peninsula.

Habit: A middle sized woody climber with scattered spines and tomentose branchlets, petioles and peduncles.

Habitat: Chiefly in evergreen forests.

Root: Tap root.

Stem: Woody, aerial, erect, climbing, branched, branchlets armed with numerous sharp recurved prickles, which help in climbing.

Leaves: Compound, opposite pinnate, pinnae 4-6 pairs, opposite leaflets 12-20 pair, 0.8-1.3 cm long, linear to oblong from an unequal sided base, stipulate, stipules large, semicordate.

Inflorescence: Terminal globose panicles, yellow and purple.

Flowers: Hermaphrodite, actinomorphic, hypogynous, complete, bracteate, flower buds purple, flower yellow, heads in dense panicles at the end of branches.

Calyx: Sepals 5, gamosepalous, campanulate, 5-toothed, valvate.

Corolla: Petals 5, gamopetalous, 5 toothed, tube longer than calyx.

Androecium: Stamens indefinite, free anthers, dithecous, dorsifixed, filaments long.

Gynoecium: Monocarpellary, ovary superior, unilocular, style long, filiform, stigma simple.

Fruit: Lomentous, pods, thick, fleshy, indehiscent.

Flowering: March - July

Floral formula: ⊕ ⚥ Br $K_{(5)}$ $C_{(5)}$ A_{α} $G_{\underline{1}}$

Uses: The fruits (pods) are used as a detergent for washing hairs, silk and woolen fabrics (Pandey 1981).

Cultivation: Through seeds and seed lings.

5. BANGHNALA

Botanical: ***Actinodaphne angustifolia*** Nees Syn. *Litsaea angustifolia* Kurz.

Burmese: Nalingyo

Assamese: Petarichawa | Kannada: Hoggodgimara | Malayalam: Malaviriji

Manipuri: Takara | Marathi: Pisa, Gulchal

Oriya: Tundambo, Jar-champa | Other venaculars: Banghnala, Baghnala

SYSTEMATIC POSITION

Bentham & Hooker (1883)	**Engler & Prantl (1899)**	**Hutchinson (1959)**
Phanerogams	Phanerogams	Angiospermae
Dicotyledones	Dicotyledoneae	Dicotyledones
Polypetalae	Archichlamydae	Lignosae
Thalmiflorae	Ranales	Laurales
Laurales	Lauraceae	Laurceae
Lauraceae		

Geographical distribution: Prevalent throughout North-east India particularly Assam, Garo and Khasi hills (Meghalaya) and Barak Drainage forest in Manipur. Also found in Chittagong (Bangladesh) and in Pegu, Yoma, Martaban and Tennaserim (Myanmar).

Habit: A middle sized evergreen, branched, dioecious tree with robust branches.

Habitat: Common in evergreen forest patches with high rainfall, high humidity and warm climate.

Root: Branched tap root.

Stem: Aerial, erect, woody, solid, cylinderical, younger parts grey and softly hairy (tomentose or pubescent), lenticels present, bark warty, rough, greyish brown, which turns to black.

Leaves: Young leaves softly tomentose, simple, coriaceous, glaucous beneath, shining above, exstipulate, 4-10 in whorls, petiole 2-2.5 cm long, length and breadth of leaves varies considerably (9-25 x 2.5-8.5 cm), lanceolate to oblanceolate or elliptic, acuminate, nerves 8-13 pairs prominent beneath, midrib and underside of secondary nerve usually hairy, margin entire.

Inflorescence: Male flowers in panicled raceme and female flowers in peduncled umbels.

Flower: Bracts 4-6, imbricate, tomentose, imbricate, pedicelllate length 0.2-0.5 cm, unisexual, hypogynous, actinomorphic, yellow.

Perianth: Tepals 6, ovate-lanceolate, polytepalous, entire, villous, valvate.

Androecium: Stamens 9-20 in male flowers, staminodes in female flowers, polyandrous, filaments villous, anther 4-celled, adnate and introse.

Gynoecium: Found in female flowers only, monocarpellary, unilocular with single ovule, placentation basal/marginal, style villous, stigma trifid and short, ovary abortive in male flowers.

Fruit: Fleshy berry, globose with 0.8 cm diameter, seated on the small cup shaped entire perianth tube, 0.5-1.0 cm long.

Seeds: Exalbuminous (Non-endospermic)

Floral formulae: Male flower: $Br._{4\text{-}6} \oplus ♂ P_6 A_{9\text{-}20} G_0$

Female flower: $Br._{4\text{-}6} \oplus ♀ P_6 A_{std\ 9} G_{\underline{1}}$

Uses: It serves as tertiary but considered as secondary host plant of muga silkworm, *Antheraea assama* (Thangavelu *et al.* 1988, Seth 1995d, Das *et al.* 2000). Besides serving as tertiary food plant of muga silkworm it is also used medicinally. Its leaf infusion is given in diabetes and uninary disorders (Sinha 1996).

Cultivation: *A. angustifolia* can be propagated through seeds and seedlings.

6. PATIHONDA

Botanical: ***Actinodaphne obovata*** Blume

Assamese: Patihonda, Pattikuta Nepali: Muslindi

SYSTEMATIC POSITION

Bentham & Hooker (1883)	Engler & Prantl (1899)	Hutchinson (1959)
Phanerogams	Phanerogams	Angiospermae
Dicotyledones	Dicotyledoneae	Dicotyledones
Polypetalae	Archichlamydae	Lignosae
Thalmiflorae	Ranales	Laurales
Laurales	Lauraceae	Laurceae
Lauraceae		

Geographical distribution: Sub-Himalayan tract from Sikkim eastwards ascending to 2000 ft. ASL in Assam, Meghalaya (Khasi hills) and Manipur (Barrak drainage forest) etc. Also found in Bangladesh and Nepal.

Habit: An evergreen, small sized, dioecious, rusty pubescent tree.

Habitat: Evergreen forest and damper portions of Sal forests with high rainfall, high humidity and warm climate.

Root: Branched tap root.

Stem: Aerial, stout, erect, woody, solid, quadrangular, greyish-brown bark exfoliating in thin papery flakes of about 0.5 cm thickness, young parts rusty pubescent.

Leaves: Simple, cauline and ramal, decussately opposite, rarely alternate, young leaves pubescent, blade 25-45 cm long, petiole 2.5-5.0 cm long, stipules absent, thin, coriaceous, elliptic–oblong or rarely obovate, obtuse–ocuminate, entire, smooth, shining above, glaucous (bluish-white) beneath, prominently nerved, 3 ribbed at or above the base, lateral nerves 22-28 on either half, prominent beneath, base obtuse (rounded).

Inflorescence: Male flowers in cluster of umbels while female flowers on panicled racemes.

Flower: Small, unisexual, pedicel 1.2 cm long, stout, bracts-4, caducous in imbricate aestivation, while in bud covered by round coriaceous scales.

Perianth: Tepals 6, petaloid, polytepalous, sub-equal, orbicular or ovate, rarely persistent, valvate.

Androecium: Found only in male flowers, stamens 6-9, arranged in two whorls, polyandrous, anthers introse and 4-celled. In females 9 staminodes present.

Gynoecium: Found in female flowers only. Monocarpellary, unilocular with single ovule, style villous, stigma trifid and placentation basal/marginal.

Fruit: Fruit ellipsoid, fleshy berry, 1.8-2.5 cm long seated on the entire cup shaped perianth tube.

Seed: Exalbuminous (Non-endospermic)

Floral formulae: Male flower : $Br_4 \oplus ♂ P_6 A_{6\text{-}9} G_0$

Female flower : $Br_4 \oplus ♀ P_6 A_{Std.\ 9} G_{\underline{1}}$

Uses: It serves as the tertiary but considered as secondary food plant of muga silkworm *Antheraea assama* (Thangavelu *et al.*1988, Seth 1995d, Das *et al.* 2000). Besides, its leaves and stamens contain three alkaloids viz., laurotetamine, N-methyl laurotetamine and actinodaphine (Sinha 1996).

Cultivation: *Actinodaphe obovata can* be propagated through seeds and seedlings.

7. BAEL

Botanical: ***Aegle marmelos*** Roxb. Syn. *A. marmelos* (L.) Correa

English: Bael Tree	Hindi: Bael, Bili, Bel	
Burmese: Okshit	French: Belindien	German: Bhelbaum
Sinhalese: Belli	Persian: Shul	
Assamese: Bel	Bengali: Bela	Gond: Mahika
Gujarati: Bil, Bili	Kannada: Bilpatre	Kol: Lohangsi
Konkani: Bello	Malayalam: Kuvalam, Kovalam	
Marathi: Bael	Oriya: Belo	Sanskrit: Bilva
Sindhi: Bila	Tamil: Vilvam, Vilva	Telugu: Maredu
Urdu: Bel		

SYSTEMATIC POSITION

Bentham & Hooker (1883)	Engler & Prantl (1899)	Hutchinson (1959)
Phanerogams	Phanerogams	Angiospermae
Dicotyledones	Dicotyledoneae	Dicotyledones
Polypetalae	Archichlamydeae	Lignosae
Disciflorae	Geraniales	Rutales

Geraniales Rutaceae Rutaceae

Rutaceae

Hindu ladies on Shivaratri and other occasions employ the sacred leaves in the worship of Lord Shiva.

Geographical distribution: Wild in the Sub-Himalayan tract and outer hills from the Jhelum eastwards, also found in south India, Singhbhum and Myanmar. Cultivated throughout India on account of its fruits and sacred nature.

Habit: A small or middle-sized deciduous tree, armed with axillary, straight, sharp spines (Fig. 3).

Habitat: Tropical - sub-tropical - sub-temperate humid habitat with varying types of soil and climatic conditions.

Root: Tap and branched.

Stem: Erect, aerial, branched, woody, cylindrical, solid, green when young, grey when old, armed with spines, gland dotted, usually strongly aromatic when fresh cut, wood yellowish white, close grained, without heartwood.

Leaves: Alternate, compound, trifoliate, sometimes pentafoliate, ovate – lanceolate, crenate, exstipulate, petiolate, gland dotted, venation reticulate unicostate.

Inflorescence: Short lateral panicles.

Flower: Bisexual, 4-5 merous, greenish white with a fine honey scent, pedicellate, regular, complete, actinomorphic, hypogynous, ebracteate.

Calyx: Sepals usually 5, flat, polysepalous, quincuncial, teeth small.

Corolla: Petals usually 5, polypetalous, imbricate, greenish white.

Androecium: Stamens numerous, 50 or even more, arranged in irregular bundles (Polyadelphous), filaments short, anthers linear.

Gynoecium: Carpels 8-20, syncarpous, ovary superior, multilocular, placentation axile, styles erect, as many as carpels, distinct, stigma capitate.

Fruit: A berry with hard rind, 4-6 inches in diameter, edible, green when young, grey or yellow on ripening. The inner contents of the fruit are mucilaginous and faintly aromatic.

Seeds: Numerous, oblong, flat, endospermic, testa densely clothed with thick fibrous hairs in a thick orange coloured sweet aromatic pulp.

Flowering: During November – December.

Fruiting: Hot season (Apri-May).

Floral formula: $\oplus$ ⚥ $K_5\ C_5\ A_\alpha\ G_{\underline{(8\text{-}20)}}$

Uses: Leaves are used as a mild laxative, for fever, catarrh or inflmmation of the mucous membrane having free discharge and for asthma. The decoction of the leaves is febrifuge and expectorant. The leaf juice is given in dropsy, constipation and jaundice. A hot poultice of the leaves is applied in opthalmia or conjunctiva with acute bronchitis and inflammation of the body. The decoction of root bark is used for intermittent fevers, melancholia and heart palpitation. The unripe fruit astringent, digestive, may be used as stomachic and also for preventing scurvy, indigestion, diarrhoea or dysentery and irritation of the elimentary canal. The ripe fruit is aromatic, cooling and mild laxative. The extract of the ripe fruit is administered in chronic constipation, dyspepsia, discomfort or pain (Kurian 1995) and used in the form of sherbet. Leaves, fruits and roots of bel contain antibiotic elements (Jain 1968). Active constituent is marmelosin which acts as a laxative, diuretic and in stong doses a cardiac depressant (Ambasta *et al.* 1986). The mucilaginous substance secreted round the seeds is used as cement and a varnish. The pulp is often used to strengthen mortar. The seeds yield a dye. The leaves are used as fodder and also for worshipping lord Shiva (Pandey 1981).

The wood is suitable for preparing charcoal and producer gas. The pulp is also used as gum and used for preparing varnish and joining the woods. The peeling of raw fruits yields a yellow dye (Jain 1968).

Cultivation: It can be propagated through seeds, seedlings, root suckers or root shoots (Brandis 1906). It coppices well hence, yield of wood may be enhanced manifold throughout the life of the trees.

The seeds are collected during April-June. The nursery is raised during May-June and seedlings are transplanted during July (Chundawat and Gautam 1993).

8. MAHARUKH

Botanical: ***Ailanthus excelsa*** Roxb.

Hindi: Maharukh	Assemese: Barkesseru	Gujarati: Ardusi
Malayalam: Matti Pongilayam	Marathi: Maharukh	Kannada: Doddamarra
Rajputana: Arua	Tamil: Perumaram	Telugu: Pedda manu
Oriya : Maha nim		

SYSTEMATIC POSITION

Bentham & Hooker (1883)	**Engler & Prantl (1899)**	**Hutchinson (1959)**
Phanerogams	Phanerogams	Angiospermae

Dicotyledones	Dicotyledoneae	Dicotyledones
Polypetalae	Archichlamydeae	Lignosae
Disciflorae	Geraniales	Rutales
Geraniales	Simarubaceae	Simarubaceae
Simarubeae		

The name "Ailanthus" is derived from "Ailanto" which means "tree of heaven", 'excelsa' is a latin word meaning 'tall'.

Geographical distribution: Indigenous in the Indian peninsula, commonly planted throughout the country especially in U.P. (Saharanpur) and Bihar.

Habit: Deciduous, moderate sized tree attaining the height of 60 – 80 ft.

Root: Tap root, branched.

Stem: Woody, wood soft and white, bark grey or white.

Leaves: Very large, compound, 30-90 cm long, leaflets numerous (8-14 pairs) on long petioles, nearly opposite, very unequal at the base, glandular – hairy, broadly falcate – lanceolate, deeply serrated, often lobed.

Inflorescence: Axillary panicles ample shorter than leaves, lax, branched.

Flower: Polygamous, small, yellow, bracteolate, pentamerous, actinomorphic, hypogynous, pedicels long.

Calyx: Sepals 5- fid, small, lobes equal, imbricate.

Corolla: Petals 5, glabrous, ovate – lanceolate, reflexed, free, valvate, yellow, spreading, edges bent inwards.

Androecium: Stamens 10 in male, 2-3 in bisexual and nil in female flowers, filaments half the length of anthers, filiform, without scales, glabrous.

Gynoecium: Carpels 2- 5 partite, styles connate, ovule one in each cell, semi anatropous.

Fruit: 1-5 seeded Samaras, 1.0 x 1.2 cm across, strongly veined, blunt or point at both ends, copper red, once or twice twisted at the base, wings very large and membranous.

Seeds: Pendulous, endospermic.

Flowering: April – May.

Sprouting: March - April.

Floral formulae: Male flower: ⊕ ♂ $K_{(5)}\ C_5\ A_{10}\ G_0$

Bisexual flower: ⊕ ⚥ $K_{(5)}\ C_5\ A_{2\text{-}3}\ G_{\underline{(2\text{-}5)}}$

Female flower: ⊕ ♀ $K_{(5)}\ C_5\ A_0\ G_{\underline{(2\text{-}5)}}$

Uses: It serves as one of the secondary food plants of eri silkworm, *Philosamia ricini* (FAO 1976, Seth 1995c, Sarmah 2000).

Medicinally, leaf extract is used for washing wounds and its juice with coconut, milk or gur (jaggery) helps in relieving post delivery pains in tribal women in Madhya Pradesh (Jabalpur) (Singh 1990).

The leaves of young plants contain acrid juice due to which cattles do not touch them. Later they are considered as a good fodder for which trees are heavily lopped. It is vastly planted in Rajasthan for fodder.

The wood is suitable for making match splints and accordingly used (Sagreiya 1967).

Cultivation: It reproduces through seeds and seedlings.

9. BARKESSERU

Botanical: ***Ailanthus glandulosa.*** Desf.; Syn. *A. altissima* (Mill) Suringle

Eng: Ailanto, Tree of heaven Assamese: Barkesseru, Barpat

SYSTEMATIC POSITION

Bentham & Hooker (1883)	Engler & Prantl (1899)	Hutchinson (1959)
Phanerogams	Phanerogams	Angiospermae
Dicotyledones	Dicotyledoneae	Dicotyledones
Polypetalae	Archichlamydeae	Lignosae
Disciflorae	Geraniales	Rutales
Geraniales	Simarubaceae	Simarubaceae
Simarubeae		

Geographical distribution: Cultivated in Northern India especially in North-east. Also found in Japan and China.

Habit: A high tree.

Root: Tap root branched.

Leaves: Compound, alternate, often exceeding 30 cm, leaflets very numerous, very unequally divided by the midrib, pubescent or sub-glabrous, paler beneath, coarsely toothed at the base, otherwise entire towards apex.

Inflorescence: Axillary or terminal much branched panicles.

Flower: Small, bracteolate, actinomorphic, polygamous, pantamerous, hypogynous.

Calyx: Sepals 5, lobes unequal, imbricate.

Corolla: Petals 5, woolly tomentose inside, valvate.

Androecium: Stamens 10 (2-3 in bisexual flowers), filaments filiform, several times larger than anthers, hispid at base, exserted.

Gynoecium: Ovary 2-5 partite, styles connate, ovules 1 in each cell (locule), semi anatropous.

Fruit: 3 membranous linear – oblong samaras, 3.2 x 0.8 cm, ripe carpels 2.5-5.0 cm long.

Seed: About 0.6 x 0.2 cm, near the centre of samara.

Floral formulae: Male flower: ⊕ ♂ $K_{(5)}\ C_{5}\ A_{10}\ G_{0}$

Bisexual flower: ⊕ ⚥ $K_{(5)}\ C_{5}\ A_{2\text{-}3}\ G_{(\underline{2\text{-}5})}$

Female flower: ⊕ ♀ $K_{(5)}\ C_{5}\ A_{0}\ G_{(\underline{2\text{-}5})}$

Uses: It serves as tertiary but considered as one of the secondary food plants of eri silkworm, *Philosamia ricini* (Sarkar 1988, Seth 1995C).

Bark antispasmodic, parasiticidal, powerful depressant of nervous system and used in diarrhoea and dysentery. Root bark used in cardiac troubles, epilepsy and asthama. Fruits emmeragogue, used in eye troubles. Leaves used in seborrhoea and scabies. Wood used for construction purposes, furniture, pulp, boards and fancy articles. Flowers contain an essential oil while seeds yield a fatty oil (Ambasta *et al.* 1986).

Cultivation: Through seeds like *A. excelsa* and *A. malabarica*

10. GOGUL

Botanical: ***Ailanthus grandis*** Prain.

Hindi: Guggul — Nepalese: Gogul

Assamese: Actaluca, Rungagora — Khasi: Diang-chao

SYSTEMATIC POSITION

Bentham & Hooker (1883)	Engler & Prantl (1899)	Hutchinson (1959)
Phanerogams	Phanerogams	Angiospermae
Dicotyledones	Dicotyledoneae	Dicotyledones
Polypetalae	Archichlamydeae	Lignosae

Disciflorae	Geraniales	Rutales
Geraniales	Simarubaceae	Simarubaceae
Simarubeae		

Geographical distribution: Nepal, Assam and along Ryang and other tributaries of the Tista river in Sikkim.

Habit: A tall tree of nearly 120 – 150 ft.

Root: Branched tap root.

Leaves: Compound, alternate, leaflets entire, 5-8 pairs, 15-20 cm long.

Inflorescence: Axillary panicles, ample shorter than leaves, lax, branched.

Flower: Polygamous, small, yellow, bracteolate, pentamerous, actinomorphic, hypogynous, pedicels long.

Calyx: Sepals 5- fid, small, lobes equal, imbricate.

Corolla: Petals 5, glabrous, ovate – lanceolate, reflexed, free, valvate, yellow, spreading, edges bent inwards.

Androecium: Stamens 10 in male, 2-3 in bisexual and nil in female flowers, filaments half the length of anthers, filiform, without scales, glabrous.

Gynoecium: Carpels 2-5, partite, styles connate, ovule one in each cell, semi anatropous.

Fruit: Samara, ripe carpels 10-15 cm long, apex rounded, obtuse.

Seeds: Pendulous, endospermic.

Flowering: April – May.

Sprouting: March - April.

Floral formulae: Male flower: ⊕ ♂ $K_{(5)}\ C_5\ A_{10}\ G_0$

Bisexual flower: ⊕ ⚥ $K_{(5)}\ C_5\ A_{2\text{-}3}\ G_{\underline{(2\text{-}5)}}$

Female flower: ⊕ ♀ $K_{(5)}\ C_5\ A_0\ G_{\underline{(2\text{-}5)}}$

Uses: It serves as one of the tertiary but recognized as secondary food plants of eri silkworm, *Philosamia ricini* (Seth 1995c, Sarmah 2000). Wood used for laminated boards, matches, tea-chests and packing cases. Wood pulp used for newsprint (Ambasta *et al.* 1986).

Cultivation: It reproduces through seeds and seedlings

11. GOGGUL DHUP

Botanical: *Ailanthus tryphysa* Alston; Syn. *A. malabarica* DC.

Hindi: Guggul Dhup   Malayalam: Mattipal   Marathi: Gugul dhup

Kannada: Guggul Dhup, Hal-maddi   Tamil: Perumaram

Telugu: Pedda-manu

SYSTEMATIC POSITION

Bentham & Hooker (1883)	**Engler & Prantl (1899)**	**Hutchinson (1959)**
Phanerogams	Phanerogams	Angiospermae
Dicotyledones	Dicotyledoneae	Dicotyledones
Polypetalae	Archichlamydeae	Lignosae
Disciflorae	Geraniales	Rutales
Geraniales	Simarubaceae	Simarubaceae
Simarubeae		

Geographical distribution: Along the Western Ghats from the Konkan Southwards up to 3000 ft. ASL.

Habit: A high tree.

Habitat: Evergreen forest.

Root: Branched tap root.

Stem: Erect, branched, aerial, solid, bark thick, rough often with bright reddish grains of resin.

Leaves: Compound, very large, leaflets 5-10 pair, glabrous, entire or wavy, lanceolate, very unequal at the base, petioles long, leaflets distant, coriaceous, much lighter beneath.

Inflorescence: Axillary or terminal panicles.

Flower: As in *Ailanthus excelsa* Roxb. but rather larger, filaments glabrous, slender, much longer than anthers, filiform, exserted.

Fruits: Ripe carpels oblong, rounded at ends, samara, broadly linear 7-12 x 2 cm, reddish, not twisted.

Flowering: Feb - March.

Floral formulae: Male flower: $\oplus$ ♂ $K_{(5)}\ C_5\ A_{10}\ G_0$

Bisexual flower: $\oplus$ ⚥ $K_{(5)}\ C_5\ A_{2\text{-}3}\ G_{\underline{(2\text{-}5)}}$

Female flower: $\oplus$ ♀ $K_{(5)}\ C_5\ A_0\ G_{\underline{(2\text{-}5)}}$

Uses: Besides serving as one of the tertiary/secondary food plants of eri silkworm (Seth 1995c, Sarmah 2000), it yields a resin (matti pal), which is used as medicine in bronchitis and dysentery and as incense (Brandis 1906). Bark carminative, febrifuge and tonic. Juice used in asthma, branchitis and dysentery. Decoction given in constipation and typhoid. Leaves yield a black dye and fruits a fatty oil. Wood is used for veneers, plywood, catamarans, match boxes, splints, paper-pulp and packing cases (Ambasta *et al.* 1986).

Cultivation: It can be cultivated through seeds.

12. SIRIS

Botanical: ***Albizzia lebbek*** Benth.

Burmese: Kokko, Kokoh	French: Accacia lebbek	
English: Rain tree	Hindi: Siris, Sirsa	Persian: Darakhtejakheria
Sinhalese: Mara	Bengali: Sirish	Gujarati: Pilosarshio
Kannada: Bage, Bagemara	Malayalam: Vaga	Marathi:Sirisha, Chichola
Oria: Bodosirisi	Santhali: Chapotsiris	Sanskrit: Bhandi, Sirisha
Tamil: Kalindi vagei, vagai	Telugu: Dirasana	Urdu: Darash

SYSTEMATIC POSITION

Bentham & Hooker (1883)	Engler & Prantl (1899)	Hutchinson (1959)
Phanerogams	Phanerogams	Angiospermae
Dicotyledones	Dicotyledoneae	Dicotyledones
Polypetalae	Archichlamydeae	Lignosae
Calyciflorae	Rosales	Leguminales
Rosales	Leguminosae	Leguminosae
Leguminosae	Mimosoideae	Mimosaceae
Mimosoideae		

It is regarded as one of the "bodhi trees", hence the great emperor Ashoka planted it along the road sides after abjuring violence and upon his conversion to Buddhism during 260 B.C. to give shade and shelter (Rao 1991).

Geographical distribution: Sub-Himalayan tract ascending up to 5000 ft. ASL and both Peninsulas. Planted largely in Egypt and throughout the tropics.

Habit: A large deciduous tree growing over 20 meters height.

Habitat: Tropical moist, tropical deciduous and subtropical humid habitat with varying types of soil and climatic conditions.

Root: Tap root, branched, deep rooted in the soil.

Stem: Aerial, erect, branched, woody, solid, hard, cylinderical, nearly glabrous, bark dark grey or brownish, sapwood large, white, heartwood dark brown, shining, mottled.

Leaves: Compound, abruptly bipinnate with a large gland on the petiole above the base and one below the uppermost pair of pinnae, petiole swollen and hairy at base, pinnae 2-3, leaflets 6-9 pairs with glands between their bases, the lateral leaflets elliptic-oblong, the two terminal obovate-oblong, pale green, 1 – 1½ inch long, oblique, obtuse.

Inflorescence: Corymbose racemes, globose umbellate heads large, white, fragrant, on 2-4 inch long peduncles in fasicles of 2-4 from the upper axils.

Flower: White, hermaphrodite, actinomorphic, regular, hypogynous, complete, fragrant, 1½ inch long upto the extremity of stamens, pedicel nearly as long as calyx.

Calyx: Sepals 5, gamosepalous, campanulate, calyx half the length of corolla, green.

Corolla: Petals 5, gamopetalous, funnel shaped, white and fragrant.

Androecium: Stamens indefinite, exserted, filaments connate at the base into a short tube.

Gynoecium: Monocarpellary, ovary superior, unilocular, placentation marginal, style long filiform, stigma terminal and simple.

Fruit: Legume, pod oblong, thin, bluntly pointed, pale yellow, smooth, shining, 4-12 seeded, straw coloured, dehiscent, 8-12 x 1-2 inch.

Flowering: April – May.

Fruiting: Pods ripen in Sept. and remain on the tree for long time.

Floral formula: $\oplus$ ⚥ $K_{(5)}\ C_{(5)}\ A_{(\alpha)}\ G_{\underline{1}}$

Uses: Wood is hard and strong and is used for furniture, oil mills, railway carriages, structural work and interior fittings. The leaves and twigs are used as fodder. This is a common roadside tree and grown for shade in tea and coffee plantations (Pandey 1981, Shukla and Mishra 1979). Randhawa (1983) considered it to be a drought resistant tree suitable for arid regions. According to him *A. lebbek* breaks easily in windstorm and block the road and cause menace to the travellers life, so they should not be planted on roadsides. It is one of the nitrogen fixing trees and hence, its growth is very fast and it may be utilized as a very good soil stabilizer/ameliorator/protector.

Besides its leaves may also be used as a fodder (Chaundawat and Gautam 1993).

Medicinally the decoction of the bark and leaves in infusion are used for dysentery and diarrhoea (Kurian 1995). Leaves and seeds used in eye troubles while bark in boils (Ambasta *et al.* 1986)

Cultivation: Through seeds and seedlings. The seeds are collected during Jan-Feb. The seedlings are raised during May-June and transplanted during July-August. Maximum germination is achieved if seeds are allowed to remain on mother tree for two months after ripening, otherwise after harvesting from mother trees, duration of storage should be increased to six months for getting germination equivalent to that obtained after *in vivo* storage. Treatment with H_2SO_4 for 1-2 minutes provided maximum germination in seeds (Punam *et al.* 2002).

13. KAJU

Botanical: *Anacardium occidentale* Linn.

Burmese: Thi-hoe-thayiet French: Acajou German: Acajoubaum

English: Cashew nut Hindi: Kaju

Persian: Badami pharangi Sinhalese: Kaju

Bengali: Hijli badam Kannada: Geru bija, Konkani: Kaju

Malayalam: Andiparuppu, Godambe, Paringi mavu Marathi: Kaju

Oriya: Lankabadam Sanskrit: Kaju taka

Tamil: Myndini, Kolia mavu Telugu: Jidi mamidi

SYSTEMATIC POSITION

Bentham & Hooker (1883)	Engler & Prantl (1899)	Hutchinson (1959)
Phanerogams	Phanerogams	Angiospermae
Dicotyledones	Dicotyledoneae	Dicotyledones
Polypetalae	Archichlamydeae	Lignosae
Disciflorae	Sapindales	Sapindales
Sapindales	Anacardiaceae	Anacardiaceae
Ancardiaceae		

Geographical distribution: It is a native of West Indies and tropical America from Mexico to Peru and Brazil. Portugese introduced it in to India from

Brazil in the sixteenth century to control coastal soil erosion. Now-a-days international trade in cashew kernels and shell-oil is monopolized by India since 33% of world production comes from India (230000 t/ 700000 t). The cultivation of cashew is concentrated especially in Kerala, Tamil Nadu, Karnataka and Andhra Pradesh and to some extent in Maharashtra, Orissa, West Bengal and Assam. It is also cultivated in East Africa (Mozombique, Tanzania), Malaya and South America (Brazil).

Habit: A low sprawling evergreen tree with a gnarled or twisted trunk.

Habitat: It can grow under a wide range of climatic and soil conditions. It is well adapted to poor soils and dry sandy areas, so it is hardy and drought resistant. It can be cultivated from sea level to 1220 m ASL, but is damaged by frost.

Root: Tap and branched.

Stem: Erect, branched, woody, cylinderical, soild, trunk crooked, resin canals present in the wood, exudes yellow gum.

Leaves: Simple, alternate, thickly, leathery, glabrous, oval or obovate, entire, 10-12.5 x 5-10 cm, apex rounded, notched, petiole ¼-½ inch, secondary nerves 8-12 pairs, conspicuous.

Inflorescence: Terminal panicles, lax.

Flower: Hermaphrodite, some times unisexual, actinomorphic, pentamerous, hypogynous, yellow, small.

Calyx: Sepals 5, gamosepalous.

Corolla: Petals 5, polypetalous, linear, ⅓– ½ inch long, pale yellow streaked with pink.

Androecium: Stamens 7-10, one longer always fertile, rest often sterile, free.

Gynoecium: Monocarpellary, ovary superior, style short, ovule solitary, pendulous.

Fruit: Drupe, kidney shaped, 1 inch long, seated on a thick pear shaped, swollen, juicy hypocarp which is swollen peduncle and receptacle, reddish or yellow in colour when ripe, 2-3 inch long and known as cashew apple. The drupe (nut), single seeded with hard, grey-green pericarp/shell, shell contains acrid juice causing severe irritation resulting into painful blisters in the skin. The seeds are inedible when raw, hence they must be cooked or roasted to drive the volatile oil before it is opened or shelled.

Flowering: November- December

Fruiting: Hot season.

Floral formula: ⊕ ⚥ $K_{(5)}\ C_5\ A_{7\text{-}10}\ G_{\underline{1}}$

Uses: The nuts fallen on the ground are dried in the sun together with the apple until rattling sound is produced by the nut on shaking. These nuts are roasted in olive or peanut oil for quick and uniform roasting and higher recovery of oil. The cashew shell oil is extracted with solvents or steam. Then shellers dip their hands in mixture of lime, ash, linseed oil or castor oil to protect from remaining traces of cashew shell oil and remove the shell by hand. Subsequently, seed coat is removed and now the edible nuts are vacuum packed for export. The kernels contain approximately 5% moisture, 20% proteins, 45% fats, 25% carbohydrates, 1.5% fibres and 2.5% minerals while cashew oil is composed of 90% anacardic acid and 10% cardol.

The fleshy pear shaped receptacle is either consumed fresh or in a preserved state. "Caju wine" is made from cashew apple. The apple is also used for making jams. The phenolic nutshell oil is widely used as a medicine and an irrilant, a wood protectant, a source of resin and as a waterproofing agent. The young leaves are used as flavouring and as a medicine. A yellow gum obtained as exudates from bark is used as an adhesive. The sap from the bark provides indelible ink (Kocchar 1981). The oil obtained from shells is also used in the manufacture of varnishes, typewriter rolls, industrial flooring tiles, gums, inks, cloth paints, water proof paper, acid and alkali resistant cements, laminating resins, rubber compounding resins, cation exchange resin for use in the treatment of water required for boilers, air conditioning units, ice plants, textile industries and for treating cracks on the soles of feet (Pandey 1981).

Medicinally, the bark is alterative, astringent and used in decoction for diarrhoea, diabetes, swellings, mouth ulcers and reducing blood pressures. The infusion of leaves and bark relieves toothache, sore gums and dysentery. The juice from pericarp is diuretic and its alcohlic solution is anthelmintic. The juice of ripe fruit is used for dysentery. The root is purgative while gum from bark is an effective insecticide (Sinha 1996, Kurian 1995). The wood may be utilized as a fuel (Chundawat and Gautam 1993). The cashew nut trees serve as host plants for commercial exploitation of wild silkworm, *Cricula trifenestrata* in Indonesia. It is believed that trees infested with these insects produce more nuts. In India, these silkworms produce Amkoni or Aam resham but their commercial exploitation is yet to be achieved (Mishra 2002, 2003).

Cultivation: Propagation is chiefly by seeds but recently cultivars have been evolved in India which can be propagated by inarching, budding and other vegetative methods. The tree begins to fruit in three years but comes to full bearing in tenth year (Kocchar 1981). The seeds are collected during March-April, nursery is raised during June and the seedlings are transplanted during July. Seeds may also be directly sown during July (Chandawat and Gautam 1993).

14. SOURSOP

Botanical: ***Annona muricata*** Linn.

English: Soursop Marathi: Mamphal

SYSTEMATIC POSITION

Bentham & Hooker (1883)	Engler & Prantl (1899)	Hutchinson (1959)
Phanerogams	Phanerogams	Angiospermae
Dicotyledones	Dicotyledoneae	Dicotyledones
Polypetalae	Archichlamydeae	Lignosae
Thalmiflorae	Ranales	Annonales
Ranales	Annonanceae	Annonanceae
Annonanceae		

Geographical distribution: Cultivated widely in Assam, Bengal, Tamil Nadu, Uttar Pradesh, Madhya Pradesh and other states in India.

Habit: Trees or large shrubs, wholly glabrous, unarmed.

Habitat: Sub-tropical to sub-temperate places in various types of soils.

Root: Tap, branched.

Stem: Aerial, erect, woody, branched, cylinderical, solid, bark glabrous (smooth), grayish white, marked by numerous anastomising longitudinal ridges.

Leaves: Simple, alternate, oblong – lanceolate, glabrous, 6.0-8.0 cm long, entire exstipulate, pellucid dotted with a peculiar smell, venation reticulate unicostate.

Inflorescence: Solitary or fasicle of flowers in the leaf axils.

Flower: Bracteate, pedicellate, regular, actinomorphic, bisexual, hypogynous, greenish white, spirocyclic.

Calyx: Sepals 3, small, triangular, polyseplous, valvate, fleshy.

Corolla: Petals 6, arranged in two whorls of three each, fleshy, triangular, polypetalous, valvate.

Androecium: Stamens indefinite, polyandrous, filaments short, anthers long, dithecous, extrose, connective thick broad and projecting beyond the anther into a short sterile projection.

Gynoecium: Carpels numerous, apocarpous, superior, each ovary unilocular, placentation marginal, style simple, short, stigma inconspicuous or beak like.

Fruit: An etario of achenes. Seeds endospermic.

Floral formula: $\oplus$ ⚥ $K_3\, C_{3+3}\, A_\alpha\, G_\alpha$

Uses: Soursop is cultivated widely for its large fleshy aromatic, sour, antiscorutic fruit which yield a sap used as beverage and in preparation of jellies and the juice pulp is used for conserves (Shukla and Mishra 1979). The fruit is a good beverage and wood is used as fuel (Chundawat and Gautam 1993). Green bark purgative. Pulp consumed with wine. Leaves and flowers used in kidney troubles (Ambasta *et al.* 1986).

The economic plantations of soursop are used as host plants of wild silkworms, *Attacus atlas* Linn. for commercial production of wild silks in Indonesia (Mishra 2003).

Cultivation: Through seeds and transplanation of seedlings (Chundawat and Gautam 1993).

15. SHARIFA

Botanical: ***Annona squamosa*** Linn

English: Custard apple, Sweetsop, Sugar apple.		Hindi: Sharifa, Sitaphal
Bengali: Ata	Manipuri: Silaphal	Sanskrit: Sitaphal

SYSTEMATIC POSITION

Bentham & Hooker (1883)	**Engler & Prantl (1899)**	**Hutchinson (1959)**
Phanerogams	Phanerogams	Angiospermae
Dicotyledones	Dicotyledoneae	Dicotyledones
Polypetalae	Archichlamydeae	Lignosae
Thalmiflorae	Ranales	Annonales
Ranales	Annonanceae	Annonanceae
Annonanceae		

Geographical distribution: Indigenous to South America and West Indies yet completely domesticated over a great part of India (Brandis 1906, Kocchar 1981). It is cultivated in Assam, Bengal, Uttar Pradesh, and Maharashtra and very popular in all parts of South India (Pandey 1981, Rao 1991).

Habit: A medium sized wholly glabrous tree or large shrub, seldom growing more than 4 meter in height.

Habitat: Hot and relatively dry localities with soils of different types viz., black cotton soil, red soil, alluvial soil etc.

Root: Tap, branched.

Stem: Erect, woody, branched, deciduous, oil cavities present, glabrous, greyish white.

Leaves: Simple, alternate, ramal, exstipulate, ovate, or oblong – lanceolate, petiolate, 5-7 cm long, entire, apex acute, glabrous, venation reticulate unicostate, gland dotted, with a peculiar smell.

Inflorescence: Solitary or group of flowers arising from axils of the leaves or from old wood (cauliflorus)

Flower: Bracteate, pedicellate, regular, actinomorphic, bisexual, hypogynous, spirocyclic, greenish, on pedicels as long as flowers.

Calyx: Sepals 3, small, triangular, free, valvate, fleshy.

Corolla: Petals 6, in two whorls of 3 each, fleshy, thick, triangular, free, valvate, exterior petals lanceolate, and 2.5 cm long, 3 interior minute or sometimes wanting.

Androecium: Stamens indefinite, polyandrous, filament short, anthers long, connectives thick broad and projecting beyond the anther into a short sterile projection, dithecous, extrose.

Gynoecium: Carpels many, apocarpous, ovary sperior, each carpel unilocular, placentation marginal, style simple, stigma inconspicuous or beak like.

Fruit: Etario of achenes, yellowish-green, 5-10 cm in diameter, tuberculate, seed endospermic, oblong, brownish black.

Flowering: Hot season.

Fruiting: Autumn or cold season.

Floral formula: ⊕ ⚥ $K_3\ C_{3+3}\ A_{\alpha}\ G_{\underline{\alpha}}$

Uses: The fruit is delicious and highly nutritious. Although not commercially cultivated on a large scale, this fruit is still a very popular one in all parts of South India. The pulp of the fruit is white or cremy white, juicy, sweet and deliciously flavoured, from which jam and beverage and fermented liquor can be prepared. The wood is used as fuel (Kocchar 1981, Pandey 1981, Chandawat and Gautam 1993, Shukla and Mishra 1979). Roots are purgative and seeds are abortifacient (Sinha 1996). Seeds yield a fatty oil (Ambasta *et al.* 1986).

Cultivation: Through seeds and transplantation of seedlings (Chundawat and Gautam 1993).

16. DHAURA

Botanical: ***Anogeissus latifolia*** wall. Syn. *Conocarpus latifolia* DC.

Eng: Alexwood
Hindi: Dhaura, Dhaunta, Dhawa, Bakli
Gujarati: Zhavdo
Gond: Arma
Kannada: Dindaga, Dindal
Malayalam: Marukinchiram
Marathi: Dhaura
Tamil: Vellay Naga
Telugu: Yellamma, Chirimanu

SYSTEMATIC POSITION

Bentham and Hooker (1883)	Engler & Prantl (1899)	Hutohinson (1959)
Phanerogams	Phanerogams	Angiospermae
Dicotyledones	Dicotyledoneae	Dicotyledones
Polypetalae	Archichlamydeae	Lignasae
Calyciflorae	Myrtiflorae	Myrtales
Myrtales	Combretaceae	Combretaceae
Combretaceae		

Geographical Distribution: Sub-Himalayan tract from Ravi to Nepal and Sri Lanka ascending up to 3000 ft. ASL, in Central India and Western Peninsula up to 4000 ft. ASL on the Nilgiris. Absent in transgangetic plains, Bengal, Assam and Myanmar.

Habit: A medium sized tree usually, ocassionaly a large tree attaining the height of 80 ft.

Root: Branched tap root.

Stem: Erect, aerial, branched, cylinderical, woody, smooth.

Leaves: Simple, broadly elliptic, obtuse at both ends, pubescent when young, glabrous when fully grown, alternate or sub-opposite, blade 4.0-9.0 cm, petiole 0.6-1.8 cm long, occasionally acute but never acuminate, secondary nerves 8-14 pairs, tertiary nerves prominent on ventral side.

Inflorescence: Dense globose heads, 0.6-0.8 cm across on short peduncles, usually in axillary racemes, peduncles single or more from same axil, usually branched and very small as compared to leaves, rusty pubescent.

Flower: Bixexual, incomplete, sepaloid, actinomorphic, epigynous, small, white or gray, bracteolate.

Calyx: Sepals 5, gamosepalous, campanulate, calyx tube compressed at the base, extends beyond the ovary, subpersistent limbs 5, small and deciduous.

Corolla: Petals nil.

Androecium: Stamens 10 in two series, filaments filiform, exserted.

Gynoecium: Carpel-1, unilocular, ovary inferior, style filiform, simple, ovules 2, pendulous from top of the locule.

Fruits: Small, coriaceous, compressed, 2-winged, 0.4x0.6 cm approximately including the wings, rusty pubescent when young but almost glabrous after maturing, terminating in a beak built by the persistent calyx tube.

Flowering: May - Aug., Sometimes Nov. - Feb.

Fruiting: Next year.

Sprouting: Naked during dry seasons. Leaf fall occurs during February though foliage start becoming coppery during November. Sprouting occurs during last week of April - May.

Floral formula: ⊕ ⚥ $K_{(5)}\ C_0\ A_{5+5}\ G_{\bar{1}}$

Varieties: Three varieties of *A. latifolia* have been reported by Hooker (1878) as hereunder:

Var. *glabra:* Leaves smooth and glabrous ventrally.

Var. *villosa*: Leaves small, densely rusty pubescent on both sides, found in Mysore.

Var. *parviflora*: Leaves very small, app. 1.2 cm long, silky tomentose, found in Central India.

Uses: It serves as one of the important secondary food plants of tropical tasar silkworm, *Antheraea mylitta* D. Larvae of fourth to fifth stage feed and spin cocoons very well on Dhaura. Hence, in M.P. and other Central Indian parts larvae are first fed on Arjun or Asan and transferred to Dhaura in later stages when scarcity of primary food plants is felt. The Railey race of *A. mylitta* D. may be exclusively reared on *Anogeissus latifola* in Chattisgarh (Sengupta 1986, Thangavelu *et al.* 2000)

Medicinally, the crushed bark of Alexwood is used for curing cough and stomach pains (Singh 1990). The leaves of *A. latifolia* have great fodder value (Khanna *et al.* 1998). Timber is used for making agricultural implements and construction purposes and also as fuel wood (Maheshwari and Singh 1991, Nair and Jayakumar 1998, Khanna *et al.* 1998).

Cultivation: It propagates through seeds in nature and may be cultivated through nursery raising followed by transplanting seedlings in atleast 8'x 8' spacing.

According to Regional Forest Research Centre, Rajahmundary, soaking of seeds in cold water for 48 hrs. yields 13% peak germination and 10% overall germination (Rao *et al.* 1998).

17. SUPARI

Botanical: *Areca catechu* Linn.

Burmese: Kun-thee, kun	French: Arec	German: Arecapalme
English: Betel nut palm, The Areca palm		Hindi: Supari
Assamese: Tambúl, Guwa	Bengali: Supari, Gua	Sihnalese: Puvakka
Gujarati: Supari	Kannada: Adike	Malayalam: Adakka
Manipuri: Kwa	Marathi: Supari	Oriya: Gua
Punjabi: Supari	Sanskrit: Guvaka, Poogiphalam	
Tamil: Pakku	Telugu: Poka, Vakka	

SYSTEMATIC POSITION

Bentham & Hooker (1883)	Engler & Prantl (1899)	Hutchinson (1959)
Phanerogams	Phanerogams	Angiospermae
Monocotyledones	Monocotyledoneae	Monocotyledones
Calycinae	Principes	Corolliferae
Palmae	Palmae	Palmales
		Palmae or Palmaceae

Areca nuts have been an important article of trade since ancient times. Ancient Indian writings mention that people chewed sliced betel nut directly or with other ingredients *viz.,* coconut shavings, sweetener, clove, cardamom, fennel and other spices, wrapped in a fresh or processed betel leaf commonly referred to as 'Pan' in India, smeared with lime paste and katha from *Acacia catechu.*

Geographical distribution: A native of Malaya, betel nut palm is now cultivated mainly in South-eastern Asian cuntries viz., India, Bangladesh, Sri Lanka, Malaysia and Indonesia. In India, betel nut palm is cultivated in approximately 120000 ha mostly in coastal regions of South Bombay, Tamil Nadu, Karnataka, West Bengal and Assam.

Habit: A graceful, erect, monoecious, solitary (single stemmed) tall palm with a height of 12-30 meters with crown of leaves (Fig. 4).

Habitat: Tropical and sub-tropical coastal climate with high humidity, high temperature and saline soils.

Root: Adventitious, emerging from the base of the stem. The primary root soon dies off and is replaced by permanent adventitious roots.

Stem: Upright, tall, unbranched, marked with annular leaf scars, 0.5 meter in diameter, topped by a crown of leaves.

Leaves: 6-9, pinnately divided, feather-palm type, large, 2-3 meters long, leaf lets numerous, quite glabrous with one or several midribs, upper confluent, petiole large, stout, with broad sheathing base.

Inflorescence: Spadix, spadices much branched and below the leaves, covered within a spath, lowest spath large, glabrous, pale straw coloured, parallel veined, spathes torned in definite way with rapid growth of flower shoots.

Flower: Unisexual, actinomorphic, incomplete, hypogynous, male flowers in the upper part of spike, female flowers in lower larger parts of the inflorescence, trimerous.

Perianth: Tepals 6, in two whorls, polyphyllous, persistent, outer whorl smaller than the perianth leaves of inner whorl, male flowers yellow, 5 mm long, with six stamens in two whorls, female flowers 10 mm long, monocarpellary, ovary superior, unilocular, style short, stigma terminal.

Fruit: A fibrous drupe, ovoid, smooth, 4-6 cm long, orange or red, pericarp fleshy, mesocarp fibrous, enclosing a single, hard, grayish brown seed with a ruminate endosperm.

Floral formulae: Male flower: ⊕ ♂ P_{3+3} A_{3+3} G_0

Female flower: ⊕ ♀ P_{3+3} A_0 $G_{\underline{1}}$

Uses: The wood is used as timber and fuel wood (Chundawat and Gautam 1993). In tropical parts of the old world from Zanzibar to India and the Pacific islands, more than 400 million people chew sliced betel nut directly or with other ingredients in betel leaf. Chewing promotes salivation that becomes red in due course and improves digestion and acts as a stimulant and carminative. The slightly intoxicating effect on the Central Nervous System is due to the alkaloid arecoline (0.1-0.4%) present in betel nut. Excessive chewing may cause dental decay, pyrrhoea, alveolaris and oral cancer. Thearecoline is used extensively in veterinary medicine as an anthelmintic or a vermifuge. Other alkaloids viz., arecaidine, arecolidine, guvacine and guvacoline are also present in areca nut in very minute quantity (Kocchar 1981). The young seeds are laxative while mature seeds are good vermifuge. The powdered nuts are used as an anthelmintic and an antiseptic agent for human beings also (Kurian 1995). Paste of the nut after rubbing on the stone with water is applied to leucodermal patches. Young nuts are used in bowel complaints and offensive ulcers. Seeds are anthelmintic, nerve tonic and astringent. Bark powder is anthelmintic and usd for tapeworms in animals. Juice of the tender fruits is laxative. Burnt nuts are used in dentrifices, while decoction of roots is a cure for sore lips and employed for liver disorders (Sinha 1996).

The spathes are used to write upon, to wrap up parcels, and as covering leaf of cheroots in Myanmar (Brandis 1906). The husks are used for making hardboard and brown tanning material. The nuts are roasted, powdered and then made into tooth powder along with other substances. The beads and other fancy articles are also made from nuts while leaves are used for thatching and woven in to mats (Pandy 1981). This is also one of the principal tannin yielding plants (Kocchar 1981).

Cultivation: It may be cultivated through either seeds are seedling (Chundawat and Gautam 1993) and seedlings are transplanted during rainy season.

18. KATHAL

Botanical: ***Artocarpus heterophyllus*** Lam. syn. *A. integrifolia* Linn.

Burmese: Peingnai, Peinne French: Jacquier German: Jackbaum

English: Jack tree Hindi: Kathal, Kathar, Kanthal,

Sinhalese: Cos Assamese: Kanthal Bengali: Kanthal

Gujarati: Vanas Kannada: Halasu Konkani: Ponnoss

Kuki: Lamphong Malayalam: Chakka Manipuri: Theibong

Marathi: Phanas, Phannas Oriya: Ponoso Punjabi: Kathal

Santhali: Kanthar Tamil: Pilla, Pilapalam Telugu: Panasa

Sanskrit: Panasa Urdu: Katahal

SYSTEMATIC POSITION

Bentham & Hooker (1883)	Engler & Prantl (1899)	Benson (1957)	Hutchinson (1959)
Phanerogams	Phanerogams	Spermatophyta	Angiospermae
Dicotyledones	Dicotyledoneae	Angiospermae	Dicotyledones
Monochlamydeae	Archichlamydeae	Dicotyledonae	Lignosae
Unisexuales	Urticales	Urticales	Urticales
Urticaceae	Moraceae	Moraceae	Moraceae

Jackfruit is the largest edible fruit in the world. It is native to India (Rao 1991), yet its center of origin is supposed to be Malaya (Kocchar 1981).

Geographical distribution: Wild in dense forests along the Western Ghats up to 4000 ft. ASL and also in large and dense forests of Myanmar (Brandis 1906). It is widely grown throughout Kerala, Koraput and coastal districts of Orissa, Uttar Pradesh, West Bengal, Bihar and Andhra Pradesh.

Habit: It is an evergreen monoecious tree growing up to a height of 12-15 meter or more in a short time, glabrous, youngest shoots and mid rib with soft stiff hairs, milky juice present (Fig. 5).

Habitat: It grows well in the tropical, dense, moist forests with all types of soil.

Root: Tap root, branched, deep.

Stem: Erect, woody, branched, branches heavy, bark brown to black, rough, exfoliating in vertical and horizontal fissures, thick.

Leaves: Simple, thickly coriaceous, round or long with a blunt point (elliptic or obovate), entire, alternate, shining above, mid rib prominent beneath, blade 10-20 cm, narrowed into 1.5-2.5 cm long petiole, stipules large, early, caducous, covering the leaf bud.

Inflorescence: Unisexual heads, developed in large stipular deciduous sheaths, male stout, cylinderic, 5-15 cm long, female globose.

Flower: Small, inconspicuous, actinomorphic, tetramerous, hypogynous.

Perianth: Tepals 4, arranged in two whorls in male flowers; confluent with each other and the receptacle forming a stout tube and minute mouth in female flowers.

Androecium: Stamen 1, only in male flowers, anthers bithecous.

Gynoecium: Carpels 2, only in female flowers, often one abortive.

Fruit: Multiple, fleshy, edible, cylinderic (oblong or round), 30-75 x 15-30 cm, hanging on short stalk from the trunk and larger branches, consisting of receptacle (of inflorescence) and the fleshy perianth (of flowers) enclosing a small number of seeds, tips of perianths hard, free appearing as conical protuberances on the rind, weighing 25-30 kg, strongly scented, green when young, yellow or light brown on maturity.

Seeds: Reniform, oily, ovules solitary, pendulous from the apex, surrounded by a fleshy endosperm.

Flowering: Cold season (Dec.– Jan).

Fruiting: Spring season.

Floral formulae: Male flower: $\oplus$ ♂ $P_{2+2}\ A_1\ G_0$

Female flower: $\oplus$ ♀ $P_{(2+2)}\ A_0\ G_{\underline{1}+1\ \text{abort}}$

Varieties: The raw fruits of "oval", "round" and "raised scented" varieties are used as vegetables also. "Singapore" or "Ceylone" varieties are early bearing popular varieties since they produce fruits within 3-5 years of planting. Some varieties produce small fruits of less than one feet long, weighing 2-5 kg, some yield 1-2 feet long weighing 10-20 kg while some yield more than two feet long. In Kerala the fruits of 3 feet are common which weigh over 30 kgs.

Uses: The fruits and seeds are used as vegetable. The fruits are eaten when ripe. Delicious and tasty deep-fried chips, pappads and sweet bread are prepared out of ripe fruits in Kerala. Seeds are dried roasted and relished as a delicacy. The flakes of the ripe fruit are preserved in Sugar and honey (Rao 1991, Pandey 1981). Leaves are used as fodder and wood is used in turnery (Chundawat and Gautam 1993).

The decoction of the root is used for diarrhoea, fever and asthma. The milky latex of the tree is applied externally for the cure of ulcers, wounds, glandular swellings and insect bites, since it promotes suppuration and curing.

The unripe fruit is astringent while ripe is laxative. The swallowing of a fresh seed helps easy digestion and prevents dirrhoea after eating enough of ripe fruits. Roasted or boiled seed gives tone and vitality to the body and helps in checking abdominal pain. The ash of rachis/receptacle mixed with a pinch of lime is applied to boil for early suppuration (Kurian 1995, Sinha 1996).

Cultivation: Jackfruit may be cultivated either through seeds, seedlings or air layerings. The seeds are collected during May-June and directly sown during July-Aug. (Chundawat and Gautam 1993).

19. NEEM

Botanical: ***Azadirachta indica*** A. Juss. syn. *Melia azadirachta* Linn.; *Melia indica* Brandis.

Burmese: Bawtamaka, Tama
French: Agem lilas
German: Gemeinder
English: Neem tree, Margosa tree, Indian Lilac
Hindi: Nim, Nimb
Persian: Neeb
Sinhalese: Kohumba
Bengali: Nim
Gujarati: Limba, Limado
Kannada: Bevu, Ollebevu
Konkani: Nim
Malayalam: Vepa
Marathi: Limba, Nimbay
Oriya: Kakophalo, Nim
Punjabi: Bakam
Sanskrit: Nimbah, Nimba, Prabhadra
Santhali: Nim
Sindhi: Nimur
Tamil: Vepa
Telugu: Vepa
Urdu: Neem

SYSTEMATIC POSITION

Bentham & Hooker (1883)	Engler & Prantl (1899)	Hutchinson (1959)
Phanerogams	Phanerogams	Angiospermae
Dicotyledones	Dicotyledoneae	Dicotyledones
Polypetalae	Archichlamydeae	Lignosae
Disciflorae	Geraniales	Meliales
Geraniales	Meliaceae	Meliaceae
Meliaceae		

Neem is most virtuous tree. India has a rich heritage on Neem whose virtues are enshrined in our ancient religious classics like Vedas and Puranas, besides Ayurvedic classics. The great emperor Ashoka planted neem trees along the roadsides to give shade and shelter during 260 B.C (Rao 1991).

Geographical distribution: Neem, a native of dry forests of India, Pakistan, Malaysia, Myanmar and Thailand, now occurs in more than 30 countries. It occurs widely in the dry regions of our country and Irawadi valley from Prome upwards in Myanmar. It is cultivated and naturalized throughout Indian sub-continent between 50-1500 m ASL, especially in upper Gangetic plains and Punjab to the Jhelum but rare in west of the Sutlej.

Habit: A large glabrous evergreen popular tree growing to a height of over 20 meters with straight trunk, elegant in form, deciduous in certain climates. It coppices and pollards well and produces root suckers (Fig. 6).

Habitat: Tropical to sub-tropical dry areas with varying soil types (pH 6.2-8.5) and rainfall between 450 – 1150 mm. It can also withstand sub-zero temperature and adverse conditions occasionally (Anonymous 1985).

Root: Tap, branched, deep rooted.

Stem: Aerial, erect, straight, branched, solid woody, cylinderical, glabrous when young, rough when matured, transversely cracked with longitudinal fissures, heartwood red, hard.

Leaves: Compound, imparipinnate, petiolate, exstipulate, long and pendant, crowded at the ends of branches, young leaves pale, tender, green and shining on maturity, leaflets linear or falcate lanceolate, 7-9 pairs, often alternate or sub-opposite, oblique, with acute apex and serrated margins, venation unicostate rediculate, the odd leaflet often wanting.

Inflorescence: Axillary cyme/panicles shorter than leaves.

Flower: Pedicellate, ebracteate, complete, hermaphrodite, white, actinomorphic, fragrant, hypogynous, pentamerous.

Calyx: Sepals 5, free (slightly connate at the base), green, imbricate.

Corolla: Petals 5, free (slightly connate at the base with staminal tube) imbricate, white, fragrant, fragrance sweet.

Androecium: Stamens 10, monoadelphous, staminal tube 10 dentate, anthers free, inserted within the tube opposite its teeth, introse, basifixed, bithecous, a disc present between stamens and ovary.

Gynoecium: Carpels 3, syncarpous, ovary superior, trilocular with 1-3 ovules in each loculus, placentation axile, style single, stigma 3 lobed, capitate, fleshy.

Fruit: A drupe, fleshy, 1.2 – 2.0 cm long, green in the beginning, bright yellow when ripe, single seeded, kernel cartilaginous and colyledons plano-convex, notched at base.

Flowering: Feb. – July.

Leaf fall: Feb.-March for a brief period.

Sprouting: March - April.

Fruiting: March - August.

Floral formulae: ⊕ ⚥ K_5 C_5 $A_{(10)}$ $G_{(\underline{3})}$

Uses: The photosynthetic rate of Neem is high due to its large leaf apex and it gives out more oxygen than most trees. Hence, it is considered as a good purifier of air and serves as an excellent shade tree for avenues and parks (Randhawa 1983). The neem tree is a good soil ameliorant, increases soil fertility and water holding capacity, makes acid soil neutral and serves as a very good windbreak. (Ketkar 1999-2000).

The wood of neem makes a good timber. It is highly durable and used for construction and agricultural implements (Jain 1968). Derivatives from the leaves, stems, fruits, seeds and other parts of neem have found use in medicine, toiletries, cosmetics, pharmaceuticals, plant and animal nutrition, pest control, energy generation and several other areas, and such uses are on the increase (Rawat *et al.* 1999-2000).

The important isolated phytochemical constituents from neem known for causing potent biological effects of their preparations for agricultural and medicinal purposes are as here under (Vander nat *et al.* 1991, Thakur *et al.* 1981):

1. **Diterpenes**:Sugiol, nimbiol (bark)
2. **Triterpenes/steroids**: 24-methylene cycloartanol (heartwood), β-sitosterol and its β-D-glucoside (heartwood, leaf), cholesterol (flowers), stigmasterol (leaf).

3. **Limonoids**: Meliantriol (seed oil), nimbocinone (fresh winter leaves), nimbindin (seed oil), azadirachtin (fresh leaves), nimbandiol (seed oil, dried leaves).
4. **Flavanols**: Nimatone (heartwood), kaempferol, quercitin and myrecetin (flowers).
5. **Flavanol glycoside**: Glycosides of kaempferol, quercitin and myrecetin (flowers) and glycosides of kaempferol and quercetin (leaves).
6. **Dihydrochalcones**: Nimbochalcin (fruits)
7. **Tannins**: Condensed tannins, gallic acid, (+)- catechin, (-)-epicatechin, (+)- callocatechin, (-) epigallocatechin (bark).
8. **Carbohydrates and proteins**: Complexes of proteins and hetero-polysaccharides, D-glucose, D-glucouronic acid, L-arabinose, L-fucose, mannose, xylose, rhamnose and amino acids.
9. **Sulphurous compounds**: Cyclic tri and tetrasulfides (leaves).

Medicinally, decoction of the neem bark is antiperiodic, invigorating and bitter, tonic, astringent used to expel worms from the body, burning sensation near the heart, fever (malaria), cough, ulcers, inflammations and leprosy. Root bark is alterative. The infusion of leaves is used as an insecticide, for eye problems, inflammation of the liver and skin disease, eczema and rheumatism. Ripe or unripe fruit is used as a purgative, to expel worms emolient, anthelmintic and also for urinary problems, tumours, piles and toothache (Chopra *et al*. 1956, Jain 1968, Kurian 1995 and Sinha 1996).

Recently, neem extracts have been found to be hypoglycaemic, anti-inflammatory, antipyretic and spermicidal (vaginal contraceptive). Isolated components like ninbidin has been found to be active against various skin diseases, nimbidin and sodium nimbidinate caused slowing of heart rate while nimbidol showed esterogenic and antipyretic activities. Azadirachtins and other isolates showed antifeedant activity (Thakur *et al*. 1981, Rembold 1989, Sharma and Saxena 1959). Hence, neem appears to have potential for inducing male infertility, vaginal and oral contraception and prevention of implantation and insecticides (Evans 1989).

Sometimes, sap oozing from neem stem in the hot months can be collected and stored. The tree also yields an amber coloured gum, which is used medicinally. The seeds yield the famous "margosa oil" with disagreeable garlic like flavour known to be effective in treatment of leprosy and skin diseases, besides being useful in the manufacture of soap. The margosa oil is used for treatment of 'mange' in dogs. The oil cake is considered as a useful biofertilizer and biopesticide. It reduces the incidence of citrus canker and citrus leaf minor. Dried leaves can ward off insects when they are put in clothes, drawers and cupboards; leaves in poultice form are used for healing of wounds. Green twigs are used as tooth brushes for cleaning teeth and mouth, particularly in pyrrhoea. Leaves are also used as a vegetable in Bengal (Randhawa 1983).

Leaves are considered as antiviral and used in pox. A paste of leaves with fresh curcuma is rubbed on the body before taking bath to prevent infection by pox. Leaves fried or boiled are taken as an antidote to stomach diseases during summer. Leaves are antiseptic, applied to boils as poultice and their decoction is given for ulcers and eczema. Gum is demulcent, tonic, used in catarrhal affections. Flowers are tonic and stomachic (Sinha 1996). The oil obtained from the seeds is used as an antiseptic. The gum, bark, leaves and seeds are used in snake bite (Pandey 1981).

The Ministry of Agriculture, Govt. of India, has recognized Neem as an insecticide and cleared 40 neem-based products. Vikwood Botanicals Inc., USA, has also commercialized "Azadirachtin" as "Margosan-O" pesticide (Ketkar 1999-2000). Extensive work on field trials of neem-coated urea on different crops established that neem products are eminently suitable as nitrogen loss inhibitors due to their nitrofication inhibition properties. This facilitates considerable increase in fertilizer use efficiency of the plants in the soil. With neem coating, anticaking property is also imparted in urea. Thus, with the current thrust on sustainable agriculture and organic farming, neem has a great practical significance in future (Suri 1999-2000).

Cultivation: Artificial regeneration of Neem is easy and is being resorted through direct sowing of seeds or through transplantation of air layerings, cuttings, seedlings, stumps and tissue cultured plants (Rawat *et al.* 1999-2000).

(a) **Direct sowing**: Ripe fruits are collected either from the trees or by sweeping the floor underneath the tree as soon as they fall, during June-July. 2000-3000 fruits weigh a kilogram. Fresh seeds may be dibbled immediately 1-5 cm deep at a spacing of 3x3 m or in lines 3 m apart on mounds, ridges, trenches or in circular saucer shaped pits since they germinate within a week or two during rains. Seeds stored with low moisture content retain their viability for about 3 months. Stored seeds with high moisture content show 78-84% germination. Germination % declines sharply after 12 weeks (Radvankski and Wickens 1981). Depulping and cleaning the seeds improve germination (Singh 1982). Quantity of seed required for such sowing (3 x 3 m) is 3-4 kg/ha (Singh 1982, Anonymous 1981). It can tolerate salts upto 6.5 d S/m at germination stage. In case of transplants, 7-10 cm high seedlings are planted with the first rain in pit of 30 cm^3 treated with 50 gm of 10% BHC dust and soil drenched with 0.2% chlorophyriphos before planting.

(b) **Vegetative propagation**: Neem coppices well and shoot cuttings and air layering with use of IBA and IAA have proved easy and moderately successful. It can also be propagated through root suckers and stem cuttings. Stumps of two-year-old plants yield high survival rate (Troups 1921, Rao 1958, Shivaganam *et al.* 1989, Palaniswamy *et al.* 1995). Approach grafting, air layering and root suckers yielded encouraging results with 85, 53 and 32 per cent respectively. Though neem is considered to be non-producer of root suckers, disturbance to root system by way of pruning while digging a trench produces one sucker/meter of trench depending upon the size of the mother tree and the rate of growth (Nadagoudar and Nataraja 1999).

(c) **Tissue culture**: Vegetatative propagation of neem using conventional methods is not greatly successful (Narayan and Jaiswal 1985). Regeneration of shoots from adult embryos and decapitated seedlings (Rangaswamy and Promila 1972), that of plantlets from shoot tips (Nalini *et al.* 1999-2000), leaflet callus (Narayan and Jaiswal 1985), leaflet callus, cotyledons and stem segments (Drew 1993, Joshi and Thengane 1993) and nodal segments of adult tree through axillary bud proliferation and subsequent rooting (Tomar 1995, Nalini *et al.* 1999, 2000, Arya and Arya 1998) have not only been achieved but effective nursery management (Ravichamy *et al.* 1999-2000) and large scale plantation of tissue culture – neem plantlets have also been achieved (Lakshmisubramanian *et al.* 1999-2000).

20. KANTA-BANS

Botanical : *Bambusa arundnacea* Willd. Syn. *B. spinosa,* Roxb

Burmese: Kyakat-Wa | English: Thorny bamboo | Hindi: Bans, Kantabans

Assamese: Kata, Kota, Kotoha | Bengali: Ketua, Kutuasi | Gujarati: Bans

Kannada: Bidru, Dongi, Hebbidru, Bidungulu | Konkani: Kalak, Padhai

Malayalam: Mula, Illi | Manipuri: Saneibi | Sanskrit: Vansh

Marathi: Kates Tokar, Bans | Tamil: Mundtaveduru, Bongu, Mungil

Telugu: Bongu-veduru, Pente-veduru

SYSTEMATIC POSITION

Bentham & Hooker (1883)	**Engler & Prantl (1899)**	**Hutchinson (1959)**
Phanerogams	Phanerogams	Angiospermae
Monocotyledones	Monocotyledoneae	Monocotyledones

Glumaceae	Glumflorae	Graminales
Gramineae	Gramineae	Gramineae

Geographical distribution: Indigenous to both Indian Peninsulas in plains and lower hills ascending up to 3000 ft. ASL. Found throughout India but cultivated largely in Northern, Central and Southern India, Orissa, Assam, Mizoram, Meghalaya and many parts of China and Myanmar.

Habit: Densely tufted, gregarious, tall, perennial arborescant grass with culms reaching up to 16-40 meters and sharp spines at the nodes in the branchlets (Fig. 7).

Habitat: Chiefly the banks of rivers, nalas and moist valleys on alluvial soils.

Root: Adventitious, fibrous, fasicled, stilt.

Stem: Underground woody rhizome giving out tillers, the aerial stems known as culms erect, fistular (hollow), cylinderical (terete), distinctly joined, woody, internodes 30-45 cm long, 10-15 cm in diameter, walls 2.5-5.0 cm thick, the nodes in lower part bear dense half whorls of stiff almost leafless branches, the upper part of culms bear large, alternating distichous, leafy branches, bending over.

Leaves: Small, thin, less than 20.0 x 2.0 cm, divided into leaf base (sheath) and leaf blade (lamina); leaf/culm sheaths alternating, coriaceous, 22-30 cm long, shorter than full grown internodes, young sheaths hairy, orange-yellow externally, shining and prominently ribbed internally; leaf blade triangular, shorter than sheath, upper side clothed with a dense brown felt of bristly hairs, nerves 30-45, usually 33-39 on 0.7 cm, venation parallel.

Inflorescence: Numerous spikelets attached to a main axis called rachis, spikelets sessile, 1.25-2.5 cm long, glabrous, shining, each spikelet bear several florets (flowers) attached to a central stock known as rachilla, sterile pair of glumes found at the base of spikelets, glumes as a rule not ciliate, a series of florets found above the glumes, florets have lemma and palea at their bases, lemma is the outer bract of the floret, palea slightly longer than glumes, the ciliate keels prominently protruding.

Flower: Sessile, bracteate (bracts are represented by lemma and palea), hermaphrodite, zygomorphic, hypogynous, irregular.

Perianth: Tepals 3, minute, membranous, scale like. The perianth leaves known as lodicules.

Androecium: Stamens 6, dithecous anthers yellow and versatile, filaments free, slender.

Gynoecium: Monocarpellary, unilocular ovary superior, almost cylinderic, single ovuled, ovules anatropous, style glabrous, short with 3 long plumose stigmas, placentation basal.

Fruit: Berry

Seeds: Endospermic, 0.2-0.3 cm long, enclosed in the persistent glume and palea.

Floral formula: •|• ⚥ P_3 lod. A_6 $G_{\underline{(1)}}$

Bamboos flower when 60 years old and a general flowering takes place in periods of 30-32 years in one district (Brandis 1906). It is believed in Mizoram and other North-eastern states that flowering in bamboo increases rat population enormously which leads to human starvation due to destruction of cereals and grains by them in large scales.

Uses: Bamboo are poor man's timber, used as rafters, scaffolding, roofing, walling, flooring, matting, basketry, cart hoods, cordage etc. Tender shoots are eaten and when bamboo clumps flower gregariously in 20-40 years, after which they die, the seed is collected and eaten as grain. They are also used for making brides in north-east India. Just as coniferous woods used in other countries, they have comparatively long fibre suitable for making paper. Certain rayon mills use bamboos for dissolving pulp as rayon can be made from any cellulosic material without regard to the fibre length (Sagreiya 1967).

The fermented shoot is used as an antidote to the injuries of nails caused by swords or arrows. The boiled decoction of fermented shoots is used in ringworms, tumors, meningitis and alopecia. Leaves are emmenagogue used in haematemesis and given to horses in cough and cold. Bamboo manna is useful as tonic, pyrigesic, expectorant and an antidote to snakebite. Decoction of the planr is abortifacient. Young shoots applied after pounding as poultice dislodge worms from ulcers. Decoction of young buds is used for normal menstrual flow (Sinha 1996).

The fermented young bamboo shoots are cherished by the local people after cooking in North-eastern India especially Manipur. Indigenous techniques are associated with the process of preparation of two products i.e., Soibum and Soidom. While in Soibum period of fermentation varies between 12-14 days, in Soidom it is 2-3 days. Soibum can be preserved for about a year, while Soidom only for a few days. Beyond such specific periods of preservation, products become unsuitable for human consumption but may be used for the treatment of dysentery, diarrhoea and healing of tumours in cattles (Sinha 1996).

The bamboo caps of Mizoram are famous. The bamboo baskets, flower pots, trays, vases, caskets and even ornaments are famous, artistic as well as decorative articles from north-eastern India. Bamboos also find a place in cultural activities of Mizo people. "Cheraw" the most colourful and distinctive "Bamboo dance" of Mizoram involves long bamboo staves, dancing skill and alertness of mind during the dance (Pandey 1981).

Cultivation: The *Bambusa arundinacea* may be usually propagated through woody rhizomes or saplings and rarely through seeds or seedlings (Brandis 1906). Mostly seedlings raised from tissue culture methods are obtained since bamboo seeds possess very short period of viability and flowering is gregarious. About 15-20 seedlings are planted in 1 m^2 of raised nursery beds but one week before planting the nursery beds are drenched with 0.1% aldrex and 0.05% bavistin to prevent termite and fungal attack respectively. Watering should be done 2-3 times a day and nursery bed is provided with a thatch to protect the seedlings from direct sunlight. For transplantation, the seedlings should be uprooted and transplanted in 45 x 5 cm pit in the fields at 6 m x 6 m spacing with 250 seedlings/ha before the onset of monsoon. The transplanted seedlings should be watered regularly in the morning and evening. Weeding should be done as and when required. After one year, the plantation is adequately irrigated with 15 days interval. Protection against damage of rodents, grazing and browsing animals may be provided by brushwood fence.

21. UDHA BANS

Botanical: ***Bambusa vulgaris*** Schrad. Syn. *B. thouarsii,* Kunth.

Burmese: Wah French: Bombou German: Baambos

English: Feathery bamboo, Golden, Bamboo Hindi: Udha bans

Assamese: Bnah Bengali: Basini bans, Bariala

Gujarat: Toncor Kannada: Bombu Konkani: Bombu

Marathi: Vansa, Kullaka, Kalaka Manipuri: Jai-baruwa

Malayalam: Mula Tamil: Ponmungil Telugu: Bongu

SYSTEMATIC POSITION

Bentham & Hooker (1883)	Engler & Prantl (1899)	Hutchinson (1959)
Phanerogams	Phanerogams	Angiospermae
Monocotyledones	Monocolyledoneae	Monocotyledones
Glumaceae	Glumiflorae	Glumiflorae
Gramineae	Gramineae	Graminiales
		Gramineae

Geographical distribution: Its home is uncertain. It is wild in Kerala forests but cultivated in hotter parts of India and North-eastern region. Usually planted in gardens as an ornamental plant.

Habit: Gregarious, tufted but clumps not dense, often polished and striped green and yellow, 7-17 meters high.

Habitat: Tropical moist conditions with alluvial, black, red or hilly soils.

Root: Adventitious, fibrous, fasicled, stilt.

Stem: From underground stem (rhizome) new aerial, erect, fistular shoots (culms) are put forth, internodes 25-45 cm long, 7.5-12.5 cm in diameter, culm sheaths pubescent outside, pubescence brown, when young streaked with green and yellow stripes, 15-25 cm long, top concavely truncate, the leafy branches formed in cluster at each node.

Leaves: Simple, thin, divided into leaf sheath and blade; leaf sheeth appressed, white hairy, leaf blade triangular with falcate ciliate auricles at the base, 30-40 X 2.5-5.0 cm, hairy below when young, nerves 21-39 on 0.7 cm, transverse veins visible on the lower suface.

Inflorescence: Spikelets compressed, 0.6-1.6 x 0.5 cm, 6-10-fid.

Flower: Flowering seldom, flowers sessile, bracts divided into lemma and thin palea, hermaphrodite, zygomorphic, hypogynous, if flowering occurs, it is at the nodes.

Perianth: Lodicules 3, minute, membranous, scale like free.

Androecium: Stamens 6, filaments slender, free, anthers dithecous.

Gynoecium: Monocarpellary, styles long, hairy, stigmas 3, plumose, ovary superior, unilocular.

Fruit: Berry.

Floral formula: •|• ⚥ $P_{3\ \text{lod}}\ A_6\ G_{\underline{(1)}}$

Uses: The split culms are made into mats and baskets. The young buds are consumed as vegetables (Pandey 1981). Roots and shoots are emollient, diuretic and diphoretic. Bark is astringent, used in haemorrhage, excessive menstruation and nausea and vomiting (Sinha 1996).

The decoction of the stem is applied externally to inflammed joints while decoction of roots taken internally promotes the flow of urine, softens and gives relief to inflammed parts. Intake of decoction of bark prevents nausea and vomiting and cures piles. The edible young shoots strengthen and give tone to the stomach and used for treatment of respiratory diseases (Kurian 1995).

Cultivation: It may be usually propagated through nodal segments woody rhizomes culm cuttings, saplings, and occasionally or rarely through seeds and seedlings. It has also been propagated successfully through tissue culture. (Mukunthakumar *et al.* 1999). According to Ambad and Jambbale (1997), the single node cuttings of 15 cm length, pretreated with bavistin 0.1% for 15 minutes and planted horizontally in polybags exhibits induction of roots and shoots and the growth of such vegetative plantlets is faster than conventional seedlings raised from seed of fast growing *Dendrocalamus strictus* bamboo.

22. KHAIRWAL

Botanical: *Bauhinia purpurea* Linn.

Burmese: Mahahlegani

English: Purple Bauhinia, Geranium tree

Hindi: Gulabi kachnar, Khairwal, Gairal, Kaliar, Kwillar.

Bengali: Deva, Koiral, Devakanchan, Raktakanchan

Kannada: Sarul, Kanchivala   Malayalam: Chuvannamandaram

Manipuri: Chingthao, angangba

Marathi: Deva, Attamatti   Oriya: Sono

Punjabi: Vanaraja, Kolar   Tamil: Mandari

Telugu: Kanchanm, Bodanta   Santhali: Singyara   Sanskrit: Kovidhara

SYSTEMATIC POSITION

Bentham & Hooker (1883)	**Engler & Prantl (1899)**	**Hutchinson (1959)**
Phanerogams	Phanerogams	Angiospermae
Dicotyledones	Dicotyledoneae	Dicotyledones
Polypetalae	Archichlamydeae	Lignosae
Calyciflorae	Rosales	Leguminales
Rosales	Leguminosae	Caesalpiniaceae
Leguminosae	Caesalpinioideae	
Caesalpinioideae		

There are many references to this tree in the Sanskrit dramas of Kalidasa, who called it Kovidhara (Randhawa 1983).

Geographical distribution: Sub-Himalayan tract from Indus eastwards ascending upto 5000 ft. ASL. Found practically wild throughout India especially in Assam, Meghalaya, (Khasi hills), hill ranges of Central and Southern India and deciduous forests of Western Peninsula. Also found in Bangladesh. Cultivated largely in park and gardens for its beautiful rosy purple flowers.

Habit: A medium sized evergreen tree with rounded crown.

Habitat: Sub-tropical hills/foot hills as well as tropical deciduous forests. It flourishes well in high well-drained soil but easily affected by low temperatures.

Root: Tap, branched.

Stem: Erect, aerial, densely branched, crowned, cylinderical, solid, bark smooth, ashy grey or brown, heartwood brown, hard, durable.

Leaves: Simple, coriaceous, bilobed, cleft about half-way down into two rounded lobes, lobes often overlapping along the inner margins, nerves 9-11, radiating from the base. Splitted leaf down in the middle gives a hoof like appearance, which is easily recognized.

Inflorescence: Short racemes in terminal panicles.

Flower: Hermaphrodite, zygomorphic, hypogynous, deep pink or rosy purple (sometimes white in *B. triandra* Roxb), large and showy, pedicels bracteolate, 0.8-1.6 cm.

Calyx: Sepals 5, gamosepalous, calyx tube obconical, 0.7 cm long, limb twice the length of tube, cleft to the base into two reflexed segments, one emarginate with two wings, the other 3 toothed (winged).

Corolla: Petals 5, oblanceolate, clawed or pointed, free, deep pink or rosy purple, base of one petal white.

Androecium: Stamens 10, fertile 3, occasionally 4, anthers basifixed.

Gynoecium: Monocarpellary, ovary superior, unilocular, placentation marginal.

Fruit: Pod, 15-30 cm long, looking like French beans, contains about 12-15 suborbicular flat seeds.

Flowering: Sept.-Nov., profuse flowering in November.

Floral formula: •|• ⚥ $Br_1\ K_{(5)}\ C_5\ A_{3+7\ stm}\ G_{\underline{1}}$

Uses: It serves as beautiful ornamental tree. Extensively planted as avenue tree in Chandigarh and South India. The bark is used for dyeing and tanning and sometimes for fibre also. The flowers are used as vegetables or pickles. The bark of underground root is poisonous. The wood is used for agricultural implements, matches, rafters, scant lings and for building purposes (Randhawa 1983, Ambasta *et al.* 1986).

The bark extract is used in insect bite, leucorrhoea, leprosy and irregular menstruation. The roots are carminative. Bark is used in diarrhoea. Flower buds are laxative and anthelmintic (Sinha 1996).

The decoction of the root is used for expelling gases, flatulence and gripping pains from the stomach and bowels. Decoction of bark is used for washing ulcers also. The bark or root and flowers mixed with boiled rice water are used as a maturant for boils and abscesses (Kurian 1995). Leaves used as a fodder and seeds yield a non-drying fatty oil (Ambasta *et al.* 1986).

Cultivation: Seeds sown in the beginning of monsoons germinate within 4-10 days. Seedlings are transplanted during rains.

23. KACHNAR

Botanical: ***Bauhinia variegata*** Linn.

Burmese: Bwechn — French: Arbre de St.Thomas

German: St.Thomasbaum — Persian: Kachnual

Eng.: Moutain Ebony — Hindi: Kachnar, Baisakhi

Bengali: Kovidara — Gujarati: Kovin dara, Bidul, Raktakanchan

Kannada: Arisinatigie, Kanchavala

Konkani: Kanchan

Marathi: Kanchan, Raktakanchan, Kachnar

Malyalam: Kovidaiam, Chuvannamandaram

Manipuri: Chingthaoangouba

Mundari: Burju — Mizo: Vau-favang — Oria: Brorod

Sanskrit: Kovidara, Kavidhara — Santhali: Jingya

Tamil: Segapuma, Mandarai, Segapu - manchori

Telugu: Bodanta, Mandari, Devakanchanamo, Bodantham

Urdu: Kachnal

SYSTEMATIC POSITION

Bentham & Hooker (1883)	Engler & Prantl (1899)	Hutchinson (1959)
Phanerogams	Phanerogams	Angiospermae
Dicotyledones	Dicotyledoneae	Dicotyledones

Polypetalae	Archichlamydeae	Lignosae
Calyciflorae	Rosales	Leguminales
Rosales	Leguminasae	Caesalpinioideae
Leguminasae	Caesalpinioideae	
Caesalpiniaceae		

'Kalidasa' specifically mentioned about "Kavidhara" in his play 'Shakuntala' (Abhigyan Shakuntalam), among the beautiful flowering plants (Rao 1991). *Bauhinia* is named after two botanists, John and Casper Bauhin who were twin brothers, while *variegata* refers to variegated flowers.

Geographical distribution : Throughout India, indigenous, common in Sub-Himalayan tracts from Indus towards Sikkim, Assam up to 4000 ft. ASL, Western Peninsula, Chotanagpur (Jharkhand), Chattisgarh and Madhya Pradesh. Cultivated largely in Himachal Pradesh (Kangra valley).

Habit: A deciduous middle-sized tree with irregular crown, wild as well as cultivated (Fig. 8).

Root: Branched tap root.

Stem: Erect, aerial, branched, branches irregular, bark dark brown and slightly rough.

Leaves: Simple, alternate, deeply cleft from the tip 1/3 the way down resulting into two obtuse lobes, subcoriaceous, pubescence grey, as broad as long, 11-15 nerved, radiating from the top of the stalk, petioles 2.5-3.5 cm long.

Inflorescence: Few flowered raceme, lateral, shortly peduncled.

Flower: Bisexual, complete, large, bracteate, pedicellate, pedicles 0.6-1.2 cm long, zygomorphic, hypogynous, generally pentamerous.

Calyx: Sepals 5, gamosepalous, calyx tube 1.8-2.5 cm, cylinderic, as long as the ovate spathaceous 5-toothed limb, valvate.

Corolla: Petals 5, subequal, obovate or oblanceolate, 3.5-5.0 cm long, 2.5-3.0 cm broad, free, white, beautifully variegated with red and yellow (4 petals light purple-red, the 5th and larger deep purple-red). Ocassionally with purely white flowers (*B. candida* Roxb.) in which the 5th petal variegated with yellowish green or dark purple, aestivation imbricate.

Androecium: Stamens 3-5, anthers versatile, filaments free, long, filiform, dehiscence longitudinal.

Gynoecium: Carpel-1, ovary superior with a long stalk and style, stigma minute, unilocutar, ovules many, placentation marginal.

Fruit: Pods flat, 15-30 cm long, 1.8-2.5 cm broad, hard, glabrous, dehiscent, 10-15 seeded.

Seeds: Non-endospermic, embryo large.

Leaf fall: February.

Flowering: Feb. – March.

Floral formula: •|• ⚥ $K_{(5)}$ C_5 A_{3-5} $G_{\underline{1}}$

Uses: It serves as one of the secondary food plants of tropical tasar silkworms, *Antheraea mylitta* D. in forests (Seth 1995 a, Thangavelu *et al.* 2000) . The leaf and flower buds are popular as a vegetable. The timber is hard and used for fuel and for making agricultural implements. The seeds yield oil. The leaves are used as wrappers for bidis. The bark yields fibre and used for tanning and dyeing. The tree also yields a gum possessing similar properties to that of cherry gum. Flower buds pickled. Leaves are source of fodder and used for bidi manufacture (Ambasta *et al.* 1986).

It is planted as an avenue tree, which presents heaviely sight of beautiful pink, white and purple varieties. (Randhawa 1983)

Medicinally, the decoction of root is used to prevent obesity and used as a carminative or as an agent for expelling gases, flatulence and gripping pains from stomach and bowels. The decoction of bark is used as tonic and anthelmintic and also used against scrofula, tuberculosis, skin ailments, ulcers and leprosy. The infusion of leaves is used as a laxative and for the cure of diarrhoea, dysentery and piles. (Kurian 1995). Decoction of bark mixed with honey is given in leucorrhoea. Flower and floral buds are laxatives. Dried buds are used for diarrhoea, dysentery and piles. (Sinha 1996).

Cultivation: It can be propagated through seeds as well as cuttings, though cuttings root with difficulty. It grows fastly and attains 6-8 ft. height during first season and flowers during second year of its plantation. (Randhawa 1983).

24. ANNATO

Botanical: *Bixa orellana* Linn.

Burmese: Thidin	English: Annato, Arnotto	Hindi: Latkan
Bengali: Latkan	Kannada: Rongmale	Marathi: Sendri
Tamil: Japhara	Telugu: Japhara	

SYSTEMATIC POSITION

Bentham & Hooker (1883)	**Engler & Prantl (1899)**	**Hutchinson (1959)**
Phanerogams	Phanerogams	Angiospermae

Dicotyledones	Dicotyledoneae	Dicotyledones
Polypetalae	Archichlamydeae	Lignosae
Parietales	Parietales	Bixales
Bixineae	Bixaceae	Bixaceae

Peoples of the Carribean islands and tropical America stained their bodies red by using the organge-red dye Annatto during war times to frighten their enemies. Therefore, they were recognized as red Indians (Kocchar 1981).

Geographical distribution: Indigenous to tropical America and West Indies, but long cultivated and naturalized in India. Annatto is now cultivated in Mexico, Brazil, Guiana (Cayenne) and the Antilles. In India, it is cultivated in Karnataka to a small extent. However, it is found as an escape in Travancore, Coramandel and Malabar Coast and in some districts of Maharastra, Bengal, Assam and Manipur.

Habit: A small evergreen tree attaining a height of 5 meter.

Habitat: Moist locations i.e., along drainage, ditches etc (Fig. 9).

Root: Tap, branched

Stem: Aerial, erect, branched, woody, cylinderical, solid, branches few, rough.

Leaves: Simple, cordate, acuminate, stalked, alternate, sometimes with reddish veins, venation unicostate reticulate, glabrous above, slightly hairy below on maturity.

Inflorescence: Panicles terminal, spacious, erect.

Flower: Bisexual, large, pedicellate, 2.5-5.0 cm in diameter, mallow like pinkish or rarely white, actinomorphic, pentamorous, hypogynous.

Calyx: Sepals 5, polysepalous, green.

Corolla: Petals 5, polypetalous, contorted in bud, pink or rarely white.

Androecium: Stamens 5, free, anthers dithecous, filaments long.

Gynoecium: Monocarpellary, ovary superior, unilocular, ovules many, placentation parietal.

Fruit: Capsule nearly globose, reddish brown or occasionally green, clothed with soft prickles, two valved, about 5 cm long, seeds 30-50 enclosed within, seeds angular, red, about 0.5 cm in diameter, surrounded by bright scarlet pulp, the aril.

Flowering and fruiting: Rainy season.

Floral formula: $\oplus$ ⚥ $K_5\ C_5\ A_5\ G_{\underline{1}}$

Varieties: 1. White flowered with green capsules.

2. Pink flowered with red capsules.

Uses: The orange red dye, annatto is obtained between 5-6% of seeds by weight. The red pulp is soaked in hot water for several days, allowed to ferment for about a week and subsequently sedimented dye is dried and cut into cakes. The pure dyestuff may be obtained by dissolving the red pulp in organic solvent like chloroform and then by evaporating the solvent to a paste like material. Two pigments viz., bixin and orellin are found to occur in this dye.

Annatto paste is used in lipstick and edible products like butter, margarine, cheese and chocolate due to its tastelessness and non-toxic nature. This dye contains about 2% vitamin-A, hence it is used with rice usually deficient in vitamin-A. It is also used in soaps, pomades, fabrics, paints, varnishes and lacquers for imparting colours. Honeybees prefer their flowers. The plant is often grown as hedge plants. Annatto is locally used as a remedy for skin diseases and burns (Kocchar 1981). Leaves are used in snakebite. The fruit is astringent and purgative. Root bark is antiperiodic and antipyretic. Seeds are cordial, febrifuge and remedy for gonorrhoea while leaves are useful in jaundice (Sinha 1996). Also suitable for floor polishes, shoe polishes and hair oils etc. (Ambasta *et al.* 1986).

Cultivation: Through seeds, seedlings and cuttings (Kocchar 1981). The direct somatic embryogenesis and plant regeneration was also achieved from excised immature zygotic embryos (Muniswamy and Sreenath 2002).

25. SEMUL

Botanical: ***Bombax ceiba*** Linn. *B. malabaricum* D.C., *Salmalia alabarica* (D.C) Schott & Endl., *B. hepataphyllum* Cav.

Burmese: Letpan

Eng.: Silk cotton tree, Red silk cotton | Hindi: Semul, Shembal, Semur

Assamese: Dumboil	Bengali: Simul	Gujarati: Shemolo
Garo: Boichu	Kannada: Burla, Sauri	Kol: Idel
Kuki: Phumchong	Malayalam: Illavu, Pulamaram	
Manipuri: Tera	Marathi: Sayar	Mundari: Edelsong
Tamil: Illavu, Pulai	Telugu: Burgha, Buraga	Sanskrit: Shalmali

SYSTEMATIC POSITION

Bentham & Hooker (1883)	**Engler & Prantl (1899)**	**Hutchinson (1959)**
Phanerogams	Phanerogams	Angiospermae
Dicotyledones	Dicotyledoneae	Dicotyledones

Polypetalae	Archichlamydeae	Lignosae
Thalmiflorae	Malvales	Tiliales
Malvales	Bombacaceae	Bombacaceae
Malvaceae		

Geographical distribution: India, Sri Lanka, Java, Sumatra, Queensland, Myanmar and North Australia. In Indian sub-continent, it is found in sub-Himalayan tract from Indus eastwards ascending to 3500 ft. ASL, throughout hotter forest regions.

Habit: A very large upto 1600 m long, deciduous tree covered with stout hard conical prickles (Fig. 10).

Habitat: Hot moist regions with alluvial soils.

Root: Branched tap root with large buttresses at the base.

Stem: Erect, stout, branches spreading horizontally in whorls, trunk and branches prickly, cylinderical, wood white, very soft turning dark on exposure, branches in whorls of 5-7.

Leaves: Palmately compound, stipulate, petiolate, leaflets 5-7, lanceolate, glabrous, 10-20 cm. long, entire, base acute, apex mucronate or cuspidate, common petiole as long as or larger than leaflets, secondary petioles 2.5-3.0 cm long, stipules small, caducous.

Inflorescence: Peduncles axillary or sub terminal, solitary or clustered, flowers numerous, fascicled or congregated at the ends of branchlets.

Flower: Complete, large, regular, hypogynous, actinomorphic, appearing before the leaves, large, scarlet, rarely white.

Calyx: Sepals 5-7 lobed or truncate, leathery, gamosepalous, cup shaped, 2.5-3.0 cm long, smooth externally, silky tomentose internally, irregularly cleft ultimately, deciduous, valvate.

Corolla: Petals 5, red or white, 5.0-8.0 cm long, oblong, fleshy, recurved, twice the length of stamens, stellate and tomentose both sides, deciduous, free.

Androecium: Stamens pentadelphous, bundles opposite the petals and divided above into numerous filaments, filaments about 70, much longer than staminal coloumn, the 5 inner most stout, forked at the top and each branch bearing reniform, long, twisted and single celled (monothecous) anther, some (10) intermediate ones simple, the outer ones numerous united at base in 5 clusters.

Gynoecium: Carpels 5, styles longer than stamens, clavate, stigmas 5, syncarpous, ovary superior, ovules numerous in each locule, placentation axile.

Fruits: Capsule oblong, hard, woody, 10-15 cm long, loculicidal, 5 valved, thickly clothed with long silky hairs inside in which the seeds are embedded in dense wool.

Seeds: Numerous, glabrous, obovate.

Leaf fall: Dec. - April; **Flowering**: Feb. - March; **Fruiting**: April - May

Floral formula : ⊕ ⚥ $K_{(5\text{-}7)}\ C_5\ A_{5\ (\text{branched})\ +10+5\ (\text{branched})}\ G_{\underline{(5)}}$

Uses: It serves as one of the secondary food plants of tropical tasar silkworm, *Antheraea mylitta* D (Seth 1995 a, Thangavelu *et al.* 2000, Naqvi and Srivastava 2003). It is cultivated in Punjab, Uttar Pradesh, the Eastern and Western Peninsulas for obtaining silky floss from the wall of the fruit, which is used for stuffing pillows, and cushions, mattresses etc. can substitute Kapok, (*B. pentandra*) (Kocchar 1981). Leaves are used as fodder (Singh 1990). Buds of *B. ceiba* are used as vegetables (Shukla and Mishra 1979).

B. ceiba has been extensively used medicinally also. Intake of flower causes sterility hence can be used for birth control (Dash 1993). In snakebite flower and young fruits are crushed and applied as an antidote. In female diseases, bark extract along with honey is prescribed. Flowers are good in colitis, whereas fruits are used in calculus affection and ulceration of bladder and kidney. Fruits are also used as stimulant, expectorant and antidote in snakebite. In diarrhoea, dysentery and menorrhagia gum is used which is demulcent, haemostatic and astringent (Sinha 1996). Bark paste is used against dysentery (Singh 1990). In sexual impotency roots are used as a good medicine since they are stimulant, tonic and aphrodisiac. The roots (Musali) are powdered along with the roots of *Grewia hirsuta* Vahl. and fried with ghee, wheat flour and jaggery (gur) and taken in body pain and for regaining vitality in Bundelkhand (U.P) (Maheshwari and Singh 1991). It serves as most important matchwood of India and also used for tea chests and as rough planking (Sagreiya 1967). It is bad fuelwood. The seeds yield 22.3% oil which may be exploited commercially for production of edible oil (Jain *et al.* 1990).

Cultivation: It can be cultivated through seeds as well as by raising of seedlings in nursery and their subsequent transplantation. Besides, it can also be propagated easily through raising of air layers (Venkatesh *et al.* 1978).

26. TAR

Botanical: ***Borassus flabellifer*** Linn.; Syn. *B. flabelliformis,* Roxb.

Burmese: Tan French: Cocotier de mer German: Delebpalme

Sinhalese: Thal

English: Palmyra Palm
Hindi: Tar, Tal
Bengali: Tal
Gurajati: Tad
Kannada: Tale Mara
Konkani: Todmadd
Malayalam: Pana
Manipuri: Kona
Marathi: Tad, Tamar
Oriya: Tala
Punjabi: Tar
Sanskrit: Trinaraja, Tal
Tamil: Panai
Telugu: Tadichettu, Potu , Penti Tadi (female tree), Tadi (male tree)

SYSTEMATIC POSITION

Bentham & Hooker (1883)	**Engler & Prantl (1899)**	**Hutchinson (1959)**
Phanerogams	Phanerogams	Angiospermae
Monocotyledones	Monocotyledoneae	Monocoledones
Calycinae	Principes	Corolliferae
Palmae	Palmae	Palmales
		Palmae or Palmaceae

Geographical distribution: It is a native of tropical Africa and cultivated and self sown throughout India and Myanmar. In India, Tar trees are found along the coastal areas of Bengal, Bihar and Western and Eastern Peninsulas.

Habit: A tall tree growing up to 20-30 meters height with a straight woody, unbranched trunk bearing a crown of leaves and circumference ring marked with the bases of perished leaves (Fig. 11).

Habitat: Coastal areas.

Root: Adventitious, emerging from the base of the stem, rootlets long and dense.

Stem: Upright, tall, unbranched, attains 100 ft height and 2 ft. diameter near the ground, young ones covered with dry leaves or the base of petioles, the old trunks marked with black narrow scars of the petioles.

Leaves: Palmately divided (fan palm type), 3-5 ft. in diameter, segments 60-80, shining, folded along the midrib, linear – lanceolate, petiole large, stout 60-120 cm long, semiterete, edges with hard horny spinescent serratures and broad sheathing base. The leaf blade remains entire and folded in the bud but becomes incised from the margins inwards with expansion of the folds. When leaf dies, the sheath persists for sometime and its tough fibres form dense matting around the base of the young leaves.

Inflorescence: Spadix simply branched, axillary, withering away after fruiting, polycarpic (fruiting several times in whole life), young spikes enclosed within numerous open spathes torned away with growth of flower shoot.

Male spikes: 1-3, at the end of branches, cylinderic, densely clothed with thick closely imbricating bracts, numerous, minute, unilateral spiklets concealed by the bracts, flowers successively open and come to the surface one by one.

Female spikes: Terminal at the branches of spadix.

Flower: Sessile, unisexual, dioecious (male and female plants separate), spiral, actinomorphic, incomplete, hypogynous, numerous, female flowers globose, 2.5 cm in diameter, male flowers minute, much smaller than the female flowers.

Calyx: Sepals 3, Polysepatous, persistent, imbricate.

Corolla: Petals 3, polypelalous, convolute.

Androecium: Stamens 6, arranged in two whorls (3+3), polyandrous, filaments connate with the corolla into a stalk, anthers bithecous.

Gynoecium: Carpels 3-4, syncarpous, ovary superior, trilocular.

Fruit: Subglobose, 15.0 cm in diameter, enclosed with enlarged partly fleshy perianth. Pyrenes 1-3, obcordate, fibrous outside, endosperm horny, hollow.

Flowering: March - April

Floral formulae: Male flower: Br. ⊕ ♂ $Br_2\ K_3\ C_3\ A_{3+3}\ G_0$

Female flower: Br. ⊕ ♀ $Br_3\ K_3\ C_3\ A_0\ G_{(\underline{3-4})}$

Uses: The bows of Tar and their leaves for writing upon were popular in ancient India. Tar yields several commercial products such as toddy, nira, palm gur and jaggery, fruit and fire. The cut flower stalks yield sugar and toddy (Palmwine). The fruit is roasted and eaten. The fibre is used for making brushes and brooms while leaves are used for making fans, umbrellas, hats, bags, baskets and mats. The leaf stalks are used for making writing and printing papers. The wood is hard, durable and resistant to saline water. The trunks of female plants are strong and used for making poles, pillars and posts (Brandis 1906, Pandey 1981). Seed and pulp of tender fruit, cotyledons in germinating seeds, tender shoots and young flesy roots are eaten (Ambasta *et al.* 1986). The young seedlings and unripe fruits are also edible (Shukla and Mishra 1979).

The roots are cooling and restorative. Juice of plant is diuretic, stimulant, antiphlegmatic, useful in inflammatory affections and dropsy. Pulp is domulcent and nutritious (Sinha 1996). The juice from leaf and flowering stalk is given in gastric ailments and diabetes respectively. The bloom on leaf base is used for clotting of blood in external wounds. Juice from flowering stalk mixed with boiled rice makes a good base poultice. After fermentation, it is applied on gangrenous and painless ulcers, boils and other skin problems.

The decoction of roots is used to restore physiological activities, expel worms, promotion of urine flow and check respiratory ailments (Kurian 1995).

Cultivation: Through seeds and seedlings.

27. CHAR (CHIRONJI)

Botanical: ***Buchanania latifolia*** Roxb.; Syn. *B. lanzan* Spreng., *Spondias elliplica* Rrottle.

Eng.: Cuddapah almond Hindi: Char, Achar, Piyal, Pial, Paira, Chironji

Bengali: Piyal, Chironji Gujarati: Charoli

Kannada: Nurkal, Murkala Malayalam: Munga, Perlu

Marathi: Char, Chirauli, Chironji Oriya: Charee

Punjabi: Chiroli Tamil: Morala, Sarail Telugu: Morli, Sara.

SYSTEMATIC POSITION

Bentham & Hooker (1883)	**Engler & Prantl (1899)**	**Hutchinson (1959)**
Phanerogams	Phanerogams	Angiospermae
Dicotyledones	Dicotyledoneae	Dicotyledones
Polypetalae	Archichlamydeae	Lignosae
Disciflorae	Sapindales	Sapindales
Sapindales	Anacardiaceae	Anacardiaceae
Anacardiaceae		

Geographical distribution: Commonly occurring in the deciduous forests of India and Myanmar barring dry regions of North-west India. It is extending from Kumaon ascending to 1500 ft., Avadh, Central India and Western Peninsula.

Habit: A medium sized subdeciduous tree reaching up to 50 ft.

Root: Branched tap root system.

Stem: Erect, aerial, sparingly branched, branchlets as thick as the finger, woody, cylinderical and soild, bark dark grey, rough with small quadrangular plates, divided by deep narrow cracks.

Leaves: Simple, exstipulate, alternate, entire, broadly oblong, coriaceous, hard, obtuse, usually softly hairy beneath, closely reticulate on both surfaces, blade 15-25 cm, petiole 0.6-0.8 cm long, stout and pubescent, secondary nerves 15-20 pairs, prominent, stout, nearly straight.

Inflorescence: Terminal and axillary cymose panicles, shorter than the leaves woolly or velvety, pyramidal with stout branches, small and caducous bracts.

Flower: Sessile, greenish-white, crowded, 1.2 cm across, hermaphrodite, regular (actimomorphic), pentamerous, perigynous with a glandular disc (thalmus).

Calyx: Sepal 5, gamosepalous, 5- toothed, persistent, imbricate.

Corolla: Petals 5, oblong, imbricate, recurved, free.

Androecium: Stamens 10, spreading as long as the petals, arranged in two whorls of 5 each, inserted at and outside the base of the annular disc, free, all fertile, anthers dithecous, dehiscing longitudinally by slit.

Gynoecium: Carpels 5-6, apocarpous, pentalocular, seated in the cavity of the fleshy disc, ovaries conical, pubescent, one fertile, others imperfect reduced to flaments, style short, stigma truncate, ovule single, solitary, pendulous, placentation basal.

Fruits: Drupe, black, 1.2 cm across, edible, subglobose, slightly compressed with hard bony, 2-valved stone, flesh scanty.

Seed: Non-endospermic, gibbous, acute at one end, cotyledons thick and fleshy radicle superior.

Floral formula: $\oplus$ ⚥ $K_{(5)}\ C_5\ A_{5+5}\ G_{\underline{1+4}}$

Uses: It serves as tertiary food plant but recognised as one of the secondary food plants of tropical tasar silkworm, *Antheraea mylitta* D (Yadav *et al.* 1997). Besides, its bark is used for tanning. The fruits as well as seeds are edible and a gum obtained from the stem is employed for printing cloth and dyeing. The gum is also used medicinally for diarrhoea. Fruit is also used in releiving burning of body and fever (Singh 1990). The seeds are used as a substitute for almond and are used in confectionary. Seeds also yield aromatic oil known as "Chironji oil" upto 51.8% (Jain *et al.* 1990, Shukla and Mishra 1979).

Cultivation: It can be cultivated through seeds and seedlings.

28. DHAK

Botanical: *Butea frondosa* Roxb.; Syn. *B. monosperma* (Lam.) Taubert

Burmese: Pauk

French: Erythrine monosperme

Persian: Palah

Sinhalese: Kaliya

English: Flame of the forest

Hindi: Dhak, Palas, Tesu

Bengali: Palas

Gujarati: Khakra, Kesudano-Jhar

Kannada: Muttuga, Muttukadagidda

Malayalam: Palas in samatha	Manipuri: Pang-gong	Marathi: Palas
Oriya: Kinjuko	Tamil: Parasa, palashu	
Sanskrit: Palasha, Plaksha	Santhali: Murup	
Telugu: Moduga, Palasmu,	Urdu: Palashpapra	

SYSTEMATIC POSITION

Bentham & Hooker (1883)	**Engler & Prantl (1899)**	**Hutchinson (1959)**
Phanerogams	Phanerogams	Angiospermae
Dicotyledones	Dicotyledoneae	Dicotyledones
Polypetalae	Archichlamydeae	Lignosae
Calyciflorae	Rosales	Leguminales
Rosales	Leguminosae	Papilionaceae
Leguminosae	Papilionaceae	
Papilionatae		

It is one of the sacred trees of Buddhists since it was associated with the life of Lord Gautam Buddha. Kalidasa specifically mentioned about "Plaksha" in his play 'Sakuntala' among the beautiful flowering plants (Rao 1991).

Geographical distribution: Common throughout India and Myanmar. In North-western Himalaya it is found up to 3000 ft. ASL. Mostly it is confined to drier districts.

Habit: A moderate sized deciduous, gregarious tree with crooked and irregular trunk, young shoots tawny tomentose, growing to a height of 12-15 meters (Fig. 12).

Habitat: Tropical deciduous forests.

Root: Tap, branched.

Stem: Erect, branched, woody, solid, cylinderical, crooked, irregular, bark rough, cracked, thick, grey-black, resinous.

Leaves: Compound, trifoliate, leaflets coriaceous, hard, the terminal arising from a cuneate base, obovate, obtuse often emerginate, as long as broad, 10-15 cm long, the laterals smaller, oblique-ovate, secondary and tertiary nerves prominent.

Inflorescence: Axillary or terminal racemes.

Flower: Orange, hermaphrodite, zygomorphic, pedicellate, complete, perigynous, paplionaceous, appear before the leaves, 5.0 cm long, pedicel 2.5 cm long, twice the length of calyx.

Calyx: Sepals 5, gamosepalous, united in a tubular calyx below the disc, 5 toothed, aestivation ascending imbricate, black.

Corolla: Petals 5, pubescent outside, unequal, uppermost and largest petals known as standard or vexillum acute, recurved, the two free lateral petals are known as wings or alae, the anterior pair of united petals known as keel or carina much curved, acute 5-6 times longer than calyx, encloses stamens and pistil, aestivation ascending imbricate.

Androecium: Stamens 10, free, inserted on a disc below calyx, upper stamen free, others connate beyond the middle, anthers uniform, obtuse, dithecous, dorsifixed.

Gynoecium: Monocarpellary, free, ovary superior, style bent and short, hairy, stigma simple.

Fruit: Pod, coriaceous, oblong, pendulous, lower portion flat, indehiscent, upper 1-seeded, dehiscent. Pale green turning yellowish brown, covered with silvery white hair.

Leaf shedding: February.

Flowering: The tree is covered with bright orange red flowers in March-April before the fresh leaves appear. Hence, it truly stands out like a flame in the forest.

Floral formula: •|• ⚥ $K_{(5)}\ C_{2+(2)+1}\ A_{9+1}\ G_{\underline{1}}$

Uses: It serves as the host plant for lac insect. The flowers yield a yellow dye. The leaves are used as substitute for plates, platters, cups etc. on festive occassions. The ripe fruits are edible. Dried leaves used as beedi wrappers. Young shoots yield fibre for ropes and root bark yields fibre for cordage and caulking boats (Ambasta *et al.* 1986).

The decoction of the root is used for curing night blindness, other diseases of eyes and elephantiasis. The decoction of leaves is also good for eye diseases. While decoction of bark is used as a laxative and anthelmintic and useful in fractures, dysentery, piles, hydrocoel, ulcers, tumours and menstrual disorders. The gum of the tree is useful for checking cough and stomatits or inflammation in the innerside of the mouth. The decoction of flowers is useful for curing leprosy in the early stages, gout and skin diseases. When the seeds are pounded with lemon juice, they act as a powerful rubefacient and successfully used as a cure for 'Herpes' or 'Dhobies' itch (Kurian 1995). Seeds are anthelmintic and gum is astringent, so given in diarrhoea. Leaves are tonic, while bark is used in snakebite. Flowers are antiseptic and depurative (Sinha 1996).

The red gum obtained from Palas is known as Bengal-Kino or Butea-gum. The gum contains tannins. Roots of palas are stated to be effective in high blood pressure and seeds are useful in skin diseases (Jain 1968). The gum serves as stimulant (Randhawa 1983).

The wood of Palas does not decompose early in water, so it is used for preparing the pot for fetching water from the well and also for building the shaft for keeping on the well. It is also used for building purposes and as fuel wood (Jain 1968).

The bark is used in tanning, root fibre is made into ropes and the seeds yield a clear oil (Randhawa 1983). In Manipur, Palas is considered as a very sacred tree. Its leaves are used for performing last rituals of those persons whose dead bodies are missing or their whereabouts are not known since quite a long time.

Cultivation: Through seeds or seedlings. Fresh seeds germinate easily after extraction from ripe pods. The seeds have short viability so they should be sown immediately. It is frost hardy, drought resistant and grows on saline soils. But they die during winter and hot weather. So the transplanting of new seedlings or plants may be continued for 2 or 3 years till sturdy shoot plant is obtained which can stand the vagaries of the weather (Randhawa 1983).

29. AAK

Botanical: *Calotropis gigantea* R. Br.

English: Crown Plant Madar	French: Mudar	German: Gross mudar
Persian: Khok	Sinhalese: Vara	
Hindi: Aak, Safed Aak, Madar, Akuwa		Bengali: Akanda
Gujarati: Akado	Kannada: Ekkada gida	
Konkani: Rhui	Malayalam: Yerrikku	Manipuri: Angkot
Marathi: Rui, Muda	Oriya: Arka	Punjabi: Ak
Sanskrit: Arkah, Mandara	Santhali: Ahauna	Sindhi: Byclospa
Tamil: Erikkh, Arkkam	Telugu: Nalle Jelledu	Urdu: Aak

SYSTEMATIC POSITION

Bentham & Hooker (1883)	Engler & Prantl (1899)	Hutchinson (1959)
Phanerogams	Phanerogams	Angiospermae
Dicotyledones	Dicotyledoneae	Dicotyledones

Gamopetalae	Sympetalae	Lignosae
Bicarpellatae	Contortae	Apocyanales
Gentianales	Asclepiadaceae	Asclepiadaceae
Asclepiadaceae		

Geographical distribution: It is commonly found in the Peninsula and Northern India, up to 500 m ASL and grows wild in the tropical countries of Asia. Gregarious in Singbhum.

Habit: A medium sized, perennial shrub, which grows about 3 meter high. Young branches inflorescence and under surface of leaves covered with white woody tomentum. All parts contain milky latex (Fig. 13).

Habitat: Tropical dry regions in all types of soil, especially sandy and barren soils.

Root: Tap root, branched, deep, covered with yellowish bark.

Stem: Erect, branched, glabrous, woody below and herbaceous above, tomentose, solid, cylinderical, bark pale.

Leaves: Simple, 10-20 cm long, obovate or obovate-oblong, subsessile, exstipulate, fleshy, covered with a waxy coat, opposite decussate, thick, heart shaped, cottony on the under surface, apex acute, glaucous, green, entire, base cordate or amplexicaul venation reticulate unicostate, milky latex present.

Inflorescence: Axillary umbellate cyme.

Flower: Bracteate, pedicellate, complete, actinomorphic, inodorous, bisexual, hypogynous, pentamerous, simple or compound, axillary or subterminal, white or purplish – lilac, buds ovoid.

Calyx: Sepals 5, polysepalous or connate at the base, greenish white, lanceolate, apex acute, hairy, aestivation imbricate.

Corolla: Petals 5, 1.25-2.5 cm across, purplish or white with purplish spots within, lobes spreading or reflexed, gamopetalous, inferior, aestivation twisted.

Corona: Much longer than broad compressed fleshy coronal appendages 5, hairy, adnate to and radiating from staminal tube, lower end free and incurved, apex appearing bifid with auricles.

Androecium: Stamens 5, filaments connate forming a staminal tube (monoadelphous) around the gynoecium, apex of the staminal tube much dilated, stigmatic head and anthers united together to form pentagonal gynostegium, anthers short, with dithecous broad membranous tips, each cell bears a single pollen pass, pollinium so that 5 anthers have 10 pollinia, pollinia united together through short stalks or caudicles to a distinct dark coloured dot like structure, corpusculum, which lies at an angle with the gynostegium, thus forming a translator apparatus.

Gynoecium: Carpels 2, synarpous, pistil free below and fused above, two distinct ovaries ending into styles form a pentangular stigmatic head to the side of which anthers are united, ovary superior, unilocular, ovules many, placentation marginal.

Fruit: A pair of follicles, with broadly ovate many seeds which bear tuft of silky hairs.

Floral formula: ⊕ ⚥ K_5 $C_{(5)}$ A_5 $G_{(2)}$

Uses: Silky fibre from seeds/fruits made into ropes, paper, clothes (Brandis 1906), threads and for stuffing pillow (Shukla and Mishra 1979). The stem yields a fibre and the gun powder charcoal is prepared from the wood (Pandey 1981). Bark yields white, silky strong and durable fibre for fishing nets, lines, bow strings and twine (Ambasta *et al.* 1986). The fresh leaves are used in the fomentation for swellings (Pandey 1981). The leaves heated over as naked flame are applied as a dry stimulant for abdominal pains. The root is used as poultice in ulcers and other skin diseases, while juice of the young floral buds is used for earaches (Kurian 1995).

Cultivation: It may be cultivated through seeds or seedlings.

30. RANGRURI

Botanical: *Canthium didymum* Roxb. Syn. *C. dicoccum* (*Gaertn.*) Merril, *C. umbellatum* Wight IC.

Hindi: Rangruri	Kannada: Hattera-nike	Marathi: Arsul
Santal : Garbhagoja	Tamil: Imburruttam	
Upper Burma: Myauklounga	Telugu: Nalla Balsu, Naikinna, Nakkareni	

SYSTEMATIC POSITION

Bentham & Hooker (1883)	Engler & Prantl (1899)	Hutchinson (1959)
Phanerogams	Phanerogams	Angiospermae
Dicotyledones	Dicotyledoneae	Dicotyledones
Gamopetalae	Sympetalae	Lignosae
Inferae	Rubiales	Rubiales
Rubiales	Rubiaceae	Rubiaceae
Rubiaceae		

Geographical distribution: Found in Western Ghats from Kandala southwards, Nilgiris, hills of Salem and adjoining districts, Orissa, Sikkim (up to 1500 ft. ASL), Khasia and Jayantia hills, common in Sri Lanka (up to 4000 ft. ASL), Malay Peninsula and Archipelago, South China and Myanmar.

Habit: A small sized evergreen tree or shrub.

Habitat: Evergreen forest.

Root: Tap root, branched.

Stem: Smooth, Unarmed, glabrous, branchlets often angular, wood hard, even grained.

Leaves: Simple, coriaceous, glabrous, glossy, polished (deep green) above, usually obtusely caudate - acuminate, elliptic, base acute, obtuse or even cordate, nerve axils eglandular, blade 10-15 cm long, petiole stout 0.6 cm long.

Inflorescence: Umbelliform, axillary, cymes subsessile or on a small peduncle, 0.6-2.5 cm long, occasionally puberulous.

Flower: Hermaphrodite, regular, complete, actimomorphic, bracts short or nil, pentamerous, white, fragrant, epigynous.

Calyx: Sepals 5, truncate or obscurely toothed, calyx tube shortly produced beyond ovary.

Corolla: Petals 5, sub-acute, synpetalous, corolla tube campanulate, cylinderic, 0.3-0.6 cm long, little longer than reflexed lobes, valvate, throat villous.

Androecium: Stamens 4-5, subsessile on the throat or mouth of the corolla.

Gynoecium : Bicarpellary, style stout and glabrous, stigma large, subquadrate, notched or bifid, ovules-1, pedulous in each cell, ovary inferior.

Fruits: Drupes, didymous, purple, 0.6-1.3 cm long, often beaked and elongated due to insect injury.

Flowering: Nov. - May.

Floral formula: $\oplus$ ⚥ $Br_0\ K_5\ \overparen{C_5\ A_5}\ G_{(\bar{2})}$

Uses: It serves as tertiary host plant but may be regarded as the secondary food plant of tropical tasar silkworm *Antheraea mylitta* D. (Seth 1995 a). Wood used for agricultural implements, combs, posts, rafters and toys. Bark used as a ferifuge, applied to fractures (Ambasta *et al.* 1984).

Cultivation: *Canthium didymum* propagates mainly through the seeds.

31. KUMBI

Botanical: *Careya arborea* Roxb.

Burmese: Bambwe | Eng.: Kumbi

Hindi: Kumbi, Kumb, Kumhi | Bengali: Kumb, Kumbi, Kumhi

Gujarati: Kumbi | Kannada: Kaval, Kaulu

Malayalam: Pera, pela | Marathi: Kumbia | Sanskrit: Kumbhi

Tamil: Ayama, Ayima | Telugu: Gadava, Araya Duddippa.

SYSTEMATIC POSITION

Bentham & Hooker (1883)	Engler & Prantl (1899)	Hutchinson (1959)
Phanerogams	Phanerogams	Angiospermae
Dicotyledones	Dicotyledoneae	Dicotyledones
Polypetalae	Archichlamydeae	Lignosae
Calyciflorae	Myrtiflorae	Myrtales
Myrtales	Lecythidaceae	Lecythidaceae
Myrtaceae		

Geographical distribution: Throughout India from 0 to 4000 ft. ASL, especially in U.P, Bihar, Bengal, Madhya Pradesh, Orissa and also in Myanmar.

Habit: A large deciduous tree attaining height up to 30-60 ft. (Fig. 14).

Root: Branched tap root.

Stem: Erect, aerial, branched, woody, bark rough, grey.

Leaves: Simple, alternate, crowded towards the ends of branches, obovate or oblong, membranous, narrowed into a short marginate petiole 0-2.5 cm. long, crenate, denticulate, 30 x 15 cm, obtuse or shortly acuminate, glabrous, secondary nerves prominent, 10-12 pairs.

Inflorescence: Spikes 7-20 cm long, few flowered

Flower: Bisexual, complete, regular (actinomorphic), cyclic, tetramerous, epigynous, 5.0 - 10.0 cm across, sessile, smell unpleasant, supporting bracts-3, unequal.

Calyx: Sepals 4, terete, campanulate, hardly produced above ovary, lobes 4, ovate, obtuse, pubescent, imbricated, calyx tube 2.0-2.5 cm long.

Corolla: Petals 4, white or greenish white, 2.5-5.0 cm long, elliptic, free, imbricated.

Androecium: Stamens numerous in several series, slightly connate at the base, filaments purple or red, filiform, outermost as well as inner most without anthers.

Gynoecium: Carpels 4-5, ovary crowned by an annular disc, tetralocular or pentalocular, syncarpous, style long, simple, stigma small, ovules in two rows in each cell, placentation axile.

Fruits: Globose, berry, 6 x 5 cm, green, fleshy, crowned with the calyx.

Seed: Ellipsoid, numerous, 0.6 cm long, lying in pulp, nonendospermic, embryo large, cotyledons obsolete.

Flowering: March - April when leaves are young.

Floral formula: $\oplus$ ⚥ $K_{(4)}\ C_4\ A_{\infty}\ G_{(\overline{4-5})}$

Uses: It serves as secondary food plant of tropical tasar silkworm, *Antheraea mylitta* D (Jolly *et al.* 1974, FAO 1976, Seth 1995 a). Juice of the bark with honey is useful for cough and cold. Fruit is edible (Singh 1990). Bark is used as an astringent (Ansari 1997). Timber is used for making agricultural implements (Singh 1990), gum stocks, planks, house posts and hoops. Bark can be used for cordage, paper making and as a substitute for English beech for fuses in ordnance factory (Ambasta *et al.* 1984).

Cultivation: It can be cultivated through either direct sowing of seeds or by raising nursery of seedlings for subsequent transplantation.

32. PAPEETA

Botanical: *Carica papaya* Linn.

Burmese: Timbawthi
Eng: Papaya

French: Abre a melonus
German: Mamaobaum
Hindi: Papeeta

Singhalese: Gaslabu
Bengali: Pappaiya
Gujarati: Papayi

Kannada: Parangi-mara
Konkani: Pappaye phal
Manipuri: Awathabi

Marathi: Papai
Mizo: Thingfanghma
Malayalam: Omaca

Punjabi: Arand–Kharpuja
Sindhi: Paputa
Tamil: Pappali

Telugu: Boppayi

SYSTEMATIC POSITION

Bentham & Hooker (1883)	Engler & Prantl (1899)	Hutchinson (1959)
Phanerogams	Phanerogams	Angiospermae
Dicotyledones	Dicotyledoneae	Dicotyledones
Polypetalae	Archichlamydeae	Lignosae
Calyciflorae	Parietales	Cucurbitales
Passiflorales	Caricaceae	Caricaceae
Passifloraceae		

Geographical distribution: A native of the West Indies and Central America with Brazilian origin. It was introduced to India in the 16th Century A.D by the Portugese. In India, it is cultivated chiefly in Uttar Pradesh, Punjab, Rajasthan, Gujarat, Maharashtra, Tamil Nadu and the Nilgiris. Cultivated in all tropical countries.

Habit: A small soft wooded, perennial, fast growing, unbranched, up to 25 ft height shrubby plant with milky juice (Fig. 15).

Root: Tap root.

Stem: Erect, aerial, branchless, soft, greyish, marked with fallen leaf scars.

Leaves: Glabrous, palmatifid, deeply seven lobed, 30-60 cm across on long hollow petioles forming a round tuft at the top, stipules absent.

Inflorescence: Axillary panicles.

Flower: Unisexual, usually dioecious but occasionally monoecious plants with pale yellow fragrant flowers. Male flowers in long stalked drooping clusters (panicles) but female flowers stalkless or sessile in short clusters, perigynous.

Calyx: Sepals 5, small, united (gamosepalous), valvate.

Corolla: Petals 5, tubular in male flowers, linear in female flowers, deciduous, twisted.

Androecium: Stamens 10, inserted in two rows in the mouth of corolla tube in fasicles opposite to petals, anthers dithecous.

Gynoecium: Carpels 5, ovary unilocular, inferior, free, ovules numerous attached to five parietal placentas.

Fruit: Fleshy berries resembling melons in size, oblong or spherical, indehiscent, sulcate.

Seeds: Black, globular, numerous, embedded in sweet pulp, endospermic.

Floral formulae: Male flower: $\oplus$ ♂ $K_{(5)}\ C_{(5)}\ A_{10}\ G_0$

Female flower: ⊕ ♀ $K_{(5)}$ $C_{(5)}$ A_0 $G_{\bar{5}}$

Uses: *Carica papaya* serves as one of the important secondary food plants of eri silkworm *Philosamia ricini* (FAO 1976, Sarkar 1988, Seth 1995c). Besides, it has immense economic potentialities in fruit and pharmaceutical industries.

The papaya is popular as a delicious table fruit throughout India. The flesh of ripe fruit is yellow or orange, soft and sweet. Unripe fruits are consumed as vegetables whereas ripe fruits are eaten fresh and raw and also employed in the preparations of candies marmalades and soft drinks as a dessert. It is grown commercially for papain (vegetable pepsin), the milky latex that oozes out from unripe fruit when lanced with a knife. The latex produces papain on drying. Papain clots the milk and helps in digestion of protein and hence has great demands in the international market. Over 750 kg of crude papain/ha. may be produced from 'CO5' variety (Rao 1991). The fruit pulp used as ingradient in face creams and hair shampoos. Milky latex from unripe fruits used as cosmetic to remove freckles and blemishes from skin.

"Washington" and "Honey dew" are most popular varieties. Following suitable varieties may be cultivated as here under (Rao 1991):

For South India: Coorg Honey Dew, Singapore, Ceylone and Sunrise (Solo).

For House Garden: C04, C03 (attractive and large fruits)

For high yielding and good keeping quality: Pusa delicious and Pusa majesty (Both are gyno-dioecious i.e. female and bisexual flowers borne on separate plants).

For large fruits: "Pusa Giant" weighing 2.5 – 3.5 kg/fruit.

For early maturity: "Pusa Dwarf" bears fruits weighing 1- 1.5 kg. just after attaining 30 cm height above the ground.

The medicinal uses of papaya are enormous since almost all parts of the plants have remedial properties as here under (Kurian 1995, Sinha 1996).

Roots: Used as tonic, anthelmintic and as a remedy for excessive bleeding of kidneys and dog bite.

Latex: Used as cosmetic, digestant, anthelmintic and for removal of freckles, corns and warts from skin, toothache and tooth decay.

Leaves: Used for treatment of wounds, fresh cuts, ulcers, eczema, warts, boils and help in digestion. Leaves yield an alkaloid carpine while plant yields a blood anti coagulant.

Flowers: Used for hoarseness, influenza, cough, bronchitis and inflammed trachea.

Seeds: Used as anthelmintic, antidote for cancer, remedy against tuberculosis, for cleaning of stomach proper functioning of the liver, promotion of menstrual flow and against suppression of urine.

Fruits: Used against constipation, gastritis, diabetes, asthma, jaundice, flatulence, cough, chest problems, stomach and duodenal ulcers and indigestion. It also promotes flow of urine, softens inflammed parts, helps in digestion and serves as remedy for all stomach and bowel diseases.

Cultivation: It can be very easily cultivated through seeds and seedlings in kitchen gardens since the germination is very easy. It germinates wildly also from thrown away seeds after eating the fruits.

33. KARAUNDA

Botanical: *Carissa carandas* Linn.; syn. *C.congesta* Wight Ic., *C. spinarum* A.DC, *C. diffusa* Roxb., *C. hirsuta* Roth., *C. villosa* Roxb.

Eng.: Bengal Currants	French: Calac	Persian: Karwanah
Sinhalese: Mahakaranda	Hindi: Karaunda	
Bengali: Bainchi	Gujarati: Karamadan	Kannada: Dodakavali
Konkani: Carandam	Malayalam: Kalavu,	Marathi: Karvanda
Oriya: Koronda	Kala-chedi	Tamil: Kalakkay
Telugu: Kalivi	Sanskrit: Avighna	Urdu: Karvanah

SYSTEMATIC POSITION

Bentham & Hooker (1883)	**Engler & Prantl (1899)**	**Hutchinson (1959)**
Phanerogams	Phanerogams	Angiospermae
Dicotyledones	Dicotyledoneae	Dicotyledones
Gamopetalae	Sympetalae	Lignosae
Bicarpellatae	Contortae	Apocynales
Gentianales	Apocynaceae	Apocynaceae
Apocynaceae		

Geographical distribution: Throughout the drier sandy or rocky soils of India from plains of Northern India, Bengal, Madhya Pradesh, Uttar Pradesh, Punjab to Western Peninsula. In Sub-Himalayan tracts, it is found up to 3000 ft. ASL. Also found in Sri Lanka, Myanmar, Malaysia, Java and Timur.

Habit: A gregarious, evergreen, thorny shrub or small tree.

Habitat: Forms underwood under Bamboos, Sal, Teak and *Pinus longifolia* in Kangra and springs up luxuriantly in clearances and hilly regions of India.

Root: Branched tap root.

Stem: Dichotomous, erect, aerial, rigid, spreading, thorny branches, sometimes climbing over tall trees, young shoot more or less pubescent, spines usually forked or trifid, some times 2.5-5.0 cm long.

Leaves: Simple, opposite, small, coriaceous, ovate or elliptic, glabrous or pubescent beneath, subsessile, base rounded or retuse, apex obtuse rarely mucronate, petiole 0.4-0.6 cm and blade 2.5 to 5.0 cm long, generally renewed in March.

Inflorescence: Corymbiform cymes at the end of branches with stout peduncles and few flowers.

Flower: Hermaphodite, actinomorphic, regular, complete, pentamerous, hypogynous, white sometimes with a pink tinge, scented.

Calyx: Sepals 5, lanceolate, ciliate, polysepalous, imbricate in bud, puberulous, acute.

Corolla: Petals 5, gamopetalous, corolla tube cylinderic, 0.8 cm long, glabrous or puberulous with swollen throats and lobes lanceolate, acute, about half as long as the tube, contorted in bud.

Androecium: Stamens 5, alternating with corolla lobes, inserted in the tube, filaments short; anthers lanceolate.

Gynoecium: Carpels 2, bilocular, superior, placentation axile, ovules-4.

Fruits: Berry dark purple, ellipsoid, edible 0.4 - 0.6 cm long, 2-seeded.

Flowering: January – May.

Floral formula: $\oplus$ ⚥ $K_{(5)} \; \overbrace{C_5 \; A_5} \; G_{(\underline{2})}$

Varieties/ Species: A softly pubescent form is *C. hirsuta* Roth. syn. *C. villosa* Roxb. which is found in Belgaum. The cultivated form of *C. spinarum* A. DC is *C. carandas* Linn., which is a taller shrub with larger leaves and fruits of 1.3 - 2.5 cm long, 4 or more seeded which turns black after ripening.

Uses: It serves as tertiary host plant but recognized as one of the secondary food plants of tropical tasar silkworm, *A. mylitta* D (Seth 1995 a). Besides, its fruits are eaten either raw or made as pickles. Ripe fruits sweet, edible, suitable for tarts, jellies and puddings. Medicinally decoction of roots is used as a laxative, for expelling worms from body and for strengthening and giving tone to the stomach. Unripe fruit cures biliousness or congestion of the liver and is also used for stimulating appettite, reducing fever and preventing or curing scurvy. The decoction of leaves is also used for reducing fever whereas ripe fruits are helpful in checking bleeding of internal organs (Kurian 1995, Sinha 1996). Wood used for making spoons and combs. It is planted for fencing due to its thorny nature (Ambasta *et al.* 1986).

Cultivation: It can be propagated through seedlings after germinating seeds in a bed. The one year old open nursery grown seedlings are dug and bare root of seedlings is washed to remove nursery soil and pruned for thick roots beyond 10 cm and foliage upto 75%. The remaining root system is dipped in slurry of clay and wrapped with sterilized moist Sphagum moss before storing in shade. The clay slurry may be prepared by suspending the clay soil in the water and adding Thiram at the rate of 2 gms/litre of slurry. Fresh water is sprinkled in morning and evening everyday during the period of storage. The stored seedling may be replanted at the spacing of alteast 8'x8'. The survival of seedlings is significantly higher in rainy season plantings followed by spring and autumn season plantings. The higher survival and growth of seedings in rainy and spring season plantings is due to better weather conditions. 86% survival after seven days of storage has been noticed and height and number of leaves (growth) are not affected in seedlings upto 5 days of storage, which may be due to hardy nature and tolerance to drought conditions. Aforesaid practices minimise expenditure on labour, transportation and carriage of noxious weeds, nematodes, diseases and insect pests from nursery to planting site (Misra and Jaiswal 1997). Karaunda can also be propagated vegetatively by use of IBA in air layers, which stimulates root initiation in *Carrissa carandas* easily (Misra and Singh 1990).

34. DADMARI

Botanical: *Cassia alata* Linn.

English: Ringworm Senna   Hindi: Dadmari, Dadmurdan

Bengali: Dadmari

SYSTEMATIC POSITION

Bentham & Hooker (1883)	Engler & Prantl (1899)	Hutchinson (1959)
Phanerogams	Phanerogams	Angiospermae
Dicotyledones	Dicotyledoneae	Dicotyledones
Polypetalae	Archichlamydeae	Lignosae
Calyciflorae	Rosales	Leguminales
Rosales	Leguminosae	Casesalpiniaceae
Leguminosae	Caesalpinoideae	
Caesalpinoideae		

Geographical distribution: Indigenous to Andamans. Found in Tamil Nadu, Bengal and tropics of both hemispheres.

Habit: A large shrub.

Habitat: High temperature and humid areas (Fig. 16).

Root: Tap, branched

Stem: Aerial, erect branched, branches woody, thick, finely downy, cylinderical.

Leaves: Compound, alternate, paripinnate, petiolate, leaflets 8-12 pairs, oblong from an oblique base, opposite, obtuse, 5.0-15.0 cm long, secondary nerves 15 pair, prominent beneath, stipules ariculate, rigid, persistant, 0.6 cm long.

Inflorescence: Spiciform pedunculate racemes, erect, rigid, yellow.

Flower: Pedicellate, pedicel short, yellow, buds enveloped in large caducous bracts, complete, hermaphrodite, zygomorphic, hypogynous.

Calyx: Sepals 5, polysepalous, yellowish green, imbricate.

Corolla: Petals 5, polypetalous, bright yellow, clawed with darker veins, broad-ovate, 3.5 – 4.0 cm long.

Androecium: Stamens 10, perfect 7, anthers nearly equal, free.

Gynoecium: Monocarpellary, ovary superior, unilocular, placentation marginal, style long, stigma simple.

Fruit: Pod flat, rigid, 15.0 x 1.5 cm, lomentaceous, cylinderical.

Flowering: Nov. – Jan.

Floral formula: •|• ⚥ K_5 C_5 $A_{7+5 \text{ stminode}}$ $G_{\underline{1}}$

Uses: Occasionally planted as ornamental plant. Medicinally juice of leaves is employed for skin troubles like ringworms and pustules and snakebite. Decoction of flowers and leaves is given in asthma and bronchitis. Seeds are used as vermifuge (Sinha 1996, Pandey 1981). Chrysophanic acid contents is high (Ambasta *et al.* 1986).

Cultivation: Through seeds and seedlings.

35. AMALTAS

Botanical: ***Cassia fistula*** Linn Syn. *C. rhombifolia* Roxb

Burmese: Ngohpin, Ngu French: Canefice German: Benblum

English: Indian laburnum, Golden shower

Singhalese: Aehala

Hindi: Amaltas, Kiral, Pudding pipe tree, Kilvali, Girimalah, Sinar

Assamese: Sonaru

Bengali: Sundali, Bandarlati, Amultas, Sondal

Gujarati: Garmala, Garmado Kannada: Kakkegida, Kakke, Kakkai

Kol: Hari Konkani: Kakkàyi

Malayalam: Konnai, Babava, Bhava, Kritmalam Konnai

Manipuri: Chahui Marathi: Bahava, Behava, Jumba

Oriya: Sonari, Sandari Punjabi: Amaltas, Kanihar,

Sanskrit: Suvarnka, Alash, Karangal, Rajataru

Tamil: Konnai, Appai, Tiru Kontai, Rela

Telugu: Relu, Rela, Aragvdbamu

SYSTEMATIC POSITION

Bentham & Hooker (1883)	**Engler & Prantl (1899)**	**Hutchinson (1959)**
Phanerogams	Phanerogams	Angiospermae
Dicotyledones	Dicotyledoneae	Dicotyledones
Polypetalae	Archichlamydeae	Lignosae
Calyciflorae	Rosales	Leguminales
Rosales	Leguminosae	Caesalpiniaceae
Leguminosae	Caesalpinioideae	
Caesalpinioideae		

The specific name "fistula" in Latin means pipe referring to the shape of the fruit (Pod) of this plant.

Geographical distribution: It is a native of India and found through out the deciduous forests and plains of the country up to the foothills of Himalaya. Ascending to 4000 ft. in the outer Himalaya. Also found in Pakistan, Mayanmar, Java, Vietnam and the Philippines.

Habit: A medium sized deciduous and hardy tree growing up to 10-12 meters, some times a large tree (Fig. 17).

Habitat: Tropical deciduous dry – moist evergreen forests.

Root: Tap, branched.

Stem: Erect, aerial, woody, branched, glabrous, solid, bark smooth, greenish-grey when young, turning to brown and rough in older trees, sap wood large, heart wood brick red, when fresh cut darkening on exposure, very hard, youngest shoots silky.

Leaves: Compound, alternate, paripinnate, 5.0-13.0 cm long, pulvinous at the base, petiolate, petioles 0.4-0.6 cm long, stipulate, stipules caducous, leaflets 4-8 pairs, ovate, entire, acute, glabrous, shining, deep green, venation unicostate reticulate, fresh leaves deep coppery colour with a soft downy under surface, folded and pendulous.

Inflorescence: Racemose, typical lax, long, pendulous recemes.

Flower: Pedicellate, pedicels long, pedicels 4.0-5.0 cm long, bracteate, bracts minute and caducous, hermaphrodite, zygomorphic, complete, hypogynous, yellow, 3.5-5.0 cm across, clustered.

Calyx: Sepals 5, polysepalous, more or less petaloid, yellowish green, inferior, quincuncial aestivation, odd sepals anterior.

Corolla: Petals 5, polypetalous, yellow, clawed, ascending imbricate aestivation, inferior.

Androecium: Stamens 10, polyandrous, unequal in length, 3 posterior stamens reduced to staminodes, anthers basifixed with abortive and indehiscent lobes, the longer filaments not thickened in the middle.

Gynoecium: Monocarpellary, ovary superior, unilocular, placentation marginal, style short, stigma terminal, hairy.

Fruit: Legume, 30-60 cm long pod, cylinderical, black when ripe, seeds covered with a somewhat sweetish pulp.

Leaf fall: March - April.

Flowering: April- May when nearly leafless. At some of the places the flowering may extend up to August.

Leaf sprouting: May- June.

Floral formula: •|• ⚥ $K_5\ C_5\ A_{7+3}\ G_{\underline{1}}$

Uses: The tree is very conspicuous and beautiful on flowering, hence, it may be recognized from long distance in the forests. The whole tree looks yellow due to complete leaf fall and over loading with flowers. Therefore, it is highly ornamental for parks, gardens and avenues.

The wood is hard and is used for buildings, bridges, carts and agricultural implements. Bark is known as Sumari and used for tanning. The dark brown sweetish pulp is used for flavouring tobacco in Bengal. The santhals in Chattisgarh use flowers as food (Randhawa 1983, Kocchar 1981, Jain 1968).

The dark brown sweetish pulp is strongly laxative and purgative (Brandis 1906, Randhawa 1983, Pandey 1981). Overdose is harmful and may cause loose motions, nausea and stomache (Jain 1968). The fruit pulp is also good for thoracic obstruction, chronic fever and Nephritis or inflammation of the kidney. The juice of leaves is used for treatment of rheumatic pain, facial paralysis, eczema, ringworm and swellings, while decoction of root is strong purgative and pyrigesic. The bark extract is useful in liver complaints and enlargement of spleen. The fruit pulp is externally useful in gout and rheumatism also. Root bark is useful as external application in ringworms and pustules while its extract is used in the treatment of black water fever (Kurian 1995, Sinha 1996).

The natural colourant obtained from fruit pulp has recently been successfully used for dyeing silk fabrics (Sharada Devi *et al.* 2002).

Wood used for house posts, ploughs, tool handles and wheels. Bark, known as Sunari, is used for tanning alongwith avaram bark (Ambasta *et al.* 1986).

Cultivation : Through seeds, seedlings and root suckers. The seeds are collected and sown during March-April. Germination of seeds may be hastened by boiling them in hot water for five minutes before sowing to soften the hard seed coat. Since germination is poor, more seeds should be planted. The seedlings should be transferred to baskets in the first rains and transplanted during July-August. The caterpillars should be removed during early stages of growth. Under favourable conditions, the tree flowers after five years otherwise after 10 years of planting (Randhawa 1983, Chundawat and Gautam 1993).

The germination of *C. fistula* seeds decreases with increase in storage period. However, this decrease is less in bigger sized seeds. Seed stored for one year under ambient conditions revealed that bigger seeds recorded 64% germination followed by medium sized seeds (56%). The potentiality of the bigger sized seeds extends in storage with high germination than small sized seeds.

36. SARENG

Botanical: *Castanopsis indica* A.DC. Syn. *Castanea indica* Roxb

Burmese: Thit-e English: Indian Chestnut.

Nepalese: Dalne Katus, Bank katus

Vernaculars: Dieng-ka-sut, Serang, Sareng, Sirang-asing, Hinguri, Amke-asing, Dieng-sarang, Aule-Katus, Dalne-Katus, Bank-Katus, Sil-Batana, Karachagne

SYSTEMATIC POSITION

Bentham & Hooker (1883)	Engler & Prantl (1899)	Hutchinson (1959)
Phanerogams	Phanerogams	Angiospermae

Dicotyledones	Dicotyledoneae	Dicotyledones
Monochlamydeae	Archichlamydeae	Lignosae
(or incompletae)	Fagales	Fagales
Unisexuales	Fagaceae	Fagaceae
Cupuliferae		
Quercineae		

Geographical distribution: Valleys and outer Himalayan ranges from Nepal to Arunachal Pradesh, Meghalaya (Khasi hills), Assam (Cachar), Mizoram, Tripura and Bangladesh (Chittagong) at an elevation between 1000-9000 ft. ASL. Also found in Myanmar between 2000-6500 ft. in the hills of upper Myanmar east of Iravadi river.

Habit: A medium sized or large evergreen, monoecious tree. Branchlets, petioles, underside of leaves and inflorescence rusty tomentose.

Habitat: Sub-tropical - sub-temperate humid localities.

Root: Branched tap root.

Stem: Aerial, erect, branched, woody, hard, bark silvery grey, warty, somewhat deeply fissured vertically with exfoliating scales of 10 mm thickness.

Leaves: Simple, coriaceous, sharply serrated with subulate teeth, glabrous above, finely pubescent beneath, elliptic-oblong, apex acute or acuminate, blade 10-15 x 4.0-10 cm, secondary nerves 12-20 pairs, prominent beneath, sub-parallel, mid rib depressed above and pubescent, base obtuse or cuneate, petiole 0.7-1.2 cm.

Inflorescence: Unisexual spikes often paniculate.

Male inflorescence: 10-15 cm densely brownish-tomentose, sub-erect, several, below 1-2 terminal female spikes.

Female spikes: Terminal above male spikes, solitary, densely brownish-tomentose.

Flower: Unisexual, incomplete, epigynous, clustered, 2-3 mm across, bracteolate.

Male flower: Perianth 6 lobed, gamoteplous, companulate, stamens 12.

Female flower: Solitary, densely tomentose perianths as in male flower, styles 3, conical to terete and slender, densely hairy at base, ovary inferior, trilocular, stigmas terminal and punctiform.

Flowering: Nov-Dec. in Sikkim and Feb.-March in Myanmar.

Fruit: Nuts ovoid, 1.25 x 1.0 cm, pubescent at apex around mucronate stylar remains. Cupules/involucre thin walled, globose, 2.5-4.0 cm across including branching needle like spines 5-10 mm, concealing the surface of the involucre, unequal, slender, simple, straight, hairy, the cupules completely enclosing the 3-4 fruits except persistent style and perianth, splitting irregularly into 3-4 segments at maturity.

Fruiting: October - December.

Floral formulae: Male flower: Br. ⊕ ♂ $P_{(6)}$ A_{12} G_0

Female flower: Br. ⊕ ♀ $P_{(6)}$ A_0 $G_{(\bar{3})}$

Uses: It serves as secondary food plant of temperate oak tasar silkworms (Seth 1995 b). Wood is used for shingles and construction work, bark contains 6-12 and leaves contain 10 percent tannins. Fruit kernel is edible and leaves are used for wrapping beedis (Negi and Naithani 1995).

Cultivation: It can be propagated either through direct sowing of seeds or transplantation of seedlings raised in nursery.

37. BUCKLAI

Botanical: ***Castanopsis lancaefolia*** Hickl. & Camus. Syn. *C. roxburghiana* Biswas. *Q. lancaefolia* Roxb.,

Nepalese: Patle katus

Assamese: Bucklai Garo: Shingra, Chauko Khasi: Dingsning

Lepcha: Siri

Vernaculars: Bucklai, Chakho-misri, Dieng-pathus, Dieng-rit, Dieng-snig-rit, Palte -kotus, Peltekong, Phong-rong-arong Sarong-jola, Shakma, Shi-rang, Siri

SYSTEMATIC POSITION

Bentham & Hooker (1883)	**Engler & Prantl (1899)**	**Hutchinson (1959)**
Phanerogams	Phanerogams	Angiospermae
Dicotyledones	Dicotyledoneae	Dicotyledones
Monochlamydeae	Archichlamydeae	Lignosae
(or incompletae)	Fagales	Fagales
Unisexuales	Fagaceae	Fagaceae
Cupuliferae		
Quercineae		

Geographical distribution: Nepal to Arunachal Pradesh, Assam, Meghalaya, Nagaland, Indo-China and Myanmar at an altitude between 1350-6600 ft. ASL.

Habit: A large evergreen monoecious tree with all glabrescent plant parts but inflorescence is pubescent.

Habitat: Sub-tropical to Sub-temperate localities.

Root: Branched tap root.

Stem: Aerial, erect, woody, solid, branched, bark greyish-brown with close fissures and cracks, thickness 4.0 mm.

Leaves: Simple, lanceolate or oblong - lanceolate, entire, thinly coriaceous, glabrous and shining above, lustre metallic below, apex abruptly acuminate, base acute or obtuse, secondary nerves 9-10 pairs, occassionly upto 16 pairs, prominent and arcuate, petiole 10-17 cm long.

Inflorescence: Unicellular spikes, occasionally androgynous, panicled, large, terminal.

Flower: Bracteolate, unisexual, incomplete, actinomorphic, epigynous, small.

Male flower: Solitary, puberulous, perianth segments 6, stamens 12.

Female flower: Panicled or few near apex of the spikes, styles 3, ovary inferior.

Fruit: Nut. Glans glabrous, rugose, 2.5 x 1.2 cm ovoid, crustaceous, marked with 4-6 ridges, ridges denticulate or entire, styles united and penetrating the apex of the glan, base oblique in ripe fruits. Cupules completely enclose the glans and split irregularly to set free the glans.

Uses: It is one of the secondary food plant of oak tasar silkworm viz., *Antheraea proylei, A. pernyi, A. roylei* and *A. frithii.* (Seth 1995b).

The wood is used for house building. Bark, wood and leaves contain tannin and the bird catchers employ acorns as bait (Negi and Naithani 1995).

Cultivation: It can be propagated through either direct sowing of seeds *in situ* or transplantation of seedlings.

38. CHAUKHU

Botanical: ***Castanopsis purpurella*** (Miq) Balak Syn.*Castanea purpurella* Miq., *Castanopsis hystrix* A. DC., *C. rufescens* Hook.f. & Thoms

Vernaculars: Chaukhu, Arkhala, Dieng-stap-dih, Dieng-sohstap, Sarang-baphang, Bon-belphoi,Tangji, Sethem, Sirang-asing, Kata-singra, Tshe-shing, Phong-rong-arong.

SYSTEMATIC POSITION

Bentham & Hooker (1883)	Engler & Prantl (1899)	Hutchinson (1959)
Phanerogams	Phanerogams	Angiospermae
Dicotyledones	Dicotyledoneae	Dicotyledones
Monochlamydeae	Archichlamydeae	Lignosae
(or incompletae)	Fagales	Fagales
Unisexuales	Fagaceae	Fagaceae
Cupuliferae		
Quercineae		

Geographical distribution: Nepal, Arunachal Pradesh, Assam, Meghalaya, Manipur and Mizoram. Also found in Bangladesh, Myanmar, Indo-China, South China and Taiwan at an altitude of 1800-2000 ft. ASL.

Habit: Large evergreen tree with pubescent young parts.

Root: Branched tap root.

Stem: Aerial, erect, branched, fluted usually, bark greyish cinnamon-brown, plain with horizontal wrinkles and vertical lenticels, thickness 6 mm.

Leaves: Simple, lanceolate or narrowly elliptic, entire or sparingly toothed towards apex, coriaceous, 7.5-18.0 x 2.5-7.5 cm., dark glossy green above, minutely scaly tomentose beneath, apex acuminate, base acute, obtuse or rounded, usually oblique, secondary nerves 7-9 pairs, petiole 7-14 mm long.

Inflorescence: Unisexual stout spikes in small panicles, male spikes numerous, female few.

Flower: Unisexual, incomplete, actinomorphic, epigynous, bracteolate, bracteoles pubescent, scarious, female flowers in threes.

Fruit: Solitary, 2.5-5.0 cm long, congested in spikes. Involucre splits into 3-4 parts exposing the glans; spines compound, pubescent, usually 3 fringed, covering whole nut surface, wiry, 2 mm long usually sharp pointed.

Nuts ovoid, more or less 3 angled and approximately 1.2 cm long.

Flowering: December - March

Fruiting: Cold season

Uses: It serves as one of the secondary food plants of temperate tasar silkworm *Antheraea proylei, A. pernyi, A. roylei* and A. *frithii* (Seth 1995 b).

Wood is used for making shingles, cart-shafts, axles, naves, poles, yokes and ploughs. After preservation treatment, railway sleepers of fairly good quality may be made from its timber. Nut is edible. Bark contains 11-13% and leaves 12% tannins (Negi and Naithani 1995).

Cultivation: It can be cultivated either directly by *in situ* sowing of seeds or transplantation of seedlings raised in nursery or growing in jungle.

39. JANGALISARU

Botanical: ***Casuarina equisetifolia*** Forst.

Burmese: Tinyu, Pinle-Kabwe

English: Beefwood tree | Hindi: Jangali saru, Jan

Bengali: Jau | Gujarati: Vilayati saru

Kannada: Sarve mara, Chabaku

Malayalam: Chavukku | Marathi: Suru | Tamil : Savukku

Telugu: Sarugudu, Chowku

SYSTEMATIC POSITION

Bentham & Hooker (1883)	Engler & Prantl (1899)	Hutchinson (1959)
Phanerogams	Phanerogams	Angiospermae
Dicotyledones	Dicotyledoneae	Dicotyledones
Monochlamydae	Archichlamydeae	Lignosae
Unisexuales	Verticillatae	Casuarinales
Casuarinaceae	Casuarinaceae	Casuarinaceae

Geographical distribution: Naturally occurring in Andamans, Chittagong and Tenasserim. Extensively planted in South India along coastal regions of both the Peninsulas. Also found in Australia, Pacific islands, Malay Peninsula and Archipleago.

Habit: Evergreen woody, xerophytic tree reaching more than 30 meters in height with long slender, drooping, jointed, leafless branches, appearing like a Pine or *Ephedra*.

Habitat: It thrives best near seacoast on loose sand.

Root: Tap root.

Stem: Aerial, erect, hard, woody, branched, branches drooping, green when young, the ends of branches thickly set with numerous long, slender branchlets which are deciduous, whorled and jointed, internodes short, 0.4-0.6 cm long, striated, 6-8 ribbed with fine hairs at the bottom of furrows between the ribs and stomata in the furrows only.

Leaves: Scale like, 6-8 whorled, connate basally forming a sheath about the twig and the leaf tips appearing only as teeth at each node.

Inflorescence: Male flowers in 2.0-3.0 cm long catkins like spikes, females in heads.

Flower: Monoecious, unisexual without perianth, bracteate, actinomorphic, diamerous, hypognous, male and female flowers sometimes found on the same branches, female flowers lower down in whorls while male spikes at the terminal end.

Male flowers: At the ends of branchlets, each with 2 inner big and 2 outer small obovate, serrately margined bracteoles, arranged in two whorl arising from within leaf sheaths, stamen-1, anther large, basifixed, tetramerous, dehiscing by vertical stits.

Female flowers: Capitate at tips of lateral branchlets, bracteolate, bracteoles 2, persistent, enclosing naked flower. Carpels 2, syncarpous, ovary superior, bilocular at first but sometimes unilocular, ovules two in each loculus, placentation parietal, style 1, very short, stigma bifid, linear.

Fruit: Samara, small winged, 1-seeded, enclosed by two woody bracteoles and many bracts all of which open like a 5-valved capsule at maturity. Fruits and their bracteoles aggregated into a dry woody cone like multiple fruit 2 cm in diameter. Seeds non-endospermic.

Flowering: May - November

Floral formulae: Male: ⊕ ♂ Brl_{2+2} P_0 A_1

Female: ⊕ ♀ Brl_2 P_0 $G_{\underline{(2)}}$

Uses: This plant is extensively cultivated for construction wood and fuel (fire wood) in coastal regions because it withstands frequent pruning and grows very fast. It is also raised as an hedge plant, along roadsides as an avenue tree and as an ornamental plant in plains. The production of wood is rapid 250 cubic ft/acre/annum (Brandis 1906) or 7-8 tonnes/ha/year (Sagreiya 1967). The ability of *C. equisetifolia* to grow on Nitrogen poor soils is due to its association with *Frankia*, the symbiotic N_2 fixing actinomycete which forms nodules on its roots. But this ability of N_2 fixation of *C. equicetifolia* exhibited large variations (Gauthier *et al.* 1985).

The wood is extremely hard cracks and splits but makes excellent fuelwood and known as 'iron wood'. The tree is very useful for dune stabilization and reclaiming the stretches of sand near sea and is of great significance in Madras and Bombay Presidencies for planning on saline sand dunes and wasteland along the coasts of Coromandel and North Kanara. Other trees and shrubs come up under its shetter since it prepares the ground for a more varied forest growth (Brandis 1906, Shukla and Mishra 1979). The bark of the tree is used for tanning (6-8% tannins) and cures diarrhoea and dysentery (Shukla and Mishra 1979). This can be used as a very good multipurpose tree (MPT) for windbreak and agro-forestry purposes (Chundawat and Gautam 1993).

Wood used for house posts, masts of country made crafts and rafters. Decoction of leaves used in colic while needles used to make activated carbon and tree yields a resin (Ambasta *et al.* 1986).

Cultivation: Through seeds and seedlings. The seeds may be collected during June and December and nursery may be raised during Feb-Nov. The seedlings may be transplanted during June-July (Chundawat and Gautam, 1993).

Seeds collected from 2-3 years and 7 years old tree are characterized by poor germinability and low vigour while seeds from 5 years old trees possess high germination and vigour hence seed should be collected from 5 years old trees (Kaja Maideen *et al.* 1997). Though vegetative propagation is possible, the rate of multiplication is very slow in this species (Goor and Barney 1976). Hence, tissue culture has been attempted for micropropagation and plantlets production has been achieved through indirect organogenesis (Parthiban *et al.* 1997).

Cladode cuttings collected from four different positions and kept for rooting in a polytunnel after giving a hormone treatment (2000 mg/l IBA) revealed a decreasing trend in rooting ability (%) as cuttings are collected progressively from lower (80.75%) to upper (46.15%) parts of the tree. The juvenile material responded fully to rooting whereas in mature tissues, the rooting response is reduced but not completely suppressed (Venkataramanan *et al.* 1998).

40. TOONA

Botanical : ***Cedrela toona*** Roxb. Syn. *Toona ciliata*

Burmese: Thikado, Tawtama English: Red cedar, Toon Tree

Hindi: Toona, Tun, Tuni,Tunni, Mahanim

Assam: Poma Bengali: Tun, Mahanim

Coorg: Noga Kannada: Tundu, Mandurike, Devdari

Kol: Katangai Mayalayam: Malarveppu, Mothagiri Vembu

Manipuri: Tairel, Tairen Marathi: Todu, Kuruk Sanskrit: Nandi Vriksha

Tamil: Thevathavam, Shanthanavembu, Tunumaram Telugu: Nandichettu

SYSTEMATIC POSITION

Bentham & Hooker (1883)	Engler & Prantl (1899)	Hutchinson (1959)
Phanerogams	Phanerogams	Angiospermae
Dicotyledones	Dicotyledoneae	Dicotyledones
Polypetalae	Archichlamydeae	Lignosae
Disciflorae	Geraniales	Meliales
Geraniales	Meliaceae	Meliaceae
Meliaceae		

Geographical distribution: Sub-Himalayan tracts from the Indus eastwards ascending to 5000 ft. ASL and valleys of outer Himalayas up to 1250 m ASL. Evergreen forests of the Western Ghats and other hills of Western Peninsula, Assam, Khasi hills (Meghalaya), Manipur, Myanmar in moister forests (Brandis 1906, Sagreiya 1967). Usually cultivated in Punjab, Assam, Bihar, Maharashtra and Madhya Pradesh.

Habit: A large fast growing, nearly evergreen tree (Fig. 18).

Habitat: Evergreen humid forests in almost all types of soil. Cultivated in road sides as shade tree.

Root: Tap root, branched.

Stem: Erect, aerial, branched, woody, cylinderical, straight, wood red, not hard, shining, does not split or warp, annual rings marked by wider vessles.

Leaves: Compound, imparipinnate, glabrous, leaflets opposite or nearly so, 5-10 pairs, lanceolate or ovate – lanceolate, acuminate, oblique, falcate, entire or undulate, blade 3-5 and petioles ¼ - ¾ inches long.

Inflorescence: Panicles, terminal, cymose, generally shorter than leaves, the last ramifications pubescent.

Flower: Hermaphrodite, complete, regular, actinomorphic, hypogynous, white, honey scented, 1/5 –1/4 inch long, pentamerous.

Calyx: Sepals 5, gamosepalous, ciliate, obtuse, green.

Corolla: Petals 5, polypetalous, glabrous but ciliate on edges, thickened inside, yellowish red on maturity.

Androecium: Stamens 5, inserted on 5 fleshy orange coloured hairy lobes of the disc, filaments glabrous, distinct, anthers bithecous.

Gynoecium: Carpels 5, syncarpous, ovary hairy, superior.

Fruit: Capsule, smooth, ¾ - 1.0 inch long, seeds with membranous wings at both ends.

Flowering: Feb - April.

Fruiting: May – July, after dehiscence empty capsules remain on tree for months.

Leaf fall: Cold season, gradually.

Leaf sprouting: Feb. – April.

Floral formula: ⊕ ⚥ $K_{(5)}\ C_5\ A_5\ G_{(\underline{5})}$

Uses: Cultivated along the road sids for shades and fuelwood. It is one of the nectar yielding trees (Randhawa 1983). The flowers yield a yellowish red dye (Gunari) which is used for dyeing cotton. The bark is used medicinally as astringent, tonic etc. for chronic dysentery of infants and external application for ulcers (Ambasta *et al.* 1986). The wood is red, soft, shining, used for furniture, tea chests, cigar boxes, shuttles and picking sticks, which are used in textile industry. The bark is also used as tan (Pandey 1981, Sagreiya 1967). The sp. gravity and calorific value of its fuel wood has been found to be 0.57 and 5168-5193 K cal/kg (Chundawat and Gautam 1993). Wood yields an essential oil. Wood is also used for floors, doors, windows, ceilings and plywoods (Ambasta *et al.* 1986).

Cultivation: Through seeds and seedlings. The size of seeds has no significant difference on germination. Seeds loose their germination when stored at room temperature (18-20°C), whereas at low temperature (6-8°C) they maintained 90% germination upto one year (Sharma *et al.* 2002).

41. BHEMLOTI

Botanical: ***Celastrus monosperma*** Roxb.

SYSTEMATIC POSITION

Bentham & Hooker (1883)	**Engler & Prantle (1899)**	**Hutchinson (1959)**
Phanerogams	Phanerogams	Angiospermae
Dicotyledones	Dicotyledoneae	Dicotyledones
Polypetalae	Archichlamydeae	Lignosae
Disciflorae	Sapindales	Celastrales
Celastrales	Celastraceae	Celastraceae
Celestrineae		

Geographical distribution: Sikkim, Assam, Khasi hills and Cachar ranging from 3000 - 6000 ft. ASL.

Habit: A climbing shrub.

Root: Tap root.

Stem: Branches not lenticelled, smooth.

Leaves: Simple, elliptic – lanceolate, alternate, petiole 0.8 – 1.2 cm long, blade 7.5 – 15 cm long.

Inflorescence: Cymes arranged in elongated axillary or terminal panicles.

Flower: Bisexual, actinomorphic, complete, regular, pentamerous, hypogynous, with well-developed disk.

Calyx: Sepals 5, small, free, imbricate.

Corolla: Petals 5, free, spreading, imbricate.

Androecium: Stamens 5, alternating with petals.

Gynoecium: Tricarpellary, trilocular, stigma trifid, ovary superior.

Fruit: Capsule 0.8 – 1.0 cm long, trivalved, single seeded.

Floral formula: ⊕ ⚥ K_5 C_5 A_5 G_3

Uses: It serves as tertiary but considered as one of the secondary food plants of muga silkworm, *Anthareae assama* in Assam (Thangavelu *et al.* 1988, Das *et al.* 2000).

Cultivation: It may be cultivated through seeds.

42. MALKANGNI

Botanical: ***Celastrus paniculatus*** Linn. Syn. *C. paniculata* Willd.

Burmese: Myinkoung, Nayoung   Eng.: Black oil plant
Hindi: Malkangni, Lakhan   Sinhalese: Duhudu
Bengali: Malkangni   Kannada: Kariganne, Kanaondi
Gujarati: Malkangna   Malyalam: Palulavam
Marathi: Kangni   Oriya: Katopesu
Punjabi: Sankhu   Sanskrit: Jyotishmati
Tamil: Valulavai   Telugu: Bavanji   Urdu: Malkanguni

SYSTEMATIC POSITION

Bentham & Hooker (1883)	**Engler & Prantle (1899)**	**Hutchinson (1959)**
Phanerogams	Phanerogams	Angiospermae
Dicotyledones	Dicotyledoneae	Dicotyledones
Polypetalae	Archichlamydeae	Lignosae

Disciflorae	Sapindales	Celastrales
Celastrales	Celastraceae	Celastraceae
Celestrineae		

Geographical distribution: Sub-himalayan tract from Jhelum eastward through Central India and Western Peninsula ascending upto 6000 ft. ASL. Also found in Pegu and upper Myanmar.

Habit: A large deciduous shrubby climber with stems upto 23 cm in diameter and 18 meters height.

Root: Branched tap root.

Stem: Aerial, erect, branched, woody, wood yellowish, medullary rays broad, twigs fairly smooth, reddish-brown and densely covered with small elongate whitish lenticells which are opening meant for diffusion of gases and respiration.

Leaves: Simple, alternate, serrate, acuminate, elliptic-ovate or obovate, blade 6-12 cm long, petiole 0.6-1.3 cm long.

Inflorescence: Panicled cymes, panicles terminal, drooping and 5-20 cm long.

Flowers: Bisexual, actinomorphic, complete, regular, greenish-yellow, 3.5-4.0 mm in diameter, pentamerous, disk broad, concave, hypogynous.

Calyx: Sepals 5, small, imbricate, free.

Corolla: Petals 5, free, spreading, imbricate.

Androecium: Stamens 5, inserted on the edge of disk, alternating with petals.

Gynoecium: Carpels 3, trilocular, stigma trifid, ovary superior.

Fruit: Capsule 3-valved, 1-1.3 cm in diameter, bright yellow in colour, ovules 2 in each cell.

Seed: 3-6, completely enclosed in an red or orange aril, copious albumen.

Flowering: March - July.

Floral formula: ⊕ ⚥ $K_5\ C_5\ A_5\ G_{\underline{3}}$

Uses: It serves as tertiary host plant but recognised as one of the secondary food plants of tropical tasar silkworm, *A. mylitta* D. in Central India (Seth 1995a). Besides, it has medicinal importance also. Decoction of leaves promotes menstrual flow while decoction of seeds is laxative, appetizer, causes vomiting and used as a remedy for cough, asthma, leprosy, headache, leucoderma and promotes the removal of mucous secretions from the bronchial tubes. The decoction of bark is used for abortion whereas the oil from seeds serves as nerve stimulant, brain tonic and checks rheumatic pains (Kurian 1995). The seed oil is also used in gout, paralysis and epilepsy

(Ansari 1997). This oil is edible and may be exploited commercially for production in large scale because the seeds contain 35-50% oil contents (Jain *et al.* 1990). Celastrine, an alkaloid and a black empyreumatic oil oleum Nigrum have been isolated from seeds. Bark is abortifacient, while seeds are tonic, aphrodisiac, nerve stimulant, brain tonic and used for rheumatic pains (Ambasta *et al.* 1986).

Cultivation: It can be propagated through seeds.

43. BHERA

Botanical: *Chloroxylon swietenia* DC. Syn. *Swietenia chloroxylon* Roxb.

Eng.: East Indian Satinwood
Hindi: Bhera, Bhirra, Girya
Gond: Bhira
Gujarati: Halda, Bheira, Billu, hardi
Kannada: Hurgalu
Kol: Sengel - sali
M.P.: Behra, Girya
Marathi: Halda, Bheira, Billu, Hardi
Tamil: Porasu
Telugu: Billa

SYSTEMATIC POSITION

Bentham & Hooker (1883)	Engler & Prantl (1899)	Hutchinson (1959)
Phanerogams	Phanerogams	Angiospermae
Dicotyledones	Dicotyledoneae	Dicotyledones
Polypetalae	Archichlamydeae	Lignosae
Disciflorae	Rutales	Meliales
Geraniales	Rutaceae	Meliaceae
Meliaceae		

Geographical distribution: Common in the deciduous forests of the Western Peninsula from the Konkan to the Niligiris. Northern edge of Satpura is its outer limit.

Habit: A middle sized deciduous tree, young parts, petiole and inflorescence clothed with short grey pubescence, smelling .

Root: A much branched tap root.

Stem: Erect, woody, solid, branched, wood yellow, heavy lustrous beautiful satin, inner wood darker without distinct heartwood, fragrant.

Leaves: Compound, paripinnate, 12.0-18.0 cm long, exptipulate, leaflets 10-20 pairs, opposite, sub-opposite or alternate, glabrous, pale, oblique (bases very unequal both sides), dotted on lower surface (transluscent glands present on the leaves), shortly petioled, apex obtuse, 2.5-3.0 cm long and 0.6-0.8 cm broad.

Inflorescence: Small terminal or axillary panicles, seldom as long as leaves.

Flower: Bisexual, complete, regular, small 0.6 cm across, actinomorphic, cream coloured, pentamerous, hypogynous, disk pubescent, fleshy and 10 lobed, pedicels longer than the flowers.

Calyx: Sepals 5, gamosepalous, deeply lobed, imbricate.

Corolla: Petals 5, clawed, spreading, imbricate in bud, unguiculate.

Androecium: Stamens 10, inserted at the disk outside, free, filaments filiform - subulate, alternate ones longer, anther cordate, small, apiculate, versatile.

Gynoecium: Carpels 3, ovary immersed in the disk, syncarpous, 3-celled and 3-lobed, each cell with 8 ovules, styles short, slender, glabrous, stigma small and capitate.

Fruit: Capsule, oblong, 2.5 x 1.3 cm, thick, coriaceous, 3-celled and loculicidally 3-valved.

Seeds: Exalbuminous, compressed, imbricate, oblong, winged.

Flowering: March – April.

Sprouting: May.

Floral formula: ⊕ ⚥ $K_{(5)}\ C_5\ A_{10}\ G_{(\underline{3})}$

Uses: It serves as tertiary host plant but recognised as one of the secondary food plants of tropical tasar silkworm, *Antheraea mylitta* D (Seth 1995a). It also yields good quality timber known as satinwood, which is used for making agricultural implements (Brandis 1906, Sagreiya 1967). Wood is durable, insect and fungus resistant so used for cabinet works, brush backs, panelling, poles, carts and railroad ties (Ambasta *et al.* 1986). The crushed leaves are used to cure wounds (Singh 1990). Leaves also used in rheumatism. Seeds yield a non-drying oil, wood an essential oil, tree an amber and a yellow dye (Ambasta *et al.* 1986).

Cultivation: It can be cultivated through seeds and seedlings.

44. MALLIGIRI

Botanical: ***Cinnamomum cecicodaphne*** Meissn. syn. *Cecicodaphne glauscence* Nees., *Tetranthera glaucescens* Wall. *Laurus glaucescens* Herb. Ham.

Nepali: Malligiri

Assamese: Ganasarai, Gundroi, Goudhori

Bengali: Rohu

Lepcha: Rohu

SYSTEMATIC POSITION

Bentham & Hooker (1883)	Engler & Prantl (1899)	Hutchinson (1959)
Phanerogams	Phanerogams	Angiospermae
Dicotyledones	Dicotyledoneae	Dicotyledones
Polypetalae	Archichlamydeae	Lignosae
Thalmiflorae	Ranales	Laurales
Laurales	Lauraceae	Lauraceae
Lauraceae		

Geographical distribution: Sub-Himalayan belt from Sikkim eastwards in Assam, Meghalaya (Khasi hills) and Manipur upto 4000 ft. ASL. Also found in Bhutan, Nepal and Bangla Desh.

Habit: A tree of moderate size.

Root: Branched tap root.

Stem: Erect, branched, stout, smooth and black on drying.

Leaves: Simple, alternate or sub-opposite, penninerved, elliptic, glabrous, blade 7-10 cm long, petiole 1.2-2.5 cm long, secondary nerves 4-8 pair, coriaceous, often glaucous ventrally, brown after drying, leaf buds enclosed in pubescent, large, orbicular, concave, imbricating, caducous scales.

Inflorescence: Panicles axillary, short, 2.5–4.0 cm long, crowded, peduncle slender, densely tomentose.

Flower: Bisexual, complete, actinomorphic, regular, small, few, pubescent outside as well as inside.

Perianth: Small, undifferentiated, biseriate of 6 laterally free segments, valvate, silky, persistent.

Androecium: Perfect stamens 9 or lesser, hairy, arranged in three whorls, inner most of 3 staminodes, outer eglandular and introse, middle biglandular and extrose.

Gynoecium: Moncarpellary, ovary superior, glabrous, unilocular, ovule solitary, pendulous, anatropous.

Fruit: Berry or drupe up to 3 cm long, oblong.

Floral formula: ⊕ ⚥ $P_{3+3}\, A_{3+3+3+3\,(Std.)}\, G_{\underline{1}}$

Uses: It serves as one of the tertiary but considered as one of the secondary food plants of eri and muga silkworms (Seth 1995c, Barah *et al.* 1992).

Wood strongly scented, highly valued for cabinet works and used for construction works, dugouts, planks, ploughs, rafters and yokes. It also yields an essential oil (Ambasta *et al.* 1986).

Cultivation: It can be cultivated through seeds and seedlings.

45. GANSARAI

Botanical: ***Cinnamomum glanduliferum*** Meissn. Syn. *Camphora glandulifera* Nees., *Laurus glandulifera* Wall.

Nepalese: Malligiri

Assamese: Gansarai, Gun serai

SYSTEMATIC POSITION

Bentham & Hooker (1883)	Engler & Prantl (1899)	Hutchinson (1959)
Phanerogams	Phanerogams	Angiospermae
Dicotyledones	Dicotyledoneae	Dicotyledones
Polypetalae	Archichlamydeae	Lignosae
Thalmiflorae	Ranales	Laurales
Laurales	Lauraceae	Lauraceae
Lauraceae		

Geographical distribution: Central Himalaya eastwards including Kumaon (Uttarakhand), Assam and Meghalaya (Khasi hills).

Habit: Similar to *C. cecicodaphne* Meissn. in more or less all characters

Root: Branched tap root.

Stem: Branches stout, smooth and black after drying.

Leaves: Alternate, elliptic or lanceolate, cordate – accuminate, penninerved, very variable in size and breadth, 7.5-12.5 cm long, thickly coriaceous, ventrally glaucous and brown after drying, petiole slender and 1.2-2.5 cm.

Inflorescence: Panicles axillary, longer than *C. cecicodaphne*, 5.0 cm long, peduncle long, very slender and glabrous, few flowered, nearly glabrous.

Flower: Shortly pedicelled, 2.5 mm across, sparsely pubescent outside, densely villous inside on six tapals arranged biseriately in valvate free fashion, stamens 9, arranged in three whorls, 3 staminodes arranged in inner most whorl, all stamens and staminodes hairy, ovary glabrous.

Fruit: Berry or drupe up to 3 cm long, oblong.

Floral formula: ⊕ ⚥ $P_{3+3}\ A_{3+3+3+3\ (Std.)}\ G_{\underline{1}}$

Uses: It serves as tertiary but regarded as secondary food plant of muga silkworm *Antheraea assama* and eri silkworm, *Samia Cynthia ricini* (Raja Ram and Samson 1991 a,b).

Wood is a good substitule for sassafros and contains d-camphor. Wood can be used for the same purposes as that of *C. cecidodaphne* and seeds yield a fat (Ambasta *et al.* 1986).

Cultivation: Through seeds and seedlings.

46. CHHANEJAM

Botanical: ***Cinnamomum obtusifolium*** Nees.

Burmese: Nalingyaw, Lulingyaw, Thitkyabo

Nepalese: Barasingoli

Assamese: Chhamejam, Pattichanda

Bengali: Tejpata, Ramtejpat, Kinton

Kumauni: Phatgoli

SYSTEMATIC POSITION

Bentham & Hooker (1883)	Engler & Prantl (1899)	Hutchinson (1959)
Phanerogams	Phanerogams	Angiospermae
Dicotyledones	Dicotyledoneae	Dicotyledones
Polypetalae	Archichlamydeae	Lignosae
Thalmiflorae	Ranales	Laurales
Laurales	Lauraceae	Lauraceae
Lauraceae		

Geographical distribution: Central and Eastern Himalaya ascending to 7000 ft. ASL in Nepal, Sikkim, Assam, Meghalaya (Khasi hills) and Andaman islands. Also found in Bangladesh and upper and lower Myanmar.

Habit: A large robust plant with large leaves, buds with small scales, buds and young shoots more or less silky.

Root: Branched tap root.

Stem: Aerial, erect, branched, woody.

Leaves: Simple, opposite, glabrous, thick, coriaceous, elliptic-oblong, obtuse, acute or acuminate, tripli-nerved, nerves not impressed dorsally, blade 20-27 cm long, petiole stout 1.2-1.8 cm long, glaucous ventrally.

Inflorescence: Panicles, stout, larger than leaves, subterminal and corymbiform, long peduncled, crowded.

Flower: Bisexual, complete, actinomorphic, regular, small, 1.2 cm across.

Perianth: Small, undifferentiated, biseriate, of 6 laterally free segments, valvate, subsilky pubescent, persistent.

Androecium: Perfect stamens 9 or lesser, anthers basifixed, filaments of first series eglandular with introse 4-celled anthers, the 3 inner 2-glandular at base with extrose 2-4 celled anthers, 3 or less staminodes of the inner most series cordate or sagittate, stamens sparsely hairy or glabrous.

Gynoecium: Carpel-1, ovary superior, hairy or glabrous, unilocular, ovules solitary, pendulous, anatropous.

Fruit: Berry, small ellipsoid or subglobose, 0.8-1.2 cm long succulent, persistent sepals 0.8 cm broad, rounded or narrow.

Floral formula: $\oplus$ ⚥ $P_{3+3}\,A_{3+3+3+3\,(std.)}\,G_{\underline{1}}$

Uses: It serves as one of the tertiary but considered as secondary food plants of muga silkworm, *Antheraea assama* (FAO 1976, Seth 1995d, Das *et al.* 2000). Leaves are used as spices.

Medicinally, root bark is used in dyspepsia and to end liver complaints (Sinha 1996). Wood suitable for planks and tea chests (Ambasta *et al.* 1986).

Cultivation: It can be cultivated through seeds and seedlings.

47. TEJPAT

Botanical: *Cinnamomum tamala* Nees & Eberm, Syn. *C. albiflorum* Nees.

Burmese: Thitchabo	English:Indian Cassia Lignea	Hindi: Tejpat
Assamese: Tejpat	Bengali: Tejpat	Gujarati: Tamalapatra
Manipuri: Tejpat	Sanskrit: Tamalpatra, Tamalaka	
Tamil: Talish - Pattireeya	Telugu: Talisapatri	

SYSTEMATIC POSITION

Bentham & Hooker (1883)	**Engler & Prantl (1899)**	**Hutchinson (1959)**
Phanerogams	Phanerogams	Angiospermae

Dicotyledones	Dicotyledoneae	Dicotyledones
Polypetalae	Archichlamydeae	Lignosae
Thalmiflorae	Ranales	Laurales
Laurales	Lauraceae	Lauraceae
Lauraceae		

Geographical distribution: Sub-Himalayan tracts and outer ranges ascending up to 7000 ft. ASL from Yamuna eastwards in Khasi hills (Meghalaya), Assam, West Bengal and Manipur. Also found in Bangla Desh, Sri Lanka and Bhutan.

Habit: A moderate sized 10.0-13.0 meter high, very aromatic evergreen tree, young shoots, more or less silky (Fig. 19).

Habitat: Damp ravines.

Root: Branched tap root.

Stem: Aerial, erect, woody, branched, cylinderical, bark slightly rough or smooth, thin, black or grey, aromatic.

Leaves: Simple, extipulate, elliptic, acute, entire, coriaceous, opposite or nearly so, often alternate on same old branches, 7.5 – 15.0 x 2.5-5.0 cm, petiole 1.00-1.25 cm long, young foliage pink, dark green above, light yellow below, tricostate parallel venation.

Inflorescence: Large axillary panicles, often from the axils of upper most leaves appearing terminal.

Flower: Bisexual, small, 0.5-0.7 cm in diameter, regular, hypogynous, actinomorphic, gray, silky, bracteate, bracts 3.

Perianth: Silky, unidfferentiated, biseriate, 3 in each whorl, free, valvate, persistent.

Androecium: Stamens 7, biseriate (4+3), the three inner 2-glandular at base, anthers 4-celled, staminodes 3, cordate or hastate.

Gynoecium: Monocarpellary, ovary superior, unilocular, ovules solitary, pendulous, anatarpous.

Fruit: Oval, black when ripe, succulent, 1.0-1.25 cm long, supported by the thickened pedicel and the base of perianth with short truncate teeth.

Floral formula: $\oplus$ ⚥ $Br_3\ P_{3+3}\ A_{4+3+3\ \text{staminode}}\ G_{\underline{1}}$

Uses: Leaves are carminative, used as spices and condiments. Medicinally bark and leaves of Tejpat are given in gonorrhoea, rheumatism, colic, diarrhoea, enlargement of spleen and snake bite (Sinha 1996). It is useful in disorders of digestive system viz., indigestion, vat, colic, diarrhoea and cough. It is also useful in stopping abortion, foetal bleeding, cold, fever and respiratory disorders. It is brain tonic, diuretic and good for liver and spleen (Choudhary 1997).

Bark is used to adulterate true Cinnamon. Leaves and bark yield essential oils. Leaves are used as Betle leaves in Kashmir (Ambasta *et al.* 1986).

Cultivation: Through seeds and seedlings.

48. DALCHINI

Botanical: *Cinnamoum zeylanicum* Breyn Syn. *C. verum* Presl.

Burmese: Hmanthia English: The Cinnamon Tree

Hindi: Dalchini, Darchini

Bengali: Dalchini Gujarati: Dalchini

Kannada: Dalchini, Lavangapatti, Lavanga-Chekke

Malayalam: Varana, Lavanga

Marathi: Dalchini Sanskrit: Darushila, Tamalapatra

Tamil: Kannalvangpatte, Karuva Telugu: Dakhini

SYSTEMATIC POSITION

Bentham & Hooker (1883)	Engler & Prantl (1899)	Hutchinson (1959)
Phanerogams	Phanerogams	Angiospermae
Dicotyledones	Dicotyledoneae	Dicotyledones
Polypetalae	Archichlamydeae	Lignosae
Thalmiflorae	Laurales	Laurales
Laurales	Lauraceae	Lauraceae
Lauraceae		

To counteract the strong unpleasant odour of burning flesh, Cinnamon has been employed in the oriental temples since antiquity. It is said that one-year supply of Cinnamon was buried by emperor Nero in A.D. 65 at the funeral rites of his wife Poppaea (Kocchar 1981).

Geographical distribution: Indigenous to Sri Lanka and South India, Cinnamon is cultivated in Seycelles, Malagasy Republic, China, Malaysia, Indonesia, Kenya, Tanzania, West Indies, South America and Myanmar. However, Sri

Lanka is leading exporter (60%). In India, it is cultivated in West Coast, especially Cannonore Dist. of Kerala and Western Ghats from South Kanara southward at low elevations and adjacent hill ranges.

Habit: A moderate sized evergreen tree usually attaining a height of 9-12 m, sometimes upto 18 m in Sri Lanka and South India. All parts are very aromatic. It differs from *C. tamala* chiefly by the more thickly coriaceous leaves and the more acute lobes of the much-enlarged fruiting calyx (Fig. 20).

Habitat: Sheltered places from almost sea level to the elevation of 3000 ft. ASL with average rainfall of 200-250 cm and mean temperature of 27°C. A hot and moist climate is ideal for it.

Root: Tap, branched.

Stem: Aerial, erect, woody, solid, cylinderical, bark slightly rough to smooth, grayish white.

Leaves: Simple, opposite, very thick, coriaceous, glabrous, 12.5 –17.5 cm in length, bluntly pointed, dark glossy green above and dull green below with 3 or 5 prominent parallel ribs, young foliage pink.

Inflorescence: Panicles as long as or slightly longer than leaves, branched, sometimes terminal.

Perianth: Undifferentiated, biseriate, small, 6, laterally free, gray or slightly yellow, silky, valvate, persistent.

Androecium: Stamens 7, biseriate, outer 4 and the inner 3 biglandular at base, anthers 4 celled, staminodes 3, cordate hastate.

Gynoecium: Monocarpellary, ovary superior, unilocular, ovules solitary, pendulous, anatropous.

Fruit: Berries, dark purple or blue, one seeded, ovoid, 1.0-2.0 cm long, supported by the much-enlarged perianth.

Floral formula: $\oplus$ ⚥ $Br_3\ P_{3+3}\ A_{4+3+3\ \text{staminod}}\ G_{\underline{1}}$

Uses: Two-three years after planting, when plants are 2.0-2.5 m high they are cut close to the ground following the monsoon rainfall to facilitate the peeling of the bark. Each coppicing produces new shoots for cutting in three years again. The inner dried bark known as Dalchini is used as a spice or condiment and sold as "the quills of commerce". The chips and refuse left after the collection of quills are used for extracting an essential oil.

Owing to the synthesis of cinnamic aldehyde, the use of Cinnamon as a spice has considerably declined, yet it is still used for flavouring cakes and pasteries, in beverages and as a constituent of curry powder. Slick cinnamon is an important ingradient of pickling and can be used to add flavour to blewed prunes. The bark has 0.5-1.5% essential cinnamon bark oil which contains cinnamic aldehyde (60-75%) as chief constituent. The cinnamon bark oil is used to flavour liqueurs, confectionery, pharmaceuticals, soaps and dental preparations. On the contrary, Cinnamon leaf oil has 70-95% eugenol and more preferred than clove oil for the synthesis of vasillin (Kocchar 1981). Cinnamon leaf oil is used as scents in sweets and perfumes in soaps for improving their quality (Jain 1968).

Dalchini is useful in dysentery, diarrhoea, vomiting and nausea (Jain 1968). Its flowers and fruits are edible. Bark is given in dyspepsia, cold and to check vomiting since it is astringent and carminative. Boiled extract of leaves is prescribed in cough (Sinha 1996). Cinnamon bark oil is antibiotic and antifungal, digestive, stimulant, tonic for intestine and used against flatulence (Jain 1968). Dalchini is useful for ulcers, indigestion, uterine prolapse, clearing menstruation, headache due to cold, increase in W.B.C in blood, typhoid, tonic for heart, diuretic, laxative and induces sleeping (Chaudhary 1997).

Cultivation: Through seeds, seedlings and cuttings (Kocchar 1981).

49. NALBALI

Botanical: ***Cipadessa fruticosa*** Blume Syn. *Mallea Rothii* A. Juss, *Ekbergia indica* Roxb., *Melia montana* Hb. Madr. ex. Wall. Cat.

Hindi: Nalbali — Oriya: Ranabili

Kannada: Chitundi — Tamil: Pulipan

SYSTEMATIC POSITION

Bentham & Hooker (1883)	Engler & Prantl (1899)	Hutchinson (1959)
Phanerogams	Phanerogams	Angiospermae
Dicotyledones	Dicotyledoneae	Dicotyledones
Polypetalae	Archichlamydeae	Lignosae
Disciflorae	Geraniales	Meliales
Geraniales	Meliaceae	Meliaceae
Meliaceae		

Geographical distribution: India, upper Myanmar, Sri Lanka, Java and Phillipines. In India, Chotanagpur, Orissa and Western Peninsula from Konkan southwards.

Habit: A shrub or small tree in dense moist forests.

Root: Branched tap root.

Stem: Erect, much branched, aerial, woody,

Leaves: Compound, alternate, 7-25 cm long, imparipinnate, odd pinnate, leaflets 3-5 pairs, 2.5-4.5 x 1.2-4.0 cm, shortly petiolate, entire or coarsely serrate, pubescent on the nerves underneath, ovate or elliptic, oblique at the base.

Inflorescence: Axillary panicles with long peduncles.

Flower: Complete, bisexual, regular, actionmorphic, hypogynous, white, subglobose, 0.3-0.4 cm long.

Calyx: Sepals 5, gamosepalous, small, cup shaped, 5 toothed.

Corolla: Petals 5, free, oblong, short, spreading, valvate in bud.

Androecium: Stamens 10, filaments broad, connate at base into a short staminal tube, bifid at the apex, anthers bithecous, short, inserted between the teeth. Disk shortly cup shaped adnate to staminal tube at the base.

Fruit: 5 - ribbed / furrowed, pentalocular, fleshy drupe, scarlet, 0.5-0.6 cm. across, each cell 1-2 seeded.

Seed: Endospermic.

Floral formula: ⊕ ⚥ $K_{(5)}$ C_5 $A_{(10)}$ $G_{\underline{(5)}}$

Uses: It serves as tertiary host plant but recognised as one of the secondary food plants of *Antheraea mylitta* D., tropical tasar silkworm in the forests (Seth 1995 a).

Cultivation: It propagates through seeds in nature.

50. NARIYAL

Botanical: ***Cocos nucifera*** Linn.

Burmese: On, Ohnthee English: Coconut palm French: Cocotier

German: Calappabaum Hindi: Nariyal, Narel

Persian: Drahakht, Narejile Singhalese: Polgedi

Bengali: Narikel, Dab Gujarati: Naliyer

Kannada: Tenginamar, Tengina

Konkani: Narla-maddo Malayalam: Teng, Thena Manipuri: Yubi

Marathi: Naral, Narel Sanskrit: Duraruha, Narikera, Narikela

Tamil: Tennai maram, Tenkai, Tengay

Telugu: Narikelamu, Tenkaya, Kobbari chettu Urdu: Nariyel

SYSTEMATIC POSITION

Bentham & Hooker (1883)	**Engler & Prantl (1899)**	**Hutchinson (1959)**
Phanerogams	Phanerogams	Angiospermae
Monocotyledones	Monocotyledoneae	Monocotyledones
Calycinae	Principes	Corolliferae
Palmae	Palmae	Palmales
		Palmae or Palmaceae

Once believed to be a native of the pacific coast of tropical America, now its place of origin has been proved to be somewhere in the Indo-Pacific region from where it scattered throughout the world by sea currents to the coastal regions (See Kocchar 1981).

Geographical distribution: Coconut is now cultivated in the Phillippines, Indonesia, India, Mexico, Papua New Guinea, West Malaya and Sri Lanka. While United States of America is biggest importer of coconut, India ranks third in its production. Coconut is mainly grown in Kerala, Tamil Nadu, Karnataka, Andhra Pradesh and to some extent in Goa, Diu and Daman, Andaman and Nicobar Islands, Maharashtra, Orissa, West Bengal, Lakshadweep, Pondicherry, Assam and Tripura.

Habit: A tall, handsome monoecious tree, up to 15-30 meter high and characterized by inclined trunk leaning towards the light at the edge of plantations and along the direction of the wind besides having prominent ring like leaf scars on straight, unbranched trunks (Fig. 21).

Habitat: Coastal and deltaic regions of tropical and sub-tropical countries.

Root: Adventitious, emerging from the base of the stem.

Stem: Trunk annulate, curved, rarely branching, base thickened with a mass of rootlets.

Leaves: Crowned, 20-30, large, paripinnate, 1.8-6.0 meter long, each weighing 11-15 kg, leaflets equidistant, linear lanceolate, 1.5-2.0 meters long, petiole stout.

Inflorescence: Spadix, axillary, central axis bearing up to 40 lateral branches, stout, androgynous, divided into numerous drooping spikes bearing at their base usually a single female flower with a few male flowers, the upper portion densely covered with 200 – 300 male flowers. The whole inflorescence is first enclosed by the spathe but subsequently comes out of it. Flowering occurs at 9-5 years of age.

Flowers: Unisexual, incomplete, actinomorphic, sessile, hypogynous, protandrous, females larger than the males.

Male flower: Tepals 6, small, unequal, arranged in two whorls of 3 each, outer being small, both whorls valvate, inner being 0.7 cm long, stamens 6, arranged in two whorls of 3 each, gynoecium rudimentary.

Female flower: Tepals 6, 1.5 cm in diameter, round concave, equal, arranged in two whorls of 3 each, free, valvate, tricarpellary, syncarpous, ovary superior, flower ovoid, supported by several broad bracteoles.

Fruit: Fibrous drupe, 3 cornered, usually ovoid, weighing 1.2-2.0 kg, differentiated into tough, smooth and hard exocarp, the middle thick fibrous mesocarp, constituting the coir of commerce; and the inner hard, dark brown endocarp with three ridges outside and the three eyes or depressions at the basal end. The one eye is larger and softer and contains a small embryo inside while other two smaller harder ones represent aborted carpels. The endocarp (shell) encloses a single hollow seed with a thin brown testa firmly adhering to the solid endosperm or meat – the copra of commerce. Coconut milk the liquid endosperm is filled in the cavity of the nut.

Flowering: Begins at 4-5 years of age and continues for more than 50 years. Flowering occurs throughout the year.

Fruiting: Ripening of fruits occurs 9-12 months after pollination.

Floral formulae: Male flower: $\oplus$ ♂ P_{3+3} A_{3+3} G_0

Female flower: $\oplus$ ♀ Brl_{α} P_{3+3} A_0 $G_{(\underline{3})}$

Varieties: Some dwarf varieties, which attain a height of about 12 meters, are also available.

Uses: Coconut palm more or less provides all the necessities of life, so it is nature's greatest gift to mankind, Due to its many and varied uses, it has been called as "Kalpavriksha", the tree of heaven.

The kernels are eaten raw or used in preparation of sweets, curries, chutneys and puddings. Ripe coconut is used as sacrificial or religious offerings to Gods in Hindu rituals. The coconut milk from tender nuts is highly prized as a refreshing and delicious drink. Coconut milk/water is used in tissue culture of plants due to presence of growth substances. "Toddy" obtained from unopened spadices is drunk fresh or may be boiled to produce palm sugar or jaggery. The fermented toddy is highly intoxicating beverage, but after 24 hours of fermentation it is unpalatable and hence it can be used for producing vinegar. "Coconut apple" (haustoria of the germinating nut) and palm cabbage (tender terminal bud) are eaten. Desiccated coconut is widely used in confectionery and bakery products.

Coconut oil, a white to yellowish solid, melts above 24°C and contains 44-51% lauric, 13.1-18.5% myristic, 7.5-10.5% palmitic, 0.2-0.5% caporic, 5.4-9.5% caprylic, 4.5-9.7% capric, 1.0-3.2% stearic, 0-1.5% arachidic, 5.0-8.2% oleic and 1.0-2.6% linoleic acids. Coconut oil has a very high saponification value so soap made from it lathers freely in hard as well as salt water also. The coconut oils are used for manufacture of margarine, lubricants, detergents, cosmetics (shampoos, creams), rubber, resins, hydraulic brake fluids for aeroplanes and candles. It is also used for illumination and annointing the body. The shells are used for making drinking vessels, buttons, combs, bangles and musical instruments. They are also used as fuel in corpra kilns. Powdered shells are used as filter in plastics. It yields wood tar and furfural on distillation. The leaves are used for thatching, mats, screens, baskets and other articles. "Poonac", the press cake is an excellent feed for cattle and poultry in the East. The tree trunk is used as a timber for house construction. Wood used for cabinets, buildings and as fuel.

The coir fibre obtained from mesocarp is used for manufacture of cordage particularly cables and hawsers for ships and sailing crafts, mats, matting and coarse cloth, rubberized coir for cushions, coir bags for mines, beltings for industries, packing materials, water proof, sound proof and fire proof boards and thermal insulator. It is also used as paper pulp, absorbent for nitroglycerine in making of explosives and filter for olive oil industries. While unspun fibre is used for chaulking boats, the bristle fibres are used for making brooms and brushes. The mattress fibres are used as a substitute for horsehair upholstery and used for mattresses. The by-products of coir industry viz., pith/bast and coir dust / cocopeat are used as manure and mulch respectively to improve the soil (Kocchar 1981).

Medicinally, the juice of the fruit and fresh or dry endosperm is useful in dysentery. Fruit is aphrodisiac and diuretic. Oil is useful in alopecia and loss of hair after fever and debilitating diseases. Coconut milk is cooling, laxative, diuretic and pyrigesic (Sinha 1996, Kocchar 1981). The decoction of root promotes urine flow and used in uterine diseases, liver complaints, bronchitis and dysentery. The infusion of young roots is used as gorgles for sore throat. The fresh juice of flower stalks is useful in constipation and after mixing with rice flour and gentle heating; it is applied to gangrenous and painless ulcers and to infections caused by boils. The ash of the bark and stem is antiseptic and used for scabbies and toothache. The jelly of the tender coconut is given in intestinal inflammations and sore throat in children (Kurion 1995).

Cultivation: Coconut is propagated through seeds. The germination is slow and shoot emerges from the eye of the nut after four months but the young plant is nourished by food reserves laid down in various parts of the fruit for the first nine months or more. Palm requires plenty of sunlight, temperature between 27-32°C and rainfall between 1270-2550 mm/annum. A porous soil behind beaches is most suitable for the growth of coconut trees but it may also grow in alluvial, volcanic, lateritic or clay soils, if they have good drainage and adequate soil aeration.

51. MASURI

Botanical: *Coriaria nepalensis* Wall.

English: Mussoorie Berry
Nepalese: Bhojinsi
Hindi: Masuri, Makola
Kashmiri: Balel, Tadrelu
Tankhul (Manipur): Guipan

SYSTEMATIC POSITION

Bentham & Hooker (1883)	Engler & Prantl (1899)	Hutchinson (1959)
Phanerogams	Phanerogams	Angiospermae
Dicotyledones	Dicotyledoneae	Dicotyledones
Polypetalae	Archichlamydeae	Lignosae
Disciflorae	Sapindales	Coriariales
Sapindales	Coriariaceae	Coriariaceae
Coriarieae		

Geographical distribution: Himachal Pradesh (Shimla), Uttararakhand (Mussoorie, Kumaon) of North-Western Himalayas from 3000 to 7000 ft. ASL and Sikkim upto 11000 ft. ASL. Also found in Manipur (Khangkhoi and Sirohee), Bhutan and Myanmar (Shan hills) from 4000-6000 ft. ASL.

Habit: A large shrub with long spreading branches.

Root: Branched tap root.

Stem: Erect, branched, aerial, branchlets quadrangular, arched, sarmentose and spreading, bark reddish – brown, glabrous, lower branches opposite.

Leaves: Simple, opposite, minutely serrulate, sessile, exstipulate, glabrous or puberu-lous ventrally, 2.5-5.0 x 1.8-4.0 cm, usually broadly rounded – cordate and abruptly acuminate, occasionally ovate-oblong or elliptic and acute, basal nerves 3-5.

Inflorescence: Becomes axillary or in the forks, clustered or alone, rachis glabrous or puberulous, 5-15 cm long with oblong, concave and deciduous bracteoles.

Flower: Pentamerous, hermaphrodite, small, 0.8 cm across, greenish – yellowish, hypogynous, disc absent.

Calyx: Sepals 5, free, spreading, imbricate, persistent.

Corolla: Petals 5, smaller than sepals, free, fleshy, keeled inside, thickened after flowering and embracing the fruit.

Androecium: Stamens 10, free, hypogynous, filaments short, anthers red, conspicuous, large, oblong and rough.

Gynoecium: Carpels 5, distinct, ovary free, unilocular, whorled on a fleshy receptacle, styles slender, subulate, fleshy and papillose, ovules one in each locule, pendulous.

Fruit: 5, 0.6-0.8 cm, oblong, compressed, dry little nuts closely enclosed by much enlarged purple, succulent petals which are keeled between the carpels on the inner face. Commonly called as Mussoorie berry.

Seed: Compressed, endosperm thin layered, testa membranous, cotyledons plano-convex and radicle superior.

Floral formula: ⊕ ⚥ $K_5\ C_5\ A_{5+5}\ G_{\underline{5}}$

Uses: It serves as one of the tertiary food plants of eri silkworm, *Philosamia ricini* (Seth 1995c, Sarmah 2000) and *Actias selene* Hubn. Leaves are poisonous in large doses. Fruit is eaten but causes colic (Sinha 1996). Wood rich in tannin, used for making boxes.

Cultivation: It can be cultivated through seeds as well as stem cuttings. As much as 60% rooting was observed by prolonged treatment (24 hours) of cuttings with IBA 100 ppm dose followed by IBA 200 ppm (53.37%) and 2000 ppm (56.67%) while NAA 50 ppm gave only 43.33% rooting (Chauhan *et al.* 1994).

52. SHISHAM

Botanical: *Dalbergia* sissoo Roxb.

Eng.: Sissoo Tree — Hindi: Shisham, Sissu

Bengali: Sissu — Gujarati: Sisam, Tanach — Kannada: Agaru, Biridi

Marathi: Sissu — Manipuri: Sissu — Malayalam: Iruvil

Oriya: Sisu, Simsapa — Punjabi: Tali, Shisham

Sanskrit: Shinshapa, Aguru — Tamil: Sisu-itti, gette

Telugu: Errasissu, Sinsupa

SYSTEMATIC POSITION

Bentham & Hooker (1883)	Engler & Prantl (1899)	Hutchinson (1959)
Phanerogams	Phanerogams	Angiospermae
Dicotyledones	Dicotyledoneae	Dicotyledones
Polypetalae	Archichlamydeae	Lignosae
Calyciflorae	Rosales	Leguminales
Rosales	Leguminosae	Papilionaceae
Leguminosae	Papilionaceae	
Papilionatae		

Geographical distribution: Sub-Himalayan tract and in outer valleys from Indus to Assam ascending generally to 3000 ft. and sometimes 5000 ft. ASL along river/canal banks in Bengal, Uttar Pradesh and Punjab. Also found in Baluchistan and Suleiman range.

Habit & Habitat: A moderate sized deciduous tree with a light crown chiefly occurring on alluvium. Wild as well as cultivated throughout India along canal banks and as an avenue tree on roadsides and as a shade tree in tea plantation as it is one of the few deciduous species which come into new flush of leaves during peak of summer (Fig. 22).

Root: Branched tap root.

Stem: Erect, aerial, branched, branches downy, bark grey, heartwood brown with darker veins

Leaves: Compound, imparipinnate, alternate, distichous, rachis zig-zag, stipules lanceolate, leaflets 3-5, broadly elliptic or ovate - orbicular, acuminate, entire, pubescent when young, glabrous when fully grown, 2.5-7.5 cm long.

Inflorescence: Axillary panicles with racemos-corymbose branches.

Flower: Bisexual, complete, zygomorphic, yellowish white, nearly sessile, pentamerous, papilionaceous, perigynous.

Calyx: Sepals 5, gamosepalous, campanulate, dentate, tooth subequal, odd sepal anterior and longest, green, aestivation ascending imbricate.

Corolla: Petals 5, pale yellow, polypetalous, papilionaceous, usually clawed, dissimilar, odd upper posterior large, conspicuous called as standard or vexillum is obovate, two lateral ones called wings are oblong and two anterior inner most farming a boat shape structure after fusion called keel or karina are obtuse, aestivation imbricate.

Androecium: Stamens 9, monadelphous, anthers bithecous, filaments united in a sheath splitting above.

Gynoecium: Capel 1, ovary superior, pubescent, style incurved, much shorter than ovary, stigma simple, unilocular, placentation marginal.

Fruit: Pod thin samara like, 1-4 seeded.

Seed: Non-endospermic with hard seed coat.

Flowering: March - May.

Sprouting: Summer.

Floral formula: ⊕ ⚥ $K_{(5)}\ C_{1+2+(2)}\ A_{(9)}\ G_{\underline{1}}$

Uses: It serves as one of the secondary food plants of tropical tasar silkworm, *Antheraea mylitta* D. in nature. (Jolly *et al.* 1974, Seth 1995a).

Medicinally, decoction of leaves is useful in gonorrhoea. Root is astringent, wood is alterative, useful in leprosy, boils, eruptions and to allay vomiting (Sinha 1996).

Timber is high class, used for building, furniture, carriage, carving cabinet, musical instruments, railway sleepers, charpai legs etc.

The pulp is used for making writing and printing papers.

Leaves used as a fodder. Wood is excellent fuel wood, suitable for charcoal making.

Planted in tea gardens to shade the tea bushes (Sagreiya 1967).

Cultivation: It can be easily cultivated through seeds or seedlings. Besides, it can also be successfully propagated through air layers easily (Nagpal and Sehgal 1985, Puri and Nagpal 1988). It can also be propagated through stem cuttings. According to Sharma and Pandey (1999), IBA 5000 ppm treatment to stem cuttings show significant effect on all parameters regarding rooting behaviour and *D. sissoo* showed more pronounced effect on aforesaid characters in comparision to *D. latifolia*. *D. sissoo* showed 79.33%, 89.3% and 62.33% rooting and 71.63%, 62.51% and 56.37% survival in IBA 2000, 5000 and 8000 ppm treatments respectively.

53. KALA-DHATURA

Botanical: ***Datura metel*** Linn. syn. *D. fastuosa* Linn.

Hindi: Kala-Dhatura, Sadahdhatura

Bengali: Dhutura Gujarati: Dhatura

Kannada: Madd-gunike, Ummata

Malayalam: Ummam, Ummatai
Manipuri: Shagot-hidak
Marathi: Dhotra, Dhutura
Oriya: Dhutura
Punjabi: Dhatura
Sanskrit: Dhatura, Dhustura
Tamil: Ummattari, Vellum-mattai
Telugu: Dhathuramu, Vellummatai

SYSTEMATIC POSITION

Bentham & Hooker (1883)	**Engler & Prantl (1899)**	**Hutchinson (1959)**
Phanerogams	Phanerogams	Angiospermae
Dicotyledones	Dicotyledoneae	Dicotyledones
Gamopetalae	Sympetalae	Herbaceae
Bicarpellatae	Tubiflorae	Solanales
Polemoniales	Solanaceae	Solanaceae
Solanaceae		

Fruits of Dhatura are offered to Lord Shiva alongwith Bilvapatra on the occasion of Mahashivaratri in India.

Geographical distribution: Wild in Western Himalayas up to 2500 ft. ASL and mountains of Western Deccan Peninsula. Now found throughout India wildly. Sometimes cultivated ornamentally. Also found in tropical America. It is wildly naturalized in the world. (Hooker 1883, Jain 1968).

Habit: An annual shrub or large herb reaching a height of 1.0-1.5 meters (Fig. 23).

Habitat: Abondoned waste lands outside the towns.

Root: Tap and branched.

Stem: Erect, branched, herbaceous, cylinderical, solid, glabrous.

Leaves: Simple, petiolate, existipulate, nearly opposite, ovate, entire or deeply toothed, pubescent, apex acute, base obtuse, venation reticulate unicostate.

Inflorescence: Solitary, terminal flowers.

Flower: Large, white or purple or whitish purple, solanaceous, bracteate, hermaphrodite, actinomorphic, pentamerous, pedicellate, complete, hypogynous.

Calyx: Sepals 5, gamosepalous, lobes large, campanulate, free above and connate below, green, persistent.

Corolla: Petals 5, white, light yellow or whitish purple, gamopetalous, tubular, bell shaped, twisted, 5 toothed, transplanted plants normally bear double flowers i.e. they have 10 petals arranged in two whorls of 5 each.

Androecium: Stamens 5, all fertile, filaments almost equal, alternate with petals, epipetalous, anthers bithecous.

Gynoecium: Carpels 2, syncarpous, ovary superior, tetralocular due to the formation of a false septum dividing placenta equally, placentation axile.

Fruit: Capsule, rounded, green, covered with short spines.

Seeds: Numerous, nearly smooth, pale brown, embryo curved, curvature amounting to semicircle or more.

Flowering: Autumn – winter.

Fruiting: Winter.

Floral formula: Br ⊕ ⚥ $K_{(5)}$ C_{o} $A_{(5)}$ $C_{(5)}$ $G_{(2)}$

Uses: Dried leaves, upper flowering branches and seeds are used medicinally. Leaves and seeds contain alkaloids, nicotine and daturin, especially hyoscyamine, scopolamine and atropine which produce sedation an intoxicating effect, when smoked or eaten, but continued use leads to disorder of the brain (Kocchar 1981, Jain 1968, Shukla and Mishra 1979). The poultice of leave is externally applied as a remedy on swelling of breasts caused due to increased lactation. The leaves are burnt and smoked for the treatment of asthma and bronchitis. The leaf paste along with turmeric is applied in chest pain. Leaves are applied on boils, sores and fish bites. Juice of the flower is applied in earache. Fruit juice is applied to scalp for curing dandruff and falling hairs (Sinha 1996, Jain 1968).

Cultivation: It may be cultivated through seeds or seedlings.

54. GUL MOHUR

Botanical: ***Delonix regia*** Syn. *Poinciana regia*

English: Peacock Flower, Gold Mohar

French: Fleur de paradis

Hindi: Gulmohur

Kannada: Doddaratna-gandhi

Malayalam: Alasippu

Marathi: Gulmohur

Tamil: Mayirkkondrai, Mayarum

Telugu: Ettaturavi, Shima sankesula

SYSTEMATIC POSITION

Bentham & Hooker (1883)	Engler & Prantl (1899)	Hutchinson (1959)
Phanerogams	Phanerogams	Angiospermae
Dicotyledones	Dicotyledoneae	Dicotyledones

Polypetalae	Archichlamydeae	Lignosae
Calyciflorae	Rosales	Leguminales
Rosales	Leguminosae	Caesalpiniaceae
Leguminosae	Caesalpinioideae	
Caesalpinioideae		

Geographical distribution: Although indigenous to the islands of Madagascar, it is distributed all over India (Randhawa 1983).

Habit: Nearly evergreen, open branched, spreading, irregularly shaped umbrella like handsome ornamental tree.

Habitat: Tropical-sub-tropical plains, susceptible to frost and cold, so they do not occur in cold regions.

Root: Tap, branched.

Stem: Erect, solid, branched, woody, bark grey, rough, branches spreading.

Leaves: Compound, long, bipinnate, pinnae entire, 8-20 pairs, stipules small, leaflets 20-30 pairs, green, giving feathery look to tree.

Inflorescence: Corymbose racemes.

Flower: Large, red or yellowish-red, hermaphrodite, complete, zygomorphic, hypogynous, bracts small, caducous.

Calyx: Sepals 5, gamosepalous, deeply cleft, valvate.

Corolla: Petals 5, spreading, racket shaped, wavy, clawed, scarlet, imbricate, borne in large bunches along and at the end of branches.

Androecium: Stamens 10, free, exserted, filaments hairy at the base.

Gynoecium: Moncarpellary, ovary flat, superior, unilocular, style clavate at the apex, stigma truncate, ovules many.

Fruit: Pods, 30-35 cm long, 5-7 cm broad, flat, many seeded, seeds oblong, variegated.

Flowering: April – June, July and sometimes also at the end of monsoon.

Fruiting: At the end of flowering period – end of the monsoon, first green turning to black and hard.

Floral formula: •|• ⚥ $K_{(5)}\ C_5\ A_{10}\ G_{\underline{1}}$

Uses: It is highly valued handsome avenue and ornamental tree. Looks very beautiful when planted together with Amaltas tree.

Seeds contain gum which may be used in textile and food industries (Ambasta *et al.* 1986).

Cultivation: Through seeds, seedlings or cuttings. The seeds are collected during Dec.-June and sown during April. The seedlings are transplanted during May-July (Chundawat and Gautam 1993). Seeds are soaked in hot water for 8 minutes before sowing. The young plants should be covered with thatch till they become strong enough to withstand the vagaries of nature especially frost during nights since Gulmohar is susceptible to frost.

55. BANSKABAN

Botanical: ***Dendrocalmus strictus*** Nees

Burmese: Myinwa | English: Male bamboo, Solid bamboo

Hindi: Bans, Khurd, Bans kaban

Bengali: Karail | Gond: Halpa, Wadur | Gujarati: Nakor vans

Kannada: Kiri-Bidaru | Malayalam: Kal, mungil | Manipuri: Khokwa

Marathi: Manwel, Bundi | Oriya: Salia bhanso, Salimbo bhanso

Sanskrit: Vansha | Tamil: Kal-Mungil

Telugu: Sanna Vedura, Kanka-Veduru, Sadanapa-veduru

SYSTEMATIC POSITION

Bentham & Hooker (1883)	Engler & Prantl (1899)	Hutchinson (1959)
Phanerogams	Phanerogams	Angiospermae
Monocotyledones	Monocotyledoneae	Monocotyledones
Glumaceae	Glumiflorae	Glumiflorae
Gramineae	Gramineae	Graminales
		Gramineae

Geographical distribution: In India it is found in the North-western arid regions, salt ranges, Aravali hills and Sub-Himalayan tract from Indus eastwards up to 3500 ft. In lower Myanmar, it is common on dry ridges and in Engdaing in upper Myanmar within 24°N (Brandis 1906).

Habit: Middle sized, deciduous, densely tufted, strongly thick walled with short, stiff spreading generally leaf less branches in the lower portion, 6-15 meters high.

Habitat: Deciduous forests with dry or moist climates (Fig. 24).

Root: Adventitious, fibrous, fasicled, stilt.

Stem: Culms grayish – green, often blotched, closely packed in dense clumps, strongly thick walled, the lower portion with short, stiff generally leafless branches, one of which is larger than the others, those on the upper portion slender and leafy. Internodes 25-38 cm long, 2.5-8.0 cm in diameter, almost solid in dry regions but hollow with thick walls in moist climate. Culm sheaths slightly shorter than internodes, glabrous or with yellowish-brown hairs, narrowing into the elongate-triangular leaf blade, which is hairy especially within.

Leaves: Simple, thin, divided into sheath and blade, densely clothed on both sides with deciduous hairs, linear, 10-25 cm long with 27-33 nerves on 0.7 cm.

Inflorescence: Spikelets hairy, 0.8-1.25 cm long, numerous in dense globose heads, 2.5-4.0 cm in diam. Glumes spinescent, empty, numerous, fertile 2-3.

Flower: Flowering irregular, few culms in a clump here and there or a few clumps in a locality come into flower while at times it may flower gregariously over large areas. Flowering may occur in most exposed situations in the hottest localities while the clumps growing in a sheltered and cooler region may remain green (Brandis 1906). Flowers sessile, hermaphordite, zygomorphic, hypogynous, bracts divided into lemma and palea.

Perianth: Lodicules none.

Androecium: Stamens 6, filaments free.

Gynoecium: Monocarpellary, ovary hairy in the upper part, style long, hairy, unilocular, superior.

Fruit: Caryopsis shining, hairy at apex, 0.6-0.8 cm long, ovoid, beeked, style persistent, embryo distinctly visible on the outside.

Flowering: March-April.

Fruiting: April-May.

Floral formula: •|• ⚥ $P_0 A_6 G_{(1)}$

Uses: Full grown culms are used for a variety of purposes. viz. posts, basket, walling, rafters, battens, scaffoldings, sticks, furniture, tent poles, water pipes, fishing rods, musical instruments, masts for country boats, roofing, matting etc and is therefore referred to as the poor mans timber. The above ground part is used for the preparation of household implements and making other goods for offering local deities in Mandakini Valley in Uttaranchal (Nautiyal *et al.* 2002). It is also the most important and the principal long fibred forest raw material so far used in India for making paper. Young shoots are eaten (Sagreiya 1967). Leaves are used as fodder.

The white deposit of calcium on the nodes is used as tonic, astringent and given to patients having calcium deficiency. Leaves are used as ecbolic to animals (Sinha 1996).

Cultivation: Usually through rhizomes, culm cuttings, nodal segments and saplings and rarely through seeds and seedlings. The saplings are raised during May-June and entire planting is done during July (Chundawat and Gautam 1993). Potential application of tissue culture for propagation of *D.strictus* has also been enormously investigated (Mascarenhas *et al.* 1990).

56. TENDU ABNUS

Botanical: ***Diospyros melanoxylon*** Roxb. Syn. *D. wightigana* Wall, *D. dubia* Wall., *D. exsculpta* Bedd., *D. tupru* Buch-Ham.

Eng.: The Ebonny of Northern India, Coromandel Ebony Persimmon

Hindi: Tendu Abnus, Tendu, Timburni

Gond: Tumri, Gujarati: Tamrug

Kannada: Abnasi, Bale, Tumari Malayalam: Kari

Marathi: Timburni, Tendu, Tumri

Oriya: Kendu Sanskrit: Dhirgha patraka

Tamil: Karai, Karundumbi, tumbi

Telugu: Tumiki, Beedi aakum, Tuniki aaku

SYSTEMATIC POSITION

Bentham & Hooker (1883)	**Engler & Prantl (1899)**	**Hutchinson (1959)**
Phanerogams	Phanerogams	Angiospermae
Dicotyledones	Dicotyledoneae	Dicotyledones
Gamopetalae	Sympetalae	Lignosae
Heteromerae	Ebenales	Ebenales
Ebenales	Ebenaceae	Ebenaceae
Ebenaceae		

Geographical distribution: Usually found in deciduous dry forests of the plains of Madhya Pradesh, Chattisgarh, Maharashtra, Chotanagpur (Jharkhand), Bihar, Orissa and Western Peninsula. To a lesser extent also found in Andhra Pradesh, West Bengal, Tamil Nadu, Gujarat and Rajasthan. Rarely found in Sri Lanka.

Habit: A middle sized deciduous tree attaining 50 ft. height and 6 ft. girth. Branchlets and inflorecence clothed with ferrugineous tomentum. It produces numerous root suckers around the tree.

Habitat: Tropical deciduous forests with high temperature and moderate rainfall.

Root: Branched tap root.

Stem: Aerial, erect, woody, solid, branched, bark dark grey to black, exfoliating in regular rectangular scales.

Leaves: Simple, mostly opposite, thickly coriaceous, elliptic or ovate, both sides hairy when young, hairy or glabrous on lowerside when matured, blade 7.5-30 cm, petiole 0.8 cm, secondary nerves 6-10 pairs, reticulate, tertiary nerves raised on the upper surface.

Inflorescence: Male flowers in short cymes, female flowers solitary.

Flower: Unisexual, incomplete, tetra - hexamerous, actinomorphic, hypogynous, calyx of male and female flowers dissimilar.

Male flowers: Calyx woolly, campanulate or cup shaped, gamosepalous. Corolla tomentose, gamopetalous, tubular, ferrugineous woolly outside, stamens 12-16, connate, anthers basifixed.

Female flowers: Calyx broad, edges recurved, ovary superior, 4-10 celled, single staminode may be present.

Fruit: Globose or ovoid berry, 2.5-3.0 cm across, edible, supported by enlarged thick, coriaceous calyx, yellow when ripe, calyx segments with more or less recurved edges.

Seed: 4-8, compressed, oblong, embedded in soft or viscid pulp, testa rugose, shining, albumen ruminate.

Flowering: April - May.

Fruiting: July.

Floral formulae: Male: ⊕ ♂ $K_{(4\text{-}6)}\ C_{(4\text{-}6)}\ A_{12\text{-}16}\ G_0$

Female: ⊕ ♀ $K_{(4\text{-}6)}\ C_{(4\text{-}6)}\ A_{\text{stm } 1.}\ G_{(\underline{4\text{-}10})}$

Varieties/Species: (i) ***Diospyros tupru*** Ham (Kannada: Tapa rada) : This species was separated from *D. melanoxylon* Roxb. by C.B.Clarke due to its smaller leaves, (8.5 x 5.0 cm), obtuse apex as well as base and densely fulvous woolly nature when young (quoted by Brandis 1906). Berry 1.8-2.0 cm across, globose smooth and 4 seeded.

(ii) ***Diospyros tomentosa*** Roxb. (Hindi: Tendu) : Found in Sub-himalayan tract from Ravi to Nepal, Eastern Rajasthan, Madhya Pradesh, Chattisgarh, Chotanagpur, Bihar, Orissa, Bengal, Bundelkhand (U.P.), Maharashtra (Poona) etc. Full-grown leaves exhibit impression of deep furrows of secondary and tertiary nerves on upper surface. Flowers and berry like that of *D. tupru* but rather larger. Gamble regarded that wood of *D. tomentosa* is lighter than that of *D. melanoxylon* (quoted by Brandis 1906). According to Brandis (1906) both the species are probably not distinct, coppice well and throw up root suckers in large numbers and consist of irregularly shaped masses of jet black ebony often with purple streaks in the centre of old stems.

Uses: *D. melanoxylon* and allied taxa/varieties/species serve as one of the tertiary food plants of *Antheraea mylitta* D. especially in Maharashtra and Andhra Pradesh for Bhandara and Andhra local ecoraces. However, it is recognised as secondary host plant (Jolly *et al.* 1974).

The tendu leaves are extensively collected by tribals for Bidi preparation as wrappers of tobacco, which gives employment to over 2 lakh people (Sagreiya 1967).

The berry of Tendu is eaten in small quantity. The ash of the bark mixed with coconut oil and ash of *Calotropis procera* (Ait.) R.Br. (Aak, Madar) is used on boils and blisters. The bark paste is used in wound healing (Maheswari and Singh 1991). The decoction of bark is also used for diarrhoea, dyspepsia and dysentery (Ansari 1997). Leaves are diuretic, laxative, carminative and styptic while dried flowers are used in urinary and skin troubles (Ambasta *et al.* 1986).

Cultivation: It may be propagated through direct sowing of seeds, transplantation of seedlings or root suckers.

57. SINATHA

Botanical: *Dodonaea viscosa* Linn. Syn. *D. burmanniana* DC.

Baluchi: Ghissani

Hindi: Sinatha, Kharata, Aliar

Bihari: Mehndi

Kannada: Bandari, Bundurgi

Malayalam: Unnataruvi

Marathi: Jakhmi

Oriya: Mohra

Punjabi: Sanatta, Mendru, Benmenu

SYSTEMATIC POSITION

Bentham & Hooker (1883)	Engler & Prantl (1899)	Hutchinson (1959)
Phanerogams	Phanerogams	Angiospermae
Dicotyledones	Dicotyledoneae	Dicotyledones
Polypetalae	Archichlamydeae	Lignosae
Disciflorae	Sapindales	Sapindales
Sapindales	Sapindaceae	Sapindaceae
Sapindaceae		

Geographical distribution: Common in drier regions of North-west and Central India, Deccan and on sea coast. Also found in Baluchistan, Afghanistan, Sri Lanka, Malay Peninsula and Archipelago. More or less in all tropical countries.

Habit: A gregarious evergreen shrub, rarely a small tree, reaching upto 7 meters.

Root: Branched tap root.

Stem: Erect, woody, branched, angular, young shoots mostly viscid with bright yellow resin.

Leaves: Simple, alternate, exstipulate, glabrous, lanceolate to linear, cuneate, coriaceous, shining, subsessile, margin entire or undulate, 2.5-12.5 cm long, 0.4-4.0 cm broad, midrib prominent, secondary nerves numerous.

Inflorescence: Terminal panicles short.

Flower: Regular, incomplete, actinomorphic, hypogynous, polygamous, sepals 5, imbricate companulate, oblong, 0.3 cm long, petals nil, stamens usually 8, filaments hairy, much shorter than anthers, disk inconspicuous.

Male flower: Stamens as long as sepals, disc obsolete.

Bisexual flower: Stamens shorter than sepals, carpels 3-4, ovary superior, style cylinderic, stigma bilobed, apcarpous, polylocular, ovules 2 in each locule, disc very small.

Fruit: A membranous capsule consisting of 2-4 broad wings from base to style, 1.2-1.8 cm diametre including wings, dividing ovary into as many chambers as carpels.

Seed: Opaque with a thick funicle, dark brown - black in colour, non-endospermic.

Flowering: Rainy season.

Fruiting: Cold season.

Floral formulae: Male flower : $\oplus$ ♂ $P_{(5)}\ A_8\ G_{\underline{0}}$

Female flower : $\oplus$ ♀ $P_{(5)}\ A_0\ G_{\underline{3\text{-}4}}$

Bisexual flower : $\oplus$ ⚥ $P_{(5)}\ A_8\ G_{\underline{3\text{-}4}}$

Uses: It serves as one of the tertiary food plants but recognised as secondary host plant of tropical tasar silkworm, *Antheraea mylitta* D (Seth, 1995 a).

Medicinally, leaf extract is applied to fresh injuries. Leaves are used in bath and fomentatin. Wood is carminative, slightly purgative and applied on wounds. Considered as useful in rheumatism (Sinha 1996). Leaves are also febrifuge, used on burns swellings and wounds. Bark contains tannins (5.8%). Seeds are edible and contain semi-drying oil. Wood used for tool handles, walking sticks, engraving and turnery (Ambasta *et al.* 1986).

It serves as one of the important plants for raising hedges (Rao 1991).

Cultivation: It can be cultivated through seeds. It sheds seeds profusely which should be collected and sown before the commencement of rains. Thinning is necessary to avoid overcrowding of seedlings. Sowing seeds in shallow pans and transplanting them is a cumbersome process. Planting should be avoided in water-logged areas until the soil dries up.

58. AMLA

Botanical: *Embellica officinalis* Gaertn syn. *Phyllanthus emblica* Linn., *Cicca embelica* Kurz., *P. taxifolius* Don., *P. glomeratus* Roxb., *Dichelastina nodicaulis* Hance.

Arabi: Ambalyi — Burmese: Tasha, Zibyu, Hziphyu

Eng.: Embellic Myrobalan, Indian Gooseberry, Malacca tree

Hindi: Amla, Aonla, Avala, Amlika.

Persian: Amlab, Amelah — Sinhalese: Nellika

Assamese: Amluki, Amlaki — Bengali: Amla, Amlaki — Gond: Nalli

Gujarati: Amali, Ambala — Kannada: Amalaka, Nelli — Kol: Miral

Konkani: Anvallo — Malayalam: Nelli — Manipuri: Heikru

Marathi: Avala — Mizo: Sunhlu — Rongmei: Tarlu

Sanskrit: Amalaki, Dhatri, Adiphala — Santhali: Meral

Tamil: Amalagam, Nelli — Telugu: Nelli, Usirkai, Usiriki, Amalaki

Urdu: Anwala

SYSTEMATIC POSITION

Bentham & Hooker (1883)	Engler & Prantl (1899)	Hutchinson (1959)
Planerogams	Phanerogams	Angiospermae
Dicotyledones	Dicotyledoneae	Dicolyledones
Monochlamydeae	Archichlamydeae	Lignosae
Incomplete	Geraniales	Euphorbiales
Unisexuales	Euphorbiaceae	Euphorpiaceae
Euphorbiaceae		

Phyllanthus has been derived from two Greek words *phullon*, which means leaf, and *anthos*, which refers to flowers. Hence, *Phyllanthus* as a whole means leaf-flower which refers to apparent bearing of flowers on leaves. *Embelica* is derived from the earlier name of the tree *Myrobalanus emblica* or *Embelica myrobolan* called by early pharmacists for its fruits. The word *Embelic* was probably derived from the Persian *Amlab* and Arabic Ambalii.

Geographical distribution: Common in deciduous forests (Wild) or cultivated in most parts of India from Jammu eastwards and southwards, except in the arid region and extreme North-Western region, ascending upto 4500 ft. altitude in the Himalaya. Also found in Sri Lanka, Myanmar, Malay Peninsula and Archipalago and China. The grafted variety from Varanasi is very handsome and produces large size fruit.

Habit: A handsome, deciduous, small-medium sized tree with smooth greenish-grey barks peeling off in concoid scales and mottled stem. Wood red and hard, medullary rays broad and conspicuous on a vertical section. It peculiarly sheds its twigs along with the leaves attached (Fig. 25).

Habitat: Tropical-subtropical/temperate climatic regions with medium to high rainfall.

Root: A branched tap root.

Stem: Aerial, woody, erect, branched, bark smooth, greenish to grey, peeling off in concoid scales, branchlets slender mostly deciduous, glabrous or finely hairy, 10-20 cm long.

Leaves: Simple, equal and symmetrically set, having the appearance of pinnate leaves, feathery, small, fine, delicate, distichous, closely set, narrowly linear – oblong, obtuse, subsessile, alternate, glabrous, approximately. 0.8-1.2 cm long, puberulous beneath when matured, imbricate when young, stipulate, stipules ovate, fimbricate, scarious, lacerate.

Inflorescence: Fasicles of small flowers in the axils of fibriate bracts on the leaf bearing branchlets usually on the naked portion below the leaves.

Flower: Unisexual, incomplete, actinomorphic, greenish-yellow, small, bracteate, regular, hypogynous, apetalous, monoecious, males numerous on short, slender pedicels, females few, subsessile, both on the same branchlet.

Male flower: Sepals 6, oblong, obtuse, stamens 3 on a short central coloumn, disc of 6 minute glands often wanting.

Female: Sepals 6, oblong, acute, disc cup shaped, edge toothed, ovary tricarpellary, superior, styles 3, bifid, connate at base.

Fruit: Fleshy, globose, 0.8-1.2 cm in diameter, with 6 obscurely vertical furrows, pale yellow, sometimes reddish on ripening comprised of 3 two seeded crustaceous cocii (6 seeded), acidic, astringent and bitter.

Flowering: March - May.

Fruiting: Nov. - Feb.

The irregular tubercular protuberances on old woody branches usually give rise to a succession of three deciduous branchlets bearing leaves and flowers. When branchlets develop on the current years shoot, they are not always deciduous and instead serve to form permanent branches. Such branchlets develop on current years shoot in the axil of dorsal bracts, which have two lateral stipules also.

Floral formulae: Male flower: ⊕ ♂ K_6 C_0 A_0 G_0

Female flower: ⊕ ♀ K_6 C_0 A_0 $G_{(3)}$

Uses: It serves as one of the tertiary food plants but recognized as one of the secondary host plants of tropical tasar silkworm, *Antheraea mylitta* D (Jelly *et al.* 1974). It also serves as host plant for lac insects *Kerria lacca* Kerr. (Sharma and Jaiswal 1997). Fruits, bark and leaves contain 28%, 8-21% and 22% tannins respectively (Ambasta *et al.* 1986).

The embelic myrobalan fruit is eaten and pickled or jam is prepared from fruits. The fruit, bark and leaves are used in tanning and dyeing (Pandey 1981). The timber is hard and red but little used as it splits badly. It yields excellent charcoal.

The fruits are richest in vitamin-C (Ascorbic acid) containing twenty times as much Vit. C as organic juice and also astringent, diuretic and laxative. They are given in diarrhoea, amoebic, dysentery, haemorrage, anaemia, gastritis, thirst (trishna roga), flatulence, gripping pains from stomach and bowels and jaundice etc. Fermented fruits are prescribed in cough and fruit juice is used in eye diseases. While flowers are said to be refrigerant, seeds are used for asthma and brochitis. The fruits are also used in constipation, bleeding gums (pyorrhoea), piles, haemoptysis (spitting of blood while coughing) and blood diseases besides being used as brain and nerve tonic. The boiled extract of leaves is known to control high blood sugar (Sinha 1996, Kurian 1995). Decoction of seeds takes care of asthma, bronchitis and leucorrhoea. Fruits also reduce fever, check vomiting, congestion of liver, leprosy, painful or difficult urination, cold in nose and nasal haemorrhages (Kurian 1995).

Amla is the most important ingredient of "chyavanaprash" which is useful in treating diseases of lungs like tuberculosis, branchitis and bronchial asthma, where the patient is emaciated and weak. The vitamin-C content of Amla does not get destroyed even after boiling for considerable period unlike many citrus fruits in which it is destroyed after exposure to heat of sun or fire. Medicated oil prepared by boiling of Bhringaraja (*Eclipta alba*) and Amla viz. Mahabbhringaraja taila is used externally for massaging the head and powder of these two drugs is also used internaly in a dose of one teaspoonful three times daily with milk against greying of hair. The powder of Bakuci mixed with the decoction of khadira and Amalaki is given in a dose of 1/2 teaspoonful twice daily to cure leucoderma. Amla powder and chyavanprash is also used for curing scurvy, menorrhogia (rakta pradara). Amla is given against loss of memory to the patient in the form of murabba, pickles and vegetables. "Triphala" powder is very popularly used for treatment of chronic constipation and stomatitis (Mukhapak) which contains Amla alongwith haritaki / Abhaya (*Terminalia chebula*) and bibhitaki (*T.belerica*) [Dash 1993, Ansari 1997].

According to Bhaishajya - ratnavali, the powder of dhatri (*Embelica officinalis*), arjuna (*Terminalia arjuna)* and abhaya (*T. chebula*) stops menstruation if taken with water for family planning (Dash 1993). Seeds of Amla have also been found to yield 16% oil (Jain *et al.* 1990).

Cultivation: Amla is commonly propagated through seeds or inarched plants. Most of the seeds are unfertile. The ripe fruits are left in the sun till they dry and the stone spits and seeds escape. The seeds should be sown in March and regularly irrigated. The growth is rapid during first few years but deteriorates subsequently (Randhawa 1983).

According to Regional Forest Research Station, Rajahmundary overnight soaking of seeds in mild hot water gives 78% germination and germination at peak is 43%. The seedlings require 20-30% culling in secondary stage (root trainer plants) to have uniformly good planting stock (Rao *et al.* 1998).

59. TUTGANTHA

Botanical: *Ephedra girardiana* Wall. Syn. *E. vulgaris* Brandis

English: Ephidrine, Yellow hemp — Vernacular: Tutgantha

Ladakhi: Tse, Teapat, Trano — Punjabi : Asmania, Budagur, Chewa

SYSTEMATIC POSITION

Bentham & Hooker (1883)	Engler & Prantl (1899)	Hutchinson (1959)
Phanerogams	Phanerogams	Gymnospermae
Gymnospermae	Gymnosperm	Gnetales
Gnetaceae	Gnetaceae	Ephedraceae

Geographical distribution: Sub-Himalayan tract between 8000-14000 ft. ASL, ascending in Sikkim up to 1650 ft. ASL. Also found in inner arid tracts. Wild in Kuram valley, Pamirs, Yark and Tibet. Cultivated on a large scale in China, India, Spain, USA, Kenya and Australia (Brandis 1906, Kocchar 1981).

Habit: A rigid, nearly erect shrub with numerous stems up to 7 feet height from a woody rootstock.

Habitat: Arid hilly regions of the temperate zones.

Root: Woody, much branched, usually creeping rootstock.

Stem: Aerial, erect, tuft of short branches with long jointed flutted, green, opposite or whorled branchlets resembling the common horsetails.

Leaves: True foliage leave lacking, a two or sometimes three-toothed sheath formed at each joint by opposite and decussately arranged scale leaves.

Flower: Dioecious, arranged in compound strobili resembling inflorescence. A perianth of two opposite, compressed membranous sepals present.

Male (staminate) strobilus: Male spikelets ovate, on short axis in axil of a leaf, 0.3-0.4 cm long, solitary or in pairs, rarely in whorls of 8; bracts thick, closely set, decussate imbricate in 2-8 opposite pairs, 1-2 pairs sterile, remaining pairs with solitary male flowers; anthers 5-8 with 2-3 lobes on a central coloumn in each flower.

Female (Ovulate) strobilus: The female strobili or spikelets, 0.6 cm long, usually solitary, occasionally in the whorls of 2-4 at the nodes of small green branchlets; supporting bracts 2-3 pairs, decussate straight, female flowers usually in pairs sometimes 3 or solitary with one ovule and single integument, prolonged into a tube, which has the functions of a style (tubilus) and enclosed in a bag which may be regarded as a perianth.

Fruit: Fleshy, consisting of succulent bracts, sweet, edible, red when ripe enclosing 1-2 ovoid, triangular seeds.

Uses: The potent alkaloids are extracted from the dried or fresh branches of the stem. The major alkaloid ephedrine (C_{10} H_{15} ON) (0.68%) constitutes nearly three-quarters of the total alkaloids (1.22%) including d- and 1-pseudo-ephidrine (iosephidrine) and ephidine. Zephrol containing ephedrine hydrochloride has been used as nasal drops to relieve nasal and sinus congestion and also for bronchial coughs. The ephedrine preparations are popularly used in asthma, hay fever, colds and night wetting of children. It counteracts over doses of depressants and causes an immediate and pronounced elevation in blood pressure since it stimulates the sympathetic nervous system when administered intravenously (Kocchar 1981, Anonymous 1962) The roots also contain a number of macrocyclic alkaloids (ephadrines) and feruloylhistamine which have hypotensive properties. Ephedras are also used as anti-inflammatory drugs and this action is ascribed to a recently isolated oxazolidone related to ephedrine (Evans 1989).

Cultivation: The seed has no obligatory resting period and usually germinates immediately. It can be grown in Northern India at an altitude of 2,500 m.

Kullu Valley is suitable for the cultivation of *E. girardiana* (Anonymous 1962).

60. LOKAT

Botanical: *Eriobotrya japonica* (Thunb.) Lindl.

English: Loquat, Japanese Medlar   Hindi: Lokat

Kannada: Lakkote   Tamil: Illakotta

SYSTEMATIC POSITION

Bentham & Hooker (1883)	Engler & Prantl (1899)	Hutchinson (1959)
Phanerogams	Phanerogams	Angiospermae
Dicotyledones	Dicotyledoneae	Dicotyledones
Polypetalae	Archichlamydeae	Lignosae
Calyciflorae	Rosales	Rosales
Rosales	Rosaceae	Rosaceae
Rosaceae		

Geographical distribution: Indigenous to China and Japan. It is largely cultivated in Uttarakhand and Punjab followed by Bihar, West Bengal, Assam and Arunachal Pradesh on a smaller scale. It is grown in most tropical and sub-tropical countries.

Habit: A middle sized evergreen tree.

Habitat: Cold humid regions with alluvial, red and black cotton and hilly soils.

Root: Tap, branched.

Stem: Aerial, erect, woody, cylinderical, branched, bark dark grey, smooth to slightly rough.

Leaves: Simple, alternate, cauline and ramal, petiolate, petiole very small so leaves are nearly sessile, broadly oblanceolate, acuminate, remotely serrate, densely woolly beneath, nerves 10-15 pairs, prominent beneath, venation unicostate reticulate.

Inflorescence: Panicles, softly woolly, with stout branches.

Flower: Hermaphrodite, complete, actinomorphic, 1.00-1.50 cm across, crowded, white, fragrant, pentamerous, perigynous.

Calyx: Sepals 5, gamosepalous, green, calyx lobes small, persistent, calyx tube adnate to ovary forming a hypanthium.

Corolla: Petals 5, broadly ovate, fragrant, white, free, arising from the rim of hypanthium.

Androecium: Stamens indefinite, all fertile, polyandrous, anthers small, dithecous.

Gynoecium: Carpels 5, ovary syncarpous, half superior, pentalocular, adnate to calyx tube, ovules 2 in each cell, styles 5, woolly below.

Fruit: Pyriform berry, edible, ovoid, intruded at the top, 2.5 – 6.5 cm long, pale yellow-organge 2-5 seeded.

Flowering: August - November.

Floral formula: ⊕ ⚥ $K_{(5)}\,C_5\,A_{\alpha}\,G_{(5)}$.

Varieties: Many varieties are cultivated in India. The colour of the fruit as well as flesh varies from pale yellow to orange along with their taste and flavour. The varieties under cultivation are golden yellow, improved golden yellow, Thames pride, large pale yellow and large Agra. California Advance and Tanaka are late varieties.

Uses: The plant is chiefly valued for its fruits, which are considered to be a sedative and relishing, nutrious and edible. The fruits is used fresh or made into jellies, jams and sauces. The plant is one of the host plants of wild silkworm *Cricula trifenestrata*, known as Aamkoni or Aam Resham in Arunachal Pradesh, Assam, North Bihar and North Bengal in fruit orchards. The Cricula silk industry is very well flourishing in Indonesia, but it has still to make a beginning in India (Mishra 2003). Flowers used as expectorant while wood is used for rulers and drawing materials (Ambasta *et al.* 1986).

Cultivation: Through seeds, seedlings, air layerings, cuttings and bud grafting.

61. NEELGERI

Botanical: *Euclyptus globulus* Labill.

Burmese: Yoe-kalit-pin German: Blauer glummibaum

Sinhalese: Eucalyptus French: Eucalyptus

English: Eucalyptus, Tasmanian, Blue gum, Eucalypat

Hindi: Neelgeri, Safeda Kannada: Neelgeri Taila

Malayalam: Karpuramaram Manipuri: Nasik

Tamil: Karpura maram Telugu: Talanoppi

SYSTEMATIC POSITION

Bentham & Hooker (1883)	Engler & Prantl (1899)	Hutchinson (1959)
Phanerogams	Phanerogams	Angiospermae
Dicotyledones	Dicotyledoneae	Dicotyledones
Polypetalae	Archichlamydeae	Lignosae
Disciflorae	Myrtiflorae	Myrtaceae
Myrtales	Myrtaceae	
Myrtaceae		

Geographical distribution: The tree is a native plant of Australia, but now extensively grown in the Nilgiris, the Anamali, the Pulney hills, Shimla hills, Shillong and other North-eastern regions. It is chiefly grown in the countries round the Mediterranean sea, Victoria and Tasmania (Brandis 1906, Pandey 1981, Kocchar 1981, Sagreiya 1967). It was planted in Nilgiris in 1843 and has grown to very large sizes.

Habit: A glabrous, evergreen lofty tree, attaining 50-100 meters height, highly aromatic young shoots and foliage glaucous-gray trunk extremely smooth and white (Fig. 26).

Habitat: Marshy places.

Root: Branched tap root.

Stem: Erect, solid, woody and cylinderical, bark smooth, white and shining.

Leaves: Simple, petiolate, exstipulate, alternate, linear-lanceolate, grey-blue, falcate, 15.0-22.0 cm long, petiole 2.5-5.0 cm long, alternate, smooth, aromatic with oil glands, venation reticulate unicostate, margin entire, apex acute, evergreen.

Inflorescence: Axillary umbells.

Flowers: Ebracteate, pedicellate, actinomorphic, hermaphrodite, epigynous, cyclic, small, regular, complete, large.

Calyx: Sepals 5, gamosepalous, green covered with a bluish white blume, broadly turbinate, warty and ribbed, reduced and fused to form a lid or cap (operculum), mouth 1.8-2.5 cm in diameter, operculam shorter than the calyx tube, aestivation valvate.

Corolla: Petals 5, gamopetalous, fused with the calyx to form a double-layered lid or cap that falls off as a whole with the opening of the flower, aestivation valvate.

Androecium: Stamens indefinite, polyandrous, inserted on the rim of the calyx tube densely packed to fold during bud condition, filaments unequal, anthers dithecous, dorsifixed.

Gynoecium: Tricarpellary, syncarpous, ovary inferior, trilocular, placentation axile.

Fruit: Dried-loculicidal capsule, angular, 1.8-2.5 cm in diameter, seeds non-endospermic, sterile seeds much narrower than the fertile seeds, seeds are winged.

Flowering: June – July.

Fruiting: December-January

Floral formula: ⊕ ⚥ $K_{(5)}\ C_{(5)}\ A_{\alpha}\ G_{(\bar{3})}$

Uses: The wood is used for fuel. The specific gravity and calorific value of wood has been found to be 0.06–0.8 and 3172-5690 k cal/kg respectively (Chaundawat and Gautam, 1993). It forms principal source of timber in Australia. The timber is used for bridges, railway sleepers, building materials, plywood industries etc. The braille paper is manufactured from its pulp (Pandey 1981). These plants are grown for afforestation in the marshy places as an anti-malarial measure or a backdrop or as a wind breakers (Pandey 1981, Rao 1991). The annual production of timber/acre at the age of 20 years on the Nilgiris was as high as 470 cubic ft. (Brandis 1906). Originally introduced for producing firewood, it is also used as cheap planking and the leaves are distilled for aromatic oil, which is used in medicines and in perfumery.

The eucalyptus oil industry is limited to India. Tannins also occur in this plant (Kocchar 1981, Pandey 1981, Sagreiya 1967). A eucalyptus tree in small house compounds is not desirable since it lowers the water table of the soil and quickly drains of the soil nutrients (Rao 1991).

The leaves are used as an ingredient for the preparation of an indigenous hair lotion (Chinghi) in Manipur. The essential oil from leaves is used in medicine in the preparation of disinfectants, expectorant and stimulant. It is an insect repellant and its tincture is given in asthma and pthisis while roots are used as purgative (Sinha 1996). The infusion of leaves is recommended for the treatment of asthma, inflammation of a gland, urinary bladder, kidneys, lining of the nose or mucous membrane with a free discharge, bronchitis, whooping cough, dysentery, tuberculosis, diabetes, fever, cold and malaria. In the inflammation of the sinus, the vapour of decoction of the leaves is breathed in. Decoction of the fresh leaves is used in inflammation of the sciatic nerve, joints and muscle pains and also for washing of wounds, sores and ulcers. The eucalyptus oil is used to reduce or eliminate fever, prevent discharge of pus from gums and for burns also (Kurian 1995).

Cultivation: It may be propagated through seeds, seedlings, air layerings and juvenile stem cuttings made from sproutings from pollarded trunks through use of root inducing hormones like IAA, NAA and IBA.

The seeds are collected during December-January and nursery is raised during Sept.-Oct. The seedlings are transplanted during July-Aug. next year (Chundawat and Gautam 1993).

Micropropagation of *E. globules* has also been achieved through tissue culture for mass multiplication using shoot top explant from selected *in vitro* seedlings (Pattanaik and Vijayakumar 1997).

62. PAYAM

Botanical: ***Evodia fraxinifolia*** Hook.f.

Nepalese: Kanukpa

Assamese: Payam   Bengal: Kanukpa

Khasi: Diengsngiyat, Dieng-borapram   Lepcha: Kanu

SYSTEMATIC POSITION

Bentham & Hooker (1883)	Engler & Prantl (1899)	Hutchinson (1959)
Phanerogams	Phanerogams	Angiospermae
Dicotyledones	Dicotyledoneae	Dicotyledones
Polypetalae	Archichlamydeae	Lignosae
Disciflorae	Geraniales	Rutales
Gerianiales	Rutaceae	Rutaceae
Rutaceae		

Geographical distribution: Sub-tropical Himalaya from Nepal to Sikkim at 4000-7000 ft. ASL and Khasi hills in Meghalaya and Assam at 3000 – 7000 ft. ASL. Also found in China.

Habit: A moderate sized much branched, densely leafy, strongly aromatic evergreen tree.

Root: Tap root branched.

Stem: Aerial, erect, much branched, branchlets thick, stout, terete.

Leaves: Compound, opposite, imparipinnate, perfectly glabrous, petioles cylinderic and pilose, leaflets 3-6 pairs, leaves 20-30 cm long, spreading, bright, green, leaflets 10-22 cm long, shortly pedicled, oblong or oblong – lanceolate, acuminate, straight or falcate, quite entire or crenulate, base rounded often oblique, nerves spreading, slender, terminal leaflet usually long petioled.

Inflorescence: Axillary or terminal, broad, brachiate cymes on short, stout, compressed peduncles finely pubescent.

Flower: Unisexual, monoecious, hypogynous, actinomorphic, bracteate, bracts minute, shortly pedicelled, greenish-white, 4-5 merous, male flowers approximately 1.2 cm across.

Calyx: Sepals 4-5, small, obtuse, imbricate, free.

Corolla: Petals 4-5, pubescent within, much longer, linear, imbricate, sessile, free.

Androecium: Stamens 4-5 in male flowers only, filaments hairy longer than petals, anthers broad, inserted at the base of disk, 4 rudimentary ovaries present in male flowers.

Gynoecium: Carpels-4, syncarpous, ovary deeply 4-lobed, tetralocular, superior, glabrous, ovules-2 in each cell (locule), stigma 4-lobed, capitate, style short.

Fruit: Red, 1.3 cm across, carpels not separating up to base, very coriaceous.

Seed: Broadly elliptic, compressed slightly with dark brown shining testa.

Flowering: May

Floral formulae: Male flower: Br ⊕ ♂ $K_{4-5}\ C_{4-5}\ A_{4-5}\ G_{(\underline{4})}$ rud.

Female flower: Br ⊕ ♀ $K_{4-5}\ C_{4-5}\ A_0\ G_{(\underline{4})}$

Uses: It serves as one of the best secondary food plants of eri silkworm, Philosamia ricini in North-east India (FAO 1976, Sarkar 1988, Seth 1995c, Sarmah 2000).

Fruits used in dysentery and chutneys. Wood used for ceiling boards, match boxes and splints, partitions, posts and tea chests. Seeds yield an oil for lamps (Ambasta *et al.* 1986).

Cultivation: It propagates through seeds and seedlings.

63. KAITHA

Botanical: ***Feronia elephantum*** Correa Syn. *F. limonia* Linn.

Burmese: Mahan, Thibin | English: Elephant or Wood apple

French: Feronier | German: Elefantehepfel

Hindi: Bilin, Kaitha, Kat bel, kowit

Persian: Kabit | Sinhalese: Divul

Bengali: Kathbel, Kait | Gujarati: Kovit, Kothi | Kannada: Bela, Byala

Konkani: Kovit	Malayalam: Vila, Villatti	Marathi: Kovit, Kavatha
Oriya: Koitho	Punjabi: Bilin	Sanskrit: Kapittha
Santhali: Kainta	Sindhi: Kavatha	Tamil: Vilva, Vilanga
Telugu: Velaga	Urdu: Kaitha	

It is supposed to be one of the favourite fruit of Lord Ganesha.

SYSTEMATIC POSITION

Bentham & Hooker (1883)	Engler & Prantl (1899)	Hutchinson (1959)
Phanerogams	Phanerogams	Angiospermae
Dicotyledones	Dicotyledoneae	Dicotyledones
Polypetalae	Archichlamydeae	Lignosae
Disciflorae	Geraniales	Rutales
Geraniales	Rutaceae	Rutaceae
Rutaceae		

Geographical distribution: Throughout India. Generally cultivated in both Peninsulas, indigenous in South India, Sri Lanka and Myanmar.

Habit: A handsome, middle-sized up to 8-meter high, glabrous, deciduous tree armed with strong, straight, axillary thorns.

Habitat: Hilly waste land.

Root: Tap and branched.

Stem: Erect, aerial, branched, woody, solid, cylinderical, green when young, gray when old, armed with spines, aromatic when fresh cut.

Leaves: Alternate, exstipulate, compound, imparipinnate, leaflets opposite, 1-4 pair, cuneate or obovate, crenate at the top, gland dotted, common petiole often narrow winged, venation reticulate unicostate.

Inflorescence: Lax panicles.

Flower: Pedicellate, dull red, generally unisexual, male and bisexual flowers frequently on the same inflorescence, pentamerous, hypogynous, ebraceate.

Calyx: Usually 5, polysepalous, aestivation quincuncial.

Corolla: Usually five, polypetalous, dull red, aestivation imbricate.

Androecium: Stamens 10-12, filaments short, subulate, from a broad villous base.

Gynoecium: Carpels indefinite, syncarpous, unilocular, a large number of ovules arranged in 5 parietal placentas.

Fruit: A berry, globose, greenish white or grey, rough, 2-3 inch in diam, like a cricket ball, rind hard, woody, nearly smooth, fragrant, mucilaginous, edible.

Seeds: Numerous, oblong, embedded in fleshy edible pulp.

Flowering: February - April.

Floral formula: ⊕ ⚥ K_5 C_5 $A_{10\text{-}12}$ $G_{(\alpha)}$

Uses: The fruit is edible and its pulp is eaten in many ways. It is sweet and sour in taste but leaves a husk throat if eaten when half ripe.

Fruit is also used for making sherbat and considered as tonic, antiscorbutic and alexipharmic. Leaves are aromatic, corminative, atringents and yield an essential oil (Ambasta *et al.* 1984).

Medicinally, leaves are mild laxative in fever. The decoction of leaves reduces or eliminates fever and removes mucous from bronchi in asthma. The juice of leaves is given against abnormal accumulation of liquid in the cellular tissue, constipation and jaundice, while warm poultice of leaves is applied in opthalmia, acute bronchitis and body inflammation. The decoction of root bark is used in intermittent fevers, menalcholia and heart palpitation. The unripe fruit strengthens and tones up the stomach, helps in digestion and cures or prevents scurvy, chronic diarrhoea and dysentery. The ripe fruit is aromatic, cooling and laxative, so its extract in water is given in chronic constipation, indigestion, discomfort and pain in stomach (Kurian 1995). The leaves of wood apple are used as fodder and a gum is obtained from the trunk and branches of the tree (Pandey 1981). Bark is used as a cosmetic (Brandis 1906).

Wood used for house building, naves of wheels, oil crushers shoe lasts, pen holders, rulers, carvings and agricultural implements (Ambasta *et al.* 1986).

Cultivation: It may be cultivated through seeds or seedlings.

64. BARGAD

Botanical: ***Ficus benghalensis*** Linn. Syn. *F. indica Roxb.*, *Urostigma bengalense* Gasp

Burmese: Pyi-nyaung English: Banyan

Hindi: Bargad, Bargat, Vat, Bar, Bor Bengali: Bot, Bar

Gujarati: Vad, Vadlo, Vor Kannada: Alada mara, Ala, Vat

Manipuri: Khongnangtaru Marathi: Wad, War, Vada Malayalam: Ala, Vatam

Sanskrit: Bahupada, Vata Tamil: Ala, Alam Telugu: Marri, Vati, Peddamarri

SYSTEMATIC POSITION

Bentham & Hooker (1883)	Engler & Prantl (1899)	Benson (1957)	Hutchinson (1959)
Phanerogams	Phanerogams	Spermatophyta	Angiospermae
Dicotyledones	Dicotyledoneae	Angiospermae	Dicotyledones
Monochlamydae	Archichlamydeae	Dicotyledonae	Lignosae
Unisexuales	Urticales	Urticales	Urticales
Urticaceae	Moraceae	Moraceae	Moraceae

Hindus regard it as one of the most important sacred trees. The Hindu ladies worship this tree on the festive occassion of "Vata Savitri" for seeking long life of their husbands. Buddhists regard this as one of the "Bodhi trees" associated with the life of Lord Buddha. The great emperor Asoka, after abjuring violence and upon his conversion to Buddhism planted this tree along the roadsides to give shade and shelter during 260 B.C.

Geographical distribution: Indigenous in the sub-Himalayan tract, the Western Peninsula and outer Himalayas as high as 4000 ft. ASL. Commonly planted by Hindus throughout India.

Habit: A large tree, branches spreading, crown 300-500 ft. in diameter supported by numerous often more than 100 root drops which afterwards become trunks. Botanical Survey of India, Calcutta has oldest Banyan tree.

Habitat: Sub-temperate – sub-tropical places.

Root: Main tap root, branched, deep, buttressed, supported by numerous aerial roots converting into trunks.

Stem: Erect, spreading, branched, crowning vastly, young shoots pubescent, bark rough, lenticellate, grey-black, aerial roots clasping as stems.

Leaves: Simple, ovate, glabrous when mature, nearly terminal on branches, mostly obtuse, base cordate or rounded, basal nerves 3-5, midrib with 1-6 pairs of secondary nerves, blade 10-20 cm long, petiole 2.5-5.0 cm long but shorter than those of *F. religiosa*.

Inflorescence: Hypanthodium. Male, female and gall flower found on same receptacle enclosed within a hypanthium, male flowers found near the mouth, female flower near the base and gall flower (female with a short style and rudimentary pistil) in the middle of the receptacles, receptacles in axillary pairs, sessile.

Flower: Unisexual, incomplete, actinomorphic, hypogynous, male flowers monoa-ndrous, female flowers monocarpellary, small, perianth very short or none.

Fruit: Globose, pubescent, 1.25 – 2.0 cm in diameter, sessile, scarlet when ripe, supported by 2-4 broad obtuse bracts.

Flowering: Feb. - March.

Fruiting: March - April.

Floral formulae: Male flowers: $\oplus$ ♂ $P_0\ A_1\ G_0$

Female flowers: $\oplus$ ♀ $P_0\ A_0\ G_{\underline{1}}$

Gall flowers: $\oplus$ ♀ $P_0\ A_0\ G_{rud}$

Uses: Besides being sacred tree, *F. bengalensis* is very important medicinally. Different parts of the plant are used for the treatment of dysentery, diarrhoea, diabetes, abscess and gonorrhoea. Leaves are antimicrobial and remove mouth ulcers. Latex is applied in rheumatism and lumbago. Young prop roots before coming in contact with soil are used for healing obstruction of urine flow and exudation of pus. Bark is used in piles (Sinha 1996). Infusion of bark tonic and astringent, used in diabetes, diarrhoea and dysentery. Leaves tonic and cooling (Ambasta *et al.* 1986).

Wood is suitable for paper pulps, furniture, well-curbs, cart-shafts and yokes. Fibre of aerial roots used to make coarse ropes (Ambasta *et al.* 1986). The wood is also used for construction purposes (Shukla and Mishra 1979). It is used as excellent avenue tree and fodder, fuel and fruits for animals and birds (Chundawat and Gautam 1993).

Cultivation: Through seeds and seedlings. The seeds are collected during March-April and nursery is raised during March-May. The seedlings are transplanted during May-June (Chundawat and Gautam 1993).

65. PUKAR

Botanical: ***Ficus benjamina*** Linn; Var-1: F. *comosa* Roxb. or *comosa*, Var-2: *F. benjamina* proper

Burmese: Nyaung-thabye   Nepali: Kabra

Assamese: Chilu-Bor   Bengali: Pukar

Malayalam: Velal, Putra-juvi   Manipuri: Khongang-bot.   Marathi: Pimpri

SYSTEMATIC POSITION

Bentham & Hooker (1883)	**Engler & Prantl (1899)**	**Benson (1957)**	**Hutchinson (1959)**
Phanerogams	Phanerogams	Spermatophyta	Angiospermae

Dicotyledones	Dicotyledoneae	Angiospermae	Dicotyledones
Monochlamyedeae	Archichlamydeae	Dicotyledonae	Lignosae
(Incompletae)	Urticales	Urticales	Urticales
Unisexuales	Moraceae	Moraceae	Moraceae
Urticaceae			

Geographical distribution: Wild in foothills of Eastern Himalaya, Assam, Chotanagpur, (Singhbhum), Andaman islands, Deccan Peninsula and Kerala (Travancore) in India up to 3000 ft. ASL. Also in Myanmar, China, Bangladesh (Chittagong).

Habit: A large perfectly glabrous monoecious tree with spreading crown and drooping branches (Fig. 27).

Root: Branched tap root.

Stem: Aerial, woody, solid, stout, glabrous, cylinderical, branched.

Leaves: Simple, alternate, broadly ovate - elliptic, thinly coriaceous, shining, abruptly acuminate, entire, base obtuse or subacute, blade 7-12 cm, nerves numerous, close, straight at right angle to mid rib, reticulate veins few, petiole 1.2-2.0 cm long, stipules lanceolate, 1.2 cm long.

Inflorescence: Hypanthodium, receptable sessile, smooth in axillary pairs with small basal bracts.

Var. **proper**: globose, red after ripening, 0.8 cm. long.

Var. **comosa**: ovoid, blood red or yellow, 2.0 cm long.

Flower: Unisexual, incomplete, minute on inner walls of receptacle.

Male Flower: Very less, staminate, pedicelled with 2 flat large sepals and a subsessile anther.

Gall Flower: Pistillate, Pedicelled with 3-4 long sepals and smooth ovary

Var. **proper:** Sepals spatulate, ovary ovoid.

Var. **comosa:** Sepals lanceolate - acuminate, ovary globose.

Female Flower: Pistillate, sessile with large style and stigma.

Var. **proper:** Sepals spatulate.

Var. **comosa:** Sepals lanceolate - acuminate.

Fruit: Achne.

Var. **proper:** 0.8 cm across, ovoid – reniform.

Var. **comosa:** 1.5-2.0 cm across, large, globose, narrowed at the base.

Floral formulae: Male flower: $\oplus$ ♂ $K_{(2)}\ P_0\ A_1\ G_0$

Gall flower: $\oplus$ ♀ K_{3-4} P_o A_o $G_{\underline{1}}$

Female flower : $\oplus$ ♀ $K_{(3-4)}$ P_o A_o $G_{\underline{1}}$

Uses: It serves as one of the tertiary but recognised as secondary food plant of tasar silkworm, *Antheraea mylitta* D (Seth 1995 a).

Medicinally, extract of leaves is useful in ulcers hence, decoction of leaves mixed with oil is applied to ulcers (Sinha 1996).

Wood suitable for match boxes and bark used for ropes (Ambasta *et al.* 1986).

Cultivation: It propagates through seeds in nature. Besides, it may also be propagated through air layerings and stem cuttings.

66. DECORATIVE-BAR

Botanical: ***Ficus elastica*** Roxb.

Burmese: Rubber pin, Nyaung Kyetpaung

English: Indian Rubber Tree, Indian Caoutchouc tree

French: Caoutchoue German: Gummibaum

Hindi: Decorative-bar Sinhalese: Rubber gasa

Assamese: Attah Bor Bengali: Bor Gujarathi: Vad

Kannada: Ala Malayalam: Ala Marathi: Vada

Sanskrit: Vata Tamil: Al Telugu: Marri

SYSTEMATIC POSITION

Bentham & Hooker (1883)	Engler & Prantl (1899)	Benson (1957)	Hutchinson (1959)
Phanerogams	Phanerogams	Spermatophyta	Angiospermae
Dicotyledones	Dicotyledoneae	Angiospermae	Dicotyledones
Monochlamydeae	Archichlamydeae	Dicotyledonae	Lignoseae
Unisexuales	Urticales	Urticales	Urticales
Urticaceae	Moraceae	Moraceae	Moraceae

Geographical distribution: Outer Himalaya from Nepal eastwards, ascending to 5000 ft., Assam, Khasi hills and upper Myanmar, South to 24°N. Cultivated in many parts of the large cities.

Habit: A glabrous, spreading tree, reaching up to 8 meters height with numerous aerial roots from the trunk and branches. Milky latex, which hardens as rubber, is present in whole plant. Germination is epiphytic (Fig. 28).

Habitat: Sub-tropical to sub-temperate moist, partially shady and warm places.

Root: Tap root, supported with aerial roots descending and developing into secondary stems.

Stem: Erect, woody, branched, sending down numerous aerial roots from the branches, trunk rough, supported by high branching buttresses, bark grey, thick.

Leaves: Simple, elliptic, abruptly and shortly acuminate, 12-25 cm long, petiolate, smooth, leathery, shining, entire, petioles 1.25-6.0 cm long, stipules pink or red and as long as the leaves, caducous, covering the leaf bud like a cap, secondary nerves numerous, nearly at right angles to mid rib.

Inflorescence: Hypanthodium, receptacles, mostly sessile, in pairs, axillary or above the scars of fallen leaves, enclosed while young by hooded involucres in hypanthium which fall off leaving a saucer shaped base under the fruit. The hypanthiums have an apical opening and hollow central cavity. Receptacles found in distal part of hypanthium are fleshy and bear male flowers at the mouth, female at the distal part and gall flowers in the middle.

Flower: Unisexual, incomplete, minute, actinomorphic, hypogynous, on inner walls of receptacles, male flowers monoandrous, stigma elongate, usually acute and feathery, perianth none or small.

Fruit: Fleshy, ovoid-oblong, multiple, sessile, greenish yellow when ripe, 1.5 cm long in hot season, seeds with curved embryo in fleshy endosperm.

Floral formulae: Male flower: $\oplus$ ♂ $K_0\, A_1\, G_0$

Gall flower: $\oplus$ ♀ $K_0\, A_0\, G_{\underline{1}}$

Female flower: $\oplus$ ♀ $K_0\, A_0\, G_{\underline{1}}$

Uses: Cultivated on a large scale for the yield of Indian Rubber from latex since 1872 in Java and since 1874 in Assam (Brandis 1906). The use of Indian rubber dates back to late eighteenth century since native people of Penang, Malaya used to carry their water and food in vessels and bags made from the hardened latex of *F. elastica* during that period. During 1770, English Chemist Joseph Priestley reported the ability of this hardened latex (Cautchouc) to erase lead pencil marks since then it came to be known as rubber and still it is used for the same purpose besides other innumerable uses in different industries.

Now a days, this tree is largely cultivated as an indoor plant. The Indian Rubber plant now has many horticultural varieties as hereunder.

1. Var. **Decora**: Leaves broader, oval shaped, thick, leathery.
2. Var. **Black prince**: Leaves grayish black.
3. Var. **Shrijvereana**: Leaves greenish, cream and grey marked.
4. Var. **Variegata**: Leaves variegated, needs more sunshine.

Medicinally, the crushed aerial or adventitious rootlets are used for healing wounds, cuts and sores. The crushed bark is used for checking bleeding of wounds. The decoction of latex is used for parasitic worms.

It is a good source of fodder (Ambasta *et al.* 1986).

Cultivation: Through seeds, seedlings and air layerings. Pots with 30-35 cm in diameter are best suited for growing seedlings/saplings of *F.elestica* (Rao 1991) as an ideal indoor plant. When cultivated as an ornamental indoor plant it requires 500-1000 foot candles light for 12 hours/day.

The tree germinates on the branches of tall forest trees, grows very slowly at first, but soon sends down aerial root-drops and when these have taken root, they begin to make progress. The tree stands on its own legs subsequently and attains a height of 100-200 ft. with a diameter of the crown of 150-200 ft. (Brandis 1906)

67. GULAR

Botanical: ***Ficus glomerata*** Roxb. Syn. *F.chittagong* Miq.

Burmese: Thapan, Yethepan	English: Cluster fig	
Hindi: Gular, Umar, Umbri	Persian: Batbar	Sinhalese: Atteeka
Bengali: Dumur	Chamba: Trimbal	Gujarati: Umar, Gular
Kannada: Athi	Konkani: Loa	Malayalam: Athi
Manipuri: Heibong	Marathi: Umbar, Umra	Merwari: Umbri
Oriya: Dimri	Sanskrit: Udumbar, Udumbra	
Santhali: Loa	Tamil: Adam, Athi	Telugu: Atti
Urdu: Gular		

SYSTEMATIC POSITION

Bentham & Hooker (1883)	Engler & Prantl (1899)	Benson (1957)	Hutchinson (1959)
Phanerogams	Phanerogams	Spermatophyto	Angiospermae
Dicotyledones	Dicotyledoneae	Angiospermae	Dicotyledones
Monochlamydeae	Archichlamydeae	Dicotyledonae	Lignosae
Unisexuales	Urticales	Urticales	Urticales
Urticaceae	Moraceae	Moraceae	Moraceae

About 260 BC, the great emperor Asoka (after abjuring violence and upon conversion to Buddhism) planted the Udumbra trees along the roadsides to give shade and shelter (Rao 1991).

Geographical distribution: Wild in the sub-Himalayan tracts and outer valleys, in ravines, on the banks of rivers and in damp places. Prevalent in Western Peninsula, Ajmer, Merwara and Mount Abu (Rajasthan), Bihar, Jharkhand (Chota Nagpur), Bengal plains, Mizoram (Khasi hills) and Manipur. Also found in lower Myanmar dry regions of the Irawadi valley and Chittagong (Bangla Desh) (Brandis 1906).

Habit: A middle sized or large tree growing up to a height of 15-18 meters, young shoots minutely hairy, never epiphytic (Fig. 29).

Habitat: Along streams or rivers.

Root: Tap root, branched.

Stem: Erect, woody, branched, bark rough, grey, thick, lenticellate spreading.

Leaves: Entire, elongate – elliptic, ovate, elliptic-lanceolate or ovate-oblong, glabrous, shiny when mature, base acute or rounded, apex tapering to a bluntish point (acuminate), 7.5-10.0 cm long, petiole 2.0-5.0 cm long, secondary nerves 6-8 pairs, the lowest pair basal.

Inflorescence: Hypanthodium. Male, female and gall flower found together on one receptacle enclosed within an enlarged hypanthium (Thalmus). The male flowers are at the mouth, female flowers at the distal end and gall flowers in the middle of the receptacles. The receptacles found in dense clusters form tubercles on the trunk and larger branches, often very large.

Flower: Unisexual, incomplete, actinomorphic, hypogynous, male flowers as a rule diandrous without pistil, gall flowers have a rudimentary pistil with 2 stamens while female flowers have only two carpels with elongate, acute feathery stigma, perianth none or small.

Fruit: Nearly globose, fleshy, 2.5-4.0 cm in diameter, clustered on leafless branches or on the old woods, rarely axillary, edible, pink or red on ripening.

Leaf shedding: Jan-April.

Leaf sprouting: Feb. - April.

Flowering: Hot season.

Fruiting: Cold season.

Floral formulae: Male flower: ⊕ ♂ P_o A_2 G_o

Gall flower: ⊕ ⚥ P_o A_2 G_{rud}

Female flower: ⊕ ♀ P_o A_2 $G_{(\underline{1})}$

Uses: It is used as an avenue tree. It yields fodder, fuel and fruit. Fruit is edible when matured. The whole plant is medicinally very useful. The decoction of bark is given for increasing lactation and for giving tone and vitality to the body, for curing asthma, for checking congestion of the liver, lencorrhoea, blood tissues, leprosy, menorrhagia, nose bleeding and for expelling intestinal worms. The decoction of the unripe fruit tones the body and checks congestion of liver, leucorrhoea and blood tissues while that of ripe fruits checks leprosy, menorrhagia, nose bleeding, expels intestinal worms and cures dry cough, laryngistis, diseases of the kidney and spleen. The decoction of leaves is given for curing bronchitis. The latex is administered in piles and diarrhoea and in combination with sesamum oil it checks cancer. External application of latex checks glandular enlargements. The sap of root is used to cure diabetes. A mixture of 40 gms of latex with cumin seeds and a pinch of sugar is used to cure gonorrhoea (Kurian 1995). Root extract is given in dysentery, latex is applied to boils and bark is given to cattles in rinder-pest disease (Sinha 1996).

Roasted fruit powder used as breakfast food. Leaves used as fodder. Wood used for well-curbs, cheap furniture, fuse box fittings and match boxes. Bark yields 14% tannins (Ambasta *et al.* 1986).

Cultivation: Through seeds, seedlings and air layerings. The seeds are collected during March–April and seedlings are raised in nursery during March-May. The seedlings are transplanted during May-June (Chundawat and Gautam 1993).

68. KAT GULAR

Botanical: ***Ficus hispida*** Linn.f. syn. *F. doemonum* Koenig, *F. oppositifolia* Roxb.

Burmese: Kadut, Kaaung	Hindi: Gobla, Kat-gular, Daduri	
Bengali: Kako dumur	Gond: Katumar	Gujarati: Jungalianjir
Kannada: Adavi-atti	Kol: Sosokera	
Malayalam: Eruma-nakku	Manipuri: Ashi-heibong	
Marathi: Dher-umbar, Bokria	Punjabi: Daduri	Sanskrit: Kakadumbura
Tamil: Chona-Atthi	Telugu: Bamari, Brahmmadi	

SYSTEMATIC POSITION

Bentham & Hooker (1883)	**Engler & Prantl (1899)**	**Benson (1957)**	**Hutchinson (1959)**
Phanerogams	Phanerogams	Spermatophyta	Angiospermae
Dicotyledones	Dicotyledoneae	Angiospermae	Dicotyledones
Monochlamyedeae	Archichlamydeae	Dicotyledonae	Lignosae
(Incompletae)	Urticales	Urticales	Urticales

Unisexuales	Moraceae	Moraceae	Moraceae
Urticaceae			

Geographical distribution: Throughout India from the Chenab eastwards excepting the arid zones of the North West, sub-Himalayan tract and outer hills ascending up to 3500 ft. ASL. Also found in Myanmar, Hongkong, North Australia, Malay islands and China.

Habit: A moderate or small sized monoecious tree, throwing up abundant root suckers.

Habitat: Sub-tropical to sub-temperate humid (moist) conditions.

Root: Branched tap root.

Stem: Aerial, woody, solid, stout, branchlets usually hollow, clothed with short stiff hairs, marked by the scars of petioles and stipules.

Leaves: Simple, usually opposite, 10-40 cm long, membranous, elongate, elliptic or obovate, entire or dentate, base rounded, cuneate or subcordate, petiolate, petiole and underside of leaves covered with short stiff hairs and upper side rough with minute scattered hairs, petiole 1.2-5.0 cm long, secondary nerves 8-10 pairs, tertiary nerves parallel and reticulate; stipules 1.2 cm, pubescent, usually in whorls of 4 on the receptacles with leafless branches.

Inflorescence: Hypanthodium, ripe receptacles hispid, yellowish with scattered bracts on the sides.

Flower: Unisexual, incomplete, minute on inner walls of receptacle, hypogynous, actinomorphic.

Male Flower: Monoandrous in the same receptacle with gall flower, very less, perianth none or small.

Gall Flower: Pistillate with hairy ovary and short style.

Female Flower: Pistillate with large style hairy and stigma cylinderic, tubular, ovary hairy, superior, found on separate receptacles.

Fruit: Fleshy, obovoid, 2.5 cm long, supported by 3 persistent basal bracts usually clustered on long leaf less panicles hanging from the old wood of trunk and branches on young plants, coppice shoots and sometimes also on old trees usually in axillary pairs.

Floral formulae: Male flower: $\oplus$ ♂ $K_0\ A_1\ G_0$

Gall flower: $\oplus$ ♀ $K_0\ A_0\ G_{\underline{1}}$

Female flower: $\oplus$ ♀ $K_0\ A_0\ G_{\underline{1}}$

Uses: It serves as tertiary but recognised as one of the secondary food plants of tropical tasar silkworm, *Antheraea mylitta* D. in Central Indian forests (Jolly *et al.* 1974).

Leaves are used in ringworms; latex on boils and seeds in dysentery.

Bark, fruit and seeds are emetic and used as purgative. The above parts also promote secretion of milk (Sinha 1996).

Fruits edible, used in jam; immuture fruits eaten in curries (Ambasta *et al.* 1986) which may cause giddiness.

Cultivation: It can be cultivated through seeds or seedlings. Attempts may also be made to propagate it through root suckers.

69. PIPAL

Botanical: ***Ficus religiosa*** Linn.

Eng: The peepal tree	Hindi: Pipal, Pipli

Burmese: Nyaung bawdi

Assamese: Borbur	Bengali: Asvattha, Ashathwa	Gonda: Ali

Kannada: Arali, Ashwath	Kol: Hissa	Malayalam: Arasu, Arayal

Manipuri: Sana-Khongnang	Marathi: Ashvatha	Tamil: Arasi

Telugu: Ravi, Ragi, Bodhi	Sanskrit: Ashvattha	Gujarati: Pipul, Jari

SYSTEMATIC POSITION

Bentham & Hooker (1883)	Engler & Prantl (1899)	Benson (1957)	Hutchinson (1959)
Phanerogams	Phanerogams	Spermatophyta	Angiospermae
Dicotyledones	Dicotyledoneae	Angiospermae	Dicotyledones
Monochlamydeae	Archichlamydeae	Dicotyledonae	Lignosae
(Incompletae)	Urticales	Urticales	Urticales
Unisexuales	Moraceae	Moraceae	Moraceae
Urticaceae			

According to Hindu mythology Pipal tree is abode of Lord Shiva since time immemorial. Even before Lord Buddha (567 BC) the Pipal tree was regarded as a sacred tree as the sacred fire drills were made from Pipal wood. Lord Buddha received enlightenment under a Pipal tree in Gaya, which was later, named as "Bodhivriksha". About 260 B.C., the great emperor Asoka (after abjuring violence and upon his conversion to Buddhism) planted the Pipal trees along the roadsides to give shade and shelter to the travellers. Prince Mahindra, son of Asoka, carried a sapling of Bodhi tree to Sri Lanka around 250 B.C. and it was planted at Anuradhapur (Rao 1991).

Geographical distribution: Wild in sub-Himalayan tracts, Bengal and Central Indian forests, cultivated universally in India and Sri Lanka, occasionally in Myanmar and rarely in the Malayan region. In outer Himalaya as high as 5000 ft ASL. Rarely in North-Western arid regions.

Habit: A large entirely glabrous tree, usually epiphytic and destructive to buildings, milky latex present, deciduous, leafless during hot season for very short period (Fig. 30).

Root: A strong branched tap root, deeply penetrating into the soil. Buttress roots present on base.

Stem: Erect, branched, spreading widely, solid, woody, cylinderical, smooth, trunk irregularly shaped.

Leaves: Simple, long petioled, coriaceous, alternate, glabrescent, shining above, broad-ovate, margin undulate, apex tailed, tail 1.5-7.5 cm long, one-third the length of whole blade, petiole 7-10 cm and blade 10-17 cm x 7-12 cm, secondary nerves about 8 pairs, prominent, distant, reticulation fine and distinct, stipules minute, ovate, acute.

Inflorescence: Hypanthodium, receptacle fleshy, sessile, in axillary pairs, smooth, depressed, spheroidal, dark purple when ripe, diam. 1.5 cm, basal bracts 3, coriaceous, broad, spreading, mouth of receptacle closed by imbricate bracts.

Flower: Minute, unisexual on the inner walls of fleshy receptacle. Male, female and gall flowers found on same receptacle.

Male Flower: Very few, often absent, found near mouth of some receptacles only, sessile, trisepalous, sepals gamosapalous, broadly ovate, anther-1, ovate-rotund, filament short.

Gall Flower (female with short style) and female flower: Shortly pedicelled or sessile, sepals-5, gamosepalous, lanceolate, carpel-1, style lateral, stigma rounded, ovule pendulous, single. Gall flowers are much more in number as compared to female flowers and many of them are without perianth.

Fruit: Sessile, fleshy, enclosing numerous minute achenes, 1.2 cm diam., dark purple with broad, coriaceous, basal bracts.

Seeds: Endosperm scanty, cotyledonous.

Floral formulae: Male flower: $\oplus$ ♂ $K_{(3)}\ A_1\ G_0$

Gall flower: $\oplus$ ♀ $K_{(5)}\ A_0\ G_{\underline{1}}$

Female flower: $\oplus$ ♀ $K_0\ A_0\ G_{0 \text{ or } \underline{1}}$

Uses: It serves as one of the tertiary but recognized as one of the secondary food plants of tropical tasar silkworm *Antherea mylitta* D. in the forest (Seth 1995 a). Besides, it can also serve as the alternative host plant for mulberry silkworm, *Bombyx mori* L. and can be used to keep the worms alive when due to some reasons mulberry is not received. However, silkworms soon suffer in growth and vigour and fall an easy prey to diseases by feeding Peepal leaves and even survivors spin flimsy cocoons if somehow, they spin the cocoons. (Ullal and Narasimhanna 1994) It is a sacred tree of Hindus as well as Buddhists.

Medicinally, bark is astringent and used in boils and gonorrhoea, fruits and seeds are laxative, seeds are cooling, leaves and young shoots are purgative. Bark infusion is useful in scabies (Sinha 1996) and ulcers. The bark is also said to be useful in rheumatic swellings (Ansari 1997). Leaves serve as fodder. Cavities in hollow ornaments are filled up by hardened latex. Bark yields a fibre and contains tannins (Ambasta *et al.* 1986).

Wood is for construction purposes (Shukla and Mishra 1979), packing cases, occassionally for felloes of wheels, spoons, bowels and suitable for match boxes (Ambasta *et al.* 1986).

Cultivation: It propagates through seeds in nature. Seeds are eaten by birds and left in faecal matter in crevices of buildings where they germinate easily and cause damage to buildings. Wild growing seedlings may easily be transplanted. It can also be successfully raised using branch cuttings of 20-25 cm. in length and 1-2 cm in diameter during Jan-Feb. This yields seedlings of 20 cm height but such plants are liable to be damaged by cattle browsing after plantation. Shyam Lal (1997) developed technology for raising long tall *Ficus religiosa* plants and recommended that branches should be 2-2.5 meters long and 4-6 cm in diameter while top of the branch should be more than 3 cm in diameter. Cow dung or soil should be put on the top of each cut portion immediately after planting and wrapped by pieces of polythene sheets. Planting should be at 30 cm spacing in first fortnight of Feb. to end of Feb. Sprouting starts from middle of March and branches grow fast and become more than 60 cm by the end of June. Such saplings may be transplanted along with some soil at transplantation sites subsequently.

70. CHILKHAN

Botanical: *Ficus retusa* Linn. Syn. *F. nitida* Thunb.

Burmese: Nyaung-ok | Hindi: Chilkhan, Kamrup | Nepali: Jamu

Bengali: Kamrup | Kannada: Pilala | Kumaon: Barri, Ajan

Malayalam: Kal atthi | Manipuri: Khognong | Marathi: Nandruk, Pilala

Tamil: Ichi, Kallichi Telugu: Yerru - Juvvi Sanskrit: Kuni

SYSTEMATIC POSITION

Bentham & Hooker (1883)	Engler & Prantl (1899)	Benson (1957)	Hutchinson (1959)
Phanerogams	Phanerogams	Spermatophyta	Angiospermae
Dicotyledones	Dicotyledoneae	Angiospermae	Dicotyledones
Monochlamydeae	Archichlamydeae	Dicotyledonae	Lignosae
(Incompletae)	Urticales	Urticales	Urticales
Unisexuales	Moraceae	Moraceae	Moraceae
Urticaceae			

Geographical distribution: Sub-Himalayan belt from Kumaon eastwards. Assam, Meghalaya (Khasi hills), Andaman Islands, Chotanagpur, Bundelkhand, Madhya Pradesh and Western Peninsula. Also, found in lower Myanmar, Bangladesh (Chittagong), Sri Lanka, China, Malay Peninsula & Archepelago.

Habit: A large, evergreen, glabrous, umbrageous tree with few slender aerial roots, often epiphytic.

Root: Branched tap root, deep penetrating.

Stem: Erect, spreading much-branched crown, bark glossy, woody, cylinderical, stout.

Leaves: Simple, alternate, green, elliptic-ovate or rhomboid (var. *nitida*), or obovate inclining to round (var. *retusa* proper), apex rounded (*retusa* proper) or shortly and bluntly acuminate, blade 5-10 cm, nerves 5-6 pairs, not much more prominent than the secondary nerves, petioles 0.6-1.2 cm, stipules lanceolate, 1.2 cm long.

Inflorescence: Hypanthodium, receptacles finely pubescent when young, sessile, small, in axillary pairs, 1.2 cm in diam., yellow or reddish, basal bracts persistent, spreading, braodly ovate, obtuse.

Flower: Unisexual, incomplete, monoecious, minute, on inner walls of receptacle.

Male Flower: Numerous, staminate, sessile or shortly pedicelled, scattered, calyx with 3 sub-spatulate sepals, stamen single, anther as long as filament, cordate-apiculate.

Gall Flower: Pistillate, shortly pedicelled or sessile, calyx with 3 broadly spatulate sepals, ovary glabrescent, style short.

Female Flower: Shortly pedicelled or sessile, pistillate, perianth much smaller than found in gall, ovary ovoid or obovoid, style short, stigma cylinderic or clavate.

Flowering: February - April.

Fruit: Achene ovoid or obovoid, yellow or reddish, sessile, 1.2 cm in diam., in axillary pairs.

Floral formulae: Male flower: ⊕ ♂ $K_{(3)}$ A_1 G_0

Gall flower: ⊕ ♀ $K_{(3)}$ A_0 $G_{\underline{1}}$

Female flower: ⊕ ♀ $K_{(3)}$ A_0 $G_{\underline{1}}$

Uses: It serves as one of the tertiary but recognised as secondary food plant of tropical tasar silkworm, *Antheraea mylitta* D (Seth 1995 a).

It makes an excellant avenue shade tree for a wide street. The wood is light reddish-grey, moderately hard and prettily mottled, used mostly as fuel (Randhawa 1983) and occassionally as furniture. Leaves lopped for fodder (Ambasta *et al.* 1986).

Bark extract is prescribed in liver complaints and dried leaves are powdered and applied in headache. Adventitious roots dried and powdered and mixed with salt are applied to a decayed or aching tooth. Root-bark and leaves are boiled in oil and applied to wounds and bruishes (Sinha 1996).

Cultivation: It can be cultivated through seeds and seedlings. It may also be propagated vegetatively through airlayerings and stem cuttings.

71. KEL

Botanical: *Ficus tjakela* Burm.f

Malayalam: Karal, Aal Marathi: Kel

SYSTEMATIC POSITION

Bentham & Hooker (1883)	**Engler & Prantl (1899)**	**Benson (1957)**	**Hutchinson (1959)**
Phanerogams	Phanerogams	Spermatophyta	Angiospermae
Dicotyledones	Dicotyledoneae	Angiospermae	Dicotyledones
Monochlamydeae	Archichlamydeae	Dicotyledonae	Lignosae
(Incompletae)	Urticales	Urticales	Urticales
Unisexuales	Moraceae	Moraceae	Moraceae
Urticaceae			

Geographical distribution: Western Peninsula and Sri Lanka upto 4000 ft.ASL.

Habit: Large, deciduous, glabrous, often cultivated tree without aerial roots.

Root: Branched deep penetrating tap root.

Stem: Erect, branched, stout, woody, bark rough dark coloured externally.

Leaves: Simple, alternate, coriaceous, very glossy above, long petioled, oval-ovate, abruptly acuminate, margin entire or slightly undulate, base rounded sometimes cordate, tertiary nerves distinct on both surfaces, young leaves enveloped in large 7-12 cm long, linear-lanceolate, thin, caducous, flaccid bud-scales, blade 12-18 cm long, nerves 7-10 pairs, petioles 4-6 cm long, slender, stipules small, ovate-lanceolate 1.2 cm long.

Inflorescence: Hypanthodium, sessile receptacles in clusters of 2-6 in axils of leaves or mostly at scars of fallen leaves, rarely very shortly peduncled, depressed, globose, whitish yellow and dotted on ripening, 0.5 cm across, with 3 broad deeply bifid basal bracts.

Flower: Unisexual, monoecious, incomplete, on inner walls of receptacle, minute, gall and female flowers alike.

Male Flower: Few, sessile, only near mouth of receptacle, calyx with 3 ovate sepals, stamen single, bigger than sepals, anther ovate, filament long.

Gall Flower & Female Flower: Sepals 3-4, style elongate, stigma cylinderic.

Fruit: Achene, obovate, 0.5 cm across, in clusters of 2-6 on woody short tubercles above scars of fallen leaves.

Floral formulae: Male flower: ⊕ ♂ $K_{(3)}\ A_1\ G_0$

Gall flower: ⊕ ♀ $K_{(3\text{-}4)}\ A_0\ G_{\underline{1}}$

Female flower: ⊕ ♀ $K_{(3\text{-}4)}\ A_0\ G_{\underline{1}}$

Uses: It serves as tertiary but recognised as secondary food plant of tropical tasar silkworms, *Antheraea mylitta* D. in jungles. (Seth 1995a)

It is a distinct and beautiful species allied to *F.infectoria* Roxb. from which it is differentiated by its clusters of receptacles in 4-6 and dry leaves which are dark greenish-grey with smooth and glistenning upper surface. The fruit is eaten by tribals in Kerala (Nair and Jayakumar 1998).

Wood is suitable for match boxes (Ambasta *et al.* 1986).

Cultivation: It can be cultivated through seeds and seedlings.

72. PIPER

Botanical: ***Ficus tsiela*** Roxb.

Hindi: Piper, Piperi, Jari

Kannada: Bili basri Marathi: Piper, Piperi Malayalam: Koyali

Sanskrit: Kaninika Tamil: Sela – Kallichi Telugu: Jadi, Juvvi

SYSTEMATIC POSITION

Bentham & Hooker (1883)	Engler & Prantl (1899)	Benson (1957)	Hutchinson (1959)
Phanerogams	Phanerogams	Spermatophyta	Angiospermae
Dicotyledones	Dicotyledoneae	Angiospermae	Dicotyledones
Monochlamydeae	Archichlamydeae	Dicotyledonae	Lignosae
(Incompletae)	Urticales	Urticales	Urticales
Unisexuales	Moraceae	Moraceae	Moraceae
Urticaceae			

Geographical distribution: Western Peninsula especially Rajasthan (Ajmer, Chanda), Maharashtra, Concan southwards, Tamil Nadu. Also in Sri Lanka. A common avenue tree in drier districts.

Habit: A large glabrous spreading tree without aerial roots. At the end of branches tufts of small leaves with very long petioles may often be seen.

Root: Branched tap root, deeply penetrating.

Stem: Erect, woody, stout, branched, crown spreading, bark smooth, greenish white.

Leaves: Simple, alternate, coriaceous, ovate, acuminate, base acute, entire with a thickened marginal nerve, blade 5-15 cm long, petiole 2.5-5.0 cm long, nerves distinct, secondary nerves 8-12 pairs, the lowest pair basal, intermediate nerves few and slender, tertiary nerve obscure, dorsal surface covered with minute raised dots, stipules ovate, acuminate 0.8-2.5 cm, leaves greenish-grey on drying, upper surface smooth and glistening.

Inflorescence: Hypanthodium, receptacles sessile, globose, paired, crowded at the end of branches, axillary or at leaf scars, smooth, purple on ripening, 1.2-1.6 cm in diam, basal bracts 3, minute, ovate.

Flowers: Unisexual, monoecious, incomplete, minute on inner walls of receptacle.

Male Flower: Few, sessile, calyx with 3 ovate, acuminate sepal longer than single stamen, anther broadly ovate, filament long, thick.

Gall Flower: Sessile or shortly pedicelled, sepals 3, ovate, shorter than achene, achenes ovate- remiform, fertile broadest, style long, stigma cylinderic.

Female Flower: Mostly sessile, sepals 3, ovate, shorter than achene, fertile, broadest, style long and stigma cylinderic.

Fruit: Achenes, purple, smooth, 1.2 cm broad, sessile, in pairs and crowded at the ends of tender branches.

Floral formulae: Male flower: ⊕ ♂ $K_{(3)}\ A_1\ G_0$

Gall flower: ⊕ ♀ $K_{(3)}\ A_0\ G_{\underline{1}}$

Female flower: ⊕ ♀ $K_{(3)}\ A_0\ G_{\underline{1}}$

Uses: It serves as tertiary but recognised as one of the secondary food plants of tropical tasar silkworm *Antheraea mylitta* D (Seth 1995 a).

Leaves serve as fodder. Wood serve as fuel and used for furniture. Bark is useful in colic (Ambasta *et al.* 1984).

It is planted as a very good shady tree in drier regions especially in Maharashtra (Randhawa 1983).

Cultivation: It can be cultivated through seeds and seedlings.

73. DIKAMALI

Botanical: *Gardenia lucida* Roxb. Syn. *G.resinifera* Roth.

Hindi: Dikamali

Gujarat: Dikamali Marathi: Dikamali Sanskrit: Jantuka

Tamil: Tikkamalli Telugu: Yerra bikki

SYSTEMATIC POSITION

Bentham & Hooker (1883)	**Engler & Prantl (1899)**	**Hutchinson (1959)**
Phanerogams	Phanerogams	Angiospermae
Dicotyledones	Dicotyledoneae	Dicotyledones
Gamopetalae	Sympetalae	Lignosae
Inferae	Rubiales	Rubiaceae
Rubiales	Rubiaceae	
Rubiaceae		

Geographical distribution: Western peninsula, common from the Konkan southwards in the drier as well as in the moist districts especially in Chandrapur and Bhandara. Also found in chittagong in Bangladesh. Absent in Myanmar.

Habit: A small deciduous tree with resinous shoots.

Habitat: Tropical dry as well as moist habitats.

Root: Branched tap root.

Stem: Aerial, erect, solid, woody, branched.

Leaves: Simple, entire, elliptic - oblong, 10-25 cm long, 5-12 cm broad, glabrous, apex obtuse or subacute, narrowed into short marginate petiole, shining, green when dry, secondary nerves 20-30 pairs, usually with bearded axils, stipules large, broadly ovate - connate (sheathing), intrapetiolar i.e.,between the petiole and the axis.

Inflorescence: Flowers solitary, axillary.

Flowers: Hermaphrodite, actimomorphic, pentamerous, regular, epigynous, complete.

Calyx: Sepals 5, gamosepalous, calyx lobes 5, linear - subulate, valvate, persistent.

Corolla: Petals 5, gamopetalous, tubular, large, pure white, fragrant, tube 2.5-5 cm long, lobes 5, obovate-oblong, valvate, contorted in bud.

Androecium: Stamens 5, epipetalous i.e, inserted at or near the throat of the corolla tube, alternating to petals, anthers bithecous, introse, dehiscing longitudinally.

Gynoecium: Bicarpellary, syncarpous, unilocular with two to many parietal placenta, ovary inferior, style single, stigma globose.

Pollinalion: Entomophillous (through insects).

Fruit: Drupe, ovoid, crowned with persistent calyx-limb, endocarp thick, woody, nearly bicelled with prominent placentas.

Seed: Small, endospermic.

Flowering: March - June.

Floral formula: $\oplus$ ⚥ $K_{(5)} \overbrace{C_{(5)} A_5} G_{(\bar{2})}$

Uses: It serves as the tertiary food plant but recognised as one of the secondary host plants of tropical tasar silkworm *Antheraea mylitta* D. in Maharashtra and Andhra Pradesh for Bhandara and Andhra local ecoraces (Jolly *et al*. 1974).

The wood is made in to combs, turney and other small articles and resin exudes from the wounds made in to the bark (Brandis 1906). The resin may be used as a carminative and stimulant and given in dyspepsia. Gum used in cutaneous diseases (Ambasta *et al.* 1986).

It also serves as an ornamental shrub or tree in the gardens due to fragrant flowers (Pandey 1981).

Cultivation: It may be cultivated through seeds or seedlings.

74. KHARPAT

Botanical: ***Garuga pinnata*** Roxb.

Burmese: Chinyok	Hindi: Kharpat, Kaikar	Nepali: Dabdabi
Assamese: Thotmola	Bengali: Jum	Gond: Gharri
Kannada: Godda	Kol: Armu	Marathi: Kudak, Kakad
Telugu: Garuga, Garugu	Sonthali: Karur	

SYSTEMATIC POSITION

Bentham & Hooker (1883)	Engler & Prantl (1899)	Hutchinson (1959)
Phanerogams	Phanerogams	Angiospermae
Dicotyledones	Dicotyledoneae	Dicotyledones
Polypetalae	Archichlamydeae	Lignosae
Disciflorae	Geraniales	Rutales
Geraniales	Burseraceae	Burseraceae
Burseraceae		

Geographical distribution: Throughout India from Yamuna eastwards to Myanmar and sourthwards to Nilgiris, sub-Himalayan tract and outer valleys ascending up to 3500 ft. ASL. Also found in Malayan Archipelago and Phillippines.

Habit: A large deciduous tree up to 50 ft. height bearing grey-pubescent young shoots and inflorescence.

Habitat: Deciduous forests in the greater part of the Western Peninsula often associated with Sal and Teak in forests.

Root: Branched tap root.

Stem: Aerial, erect, woody, hard, branched, young shoots grey-pubescent.

Leaves: Compound, imparipinnate, crowded near the ends of branches, 20-30 cm long or more, pubescent when young, glabrous when mature, leaflets 6-9 pairs, opposite-subopposite, lanceolate or ovate-lanceolate, crenate, obtusely acuminate, oblique.

Inflorescence: Axillary or terminal panicles, very large, compound, pubescent, several when terminal.

Flower: Yellow, 4-6 cm long, pubescent, pentamerous, complete, bisexual, actinomorphic, hypogynous.

Calyx: Sepals 5, gamosepalous, campanulate, 10-ribbed, 5-cleft, lined by a thin disk with a crenate margin, calyx teeth ovate, pubescent, valvate.

Corolla: Petals 5, inserted between the calyx teeth, induplicate - valvate.

Androecium: Stamens 10, equal, inserted in the tube of the calyx at the margin of the disk, as long as the petals and hairy at the filament base.

Gynoecium: Carpels 4-5, tetra - pentalocular, ovary sessile, ovoid, narrowing into the hairy style, superior, style erect, stigma capitate, 4-5-lobed, ovules 2 in each cell.

Fruit: Globose, fleshy, drupe of black colour having size of a gooseberry encloses 2, rarely more, bony, tuberculated stones bearing single seeds.

Leaf fall: During dry season.

Sprouting: April - May.

Flowering: April - June.

Floral formula: ⊕ ⚥ $K_{(5)}\ C_5\ A_{10}\ G_{\underline{(4\text{-}5)}}$

Uses: It serves as tertiary but serves as one of the secondary food plants of tropical tasar silkworm, *Antheraea mylitta* D. syn. *A.paphia* (Linn.) in the forests (Jolly *et al.* 1974, Chaturvedi 1999). Fruits are eaten raw and pickled (Brandis 1906). They possess cooling and digestive properties (Sinha 1996).

Cultivation: It may be cultivated through seeds, seedlings or cuttings. It grows readily from cuttings (Brandis 1906).

75. GAMBHAR

Botanical: *Gmelina arborea* Roxb.

Burmese: Yamane, Yemane Eng: Yemane

Hindi: Gumahar, Khambhari, Gamhar, Khamara

Nepali: Khamari, Gambari Sanskrit: Gandhari, Shriparni

Assamese: Gomari Bengali: Gamari Gujarati: Shewan

Kannada: Kuli, Shivani Malayalam: Kumbil Marathi: Shivani

Oriya: Bhodropoorni Punjabi: Kumhar, Ban Rajputana: Sewan

Tamil: Umi-thekku, Kumadi Telugu: Gummadi

SYSTEMATIC POSITION

Bentham & Hooker (1883)	Engler & Prantl (1899)	Hutchinson (1959)
Phanerogams	Phanerogams	Angiospermae
Dicotyledones	Dicotyledoneae	Dicotyledones
Gamopetalae	Sympetalae	Lignosae
Bicarpellatae	Tubiflorae	Verbenales
Lamiales	Verbenaceae	Verbenaceae
Verbenaceae		

Geographical distribution: From Chenab eastwards up to 3000 ft. ASL in Sub-Himalayan belt. Also in Western Peninsula, Aravali hills, Central India and Singbhum (Jharkhand).

Habit: A large deciduous spineless tree.

Root: Branched tap root.

Stem: Woody, solid, aerial, erect, bark smooth, pale, wood greyish white, light and strong, young shoots softly tawny tomentose.

Leaves: Simple, broad-ovate, acuminate, often with a cordate base, sub-opposite-alternate, basal nerves 3-5, petiole 7-12 cm long, blade 10-20 cm long, shining glands at the insertion of leaves present, entire.

Inflorescence: Terminal racemes.

Flower: Complete, bisexual, large, yellow, tinged with brown, actinomorphic, hypogynous, pentamerous.

Calyx: Sepals 5, gamosepalous, aestivation valvate, softly tomentose.

Corolla: Petals 5, gamopetalous, tubular, ventricose 2.5-4.0 cm long, linear-lanceolate, softly tomentose, corolla tube slender below wider upwards.

Androecium: Stamens 4, didynamous, inserted on corolla tube, alternating with lobes, the fifth posterior represented by staminode or absent, filaments free, anthers oblong, dithecous, dorsifixed, dehiscence longitudinal and introse, diverging.

Gynoecium: Carpels 4, style slender, shortly bifid, rarely 4-fid, single ovule in each chamber, placentation parietal in young but becomes axile in old ovary.

Fruit: Drupe, yellow after ripening, 2.5 cm long.

Seed: Exalbuminous, ovate and sessile, embryo straight.

Sprouting: Leaves fall during Feb-April and sprouting occurs between April – May.

Flowering: February – April before sprouting.

Floral formula: ⊕ ⚥ $K_{(5)}$ $C_{(5)}$ A_{4+1std} $G_{\underline{(4)}}$

Uses: It serves as secondary food plant of muga (*Antheraea assama*) as well as eri (*Philosamia ricini*) silkworms (Seth 1995c,d, Das *et al*. 2000, Sarmah 2000). Besides, its wood possesses good timber value and used as mortar in Kerala and for making plough parts and drums in U.P. Medicinally, leaves are demulcent, used in gonorrhoea, cough and to remove foetid discharges and worms from ulcers. Fruit decoction is used in fever and billious affections, flower in blood diseases and plant in snakebite and scorpion sting. Dasamula, an Ayurvedic preparation cosist of its root as one of the ingradients (Pandey 1981, Sinha 1996, Maheshwari and Singh 1991, Nair and Jayakumar 1998).

Wood used for furniture, carriages, carvings, axles, shafts, artificial limbs, planking, tight cooperage, printing blocks, venetian blinds, musical instruments, picture frames and lacquered boxes. Also employed in well-work, bridges, ship building, tea chest plywood, match boxes, paper making etc. (Ambasta *et al.* 1986).

Cultivation: It may be cultivated through seeds, air layerings as well as stem cuttings through use of root inducing hormones like NAA and IAA. Maximum rooting in KFRI, Peechi in stem cuttings was observed in NAA 100 ppm during April (60%) followed by IAA 100 ppm during March (50%). The duration of hormonal treatment to cuttings was 24 hr. (Surendran and Seethalakshmi 1987).

76. ANJAN

Botanical: *Hardwickia binata* Roxb.

Eng.: Anjan	Hindi: Anjan	
Gond: Chhota-dhundera		
Kannada: Karachi, Kamara	Malayalam: Aacha	Marathi: Anjan, Parsid
Sanskrit: Anjan	Tamil: Acha, Calam	Telugu: Yepi, Naryepi

SYSTEMATIC POSITION

Bentham & Hooker (1883)	Engler & Prantl (1899)	Hutchinson (1959)
Phanerogams	Phanerogams	Angiospermae
Dicotyledones	Dicotyledoneae	Dicotyledones
Polypetalae	Archichlamydeae	Lignosae
Calyciflorae	Rosales	Leguminales

Rosales	Leguminosae	Caesalpiniaceae
Leguminosae	Caesalpinioideae	
Caeslpinioideae		

Geographical distribution: Drier regions of Western Peninsula and south of Sone river (Bihar) ascending up to 3500 ft. ASL.

Habit: A gregarious tree reaching up to 100 ft. heights (Fig. 31).

Root: Branched tap root.

Stem: Erect, aerial, branched, branches slender, glabrous, drooping, heart-wood dark red, extremely hard, woody, cylinderical.

Leaves: Leaflets 2, (bipinnate) sessile, obtuse, oblique, ovate-trapezoid, 2.5-7.0 cm long, 4-5 longitudinal nerves radiating from base.

Inflorescence: Racemes few flowered, lax, in axillary and terminal panicles.

Flower: Bisexual, incomplete, regular, greenish yellow, pedicels as long as the calyx, bracteoles absent, glabrous, entire, zygomorphic, hypogynous.

Calyx: Sepals 5, rarely 4, oblong, obtuse, imbricated, sub-petaloid, persistent, 0.3-0.4 cm long.

Corolla: Nil.

Androecium: Stamens 10, rarely 8 (twice the number of sepals), included, free, anthers short, versatile, filaments filiform, dehiscence longitudinal.

Gynoecium: Carpel 1, ovary superior, sessile, unilocular, 2-ovuled, style filiform, stigma capitate, placentation marginal.

Fruit: Flat, coriaceous, oblong-lanceolate pod, thin, with parallel longitudinal veins, flexible, 5.0-7.0 cm long, the seed lies near the tip.

Seed: Non-endospermic.

Flowering: Rainy season.

Fruiting: Cold season.

Floral formula: $\oplus$ ⚥ $K_5\ C_0\ A_{5+5}\ G_{\underline{1}}$

Uses: It serves as one of the primary / secondary food plants of tropical tasar silkworm, *Antheraea mylitta* D. Belguam ecorace may exclusively be reared on Anjan. (FAO 1976, Sengupta 1986, Thangavelu *et al*. 2000).

Wood is extremely hard and heavy, used for construction, agricultural implements and building purposes (Nair and Jayakumar 1998). Wood used for carving, turning and ornamental work; naves of cart wheels, ploughs, oil mills, posts and beams, clod crushers, bridges, wells, mine props, pontoons, oars and parquet floors. Also suitable for lathe chucks, tool-handles, railway keys, tent pegs, bench screws and brake blocks (Ambasta *et al.* 1986). A resin tapped from the heartwood is used to prepare a varnish. The bark is source of fibre, which is spun into ropes and cordage. The green leaves are used as a fodder. The pods, leaves and bark contain 6%, 7% and 9% tannins respectively (Prasad *et al.* 1974, Masilamani and Vadivelu 1997). It is a fairly drought resistant tree capable of colonising dry shallow soils and rocky terrain where other species fail (Singh 1982).

Leaves used as fodder and manure.

Cultivation: It can be cultivated through directly sowing seeds or planting seedlings easily. The seeds loose their viability shortly and their whole viability is lost after 10 months of storage. Pretreatment of seeds with 200-ppm ethrel after cold water soaking prolongs the viability up to eight months with higher germination and vigour (Masilamani and Vadivelu 1997). The standard rooting media for raising seedlings in nursery has been found to be sand, silt and FYM mixture in 2:1:1 ratio (Mutha *et al.* 1997).

77. KESSERU

Botanical: ***Heteropanax fragrans*** Seem.

Burmese: Tachanza	Hindi: Kesseru, Tarla, Dinha	Nepali: Lal totilla
Assamese: Kesseru	Bengali: Guti-suna	Mundari: Rengebanam

SYSTEMATIC POSITION

Bentham & Hooker (1883)	Engler & Prantl (1899)	Hutchinson (1959)
Phanerogams	Phanerogams	Angiospermae
Dicotyledones	Dicotyledoneae	Dicotyledones
Polypetalae	Archichlamydeae	Lignosae
Calyciflorae	Umbelliflorae	Araliales
Umbellales	Araliaceae	Araliaceae
Araliaceae		

Geographical distribution: Hills and plains of Assam, Meghalaya (Khasi hills), Manipur, Tripura, Arunachal Pradesh, Bangladesh (Chittagang), Andamans and Myanmar, China etc.

Habit: A small glabrous soft wooded deciduous tree up to 60 ft. height (Fig. 32).

Habitat: Sub-Himalayan tract with high humidity, rainfall and moderate temperature. It grows well in almost all types of soil but prefers sandy loam soils.

Root: Branched tap root.

Stem: Erect, woody, aerial, unarmed, glabrous, branched.

Leaves: Compound, tripinnate, alternate, leaflets, opposite, ovate, elliplic or elliptic – lanceolate, 7-18 cm long, glabrous, stipules absent, petioles long.

Inflorescence: Elongated panicled racemes.

Flower: Unisexual or bisexual, regular, pentamerous, epigyous, yellow, fragrant, male flowers sessile, or nearly so in globose heads, bisexual flowers usually at the ends of racemes in umbels.

Calyx: Sepals 5, adnate to ovary, limbs short, entire, united.

Corolla: Petals 5, free, valvate, triangular, caducous.

Androecium: Stamens 5, filaments inflexed in bud, anthers didynamous, versatile.

Gynoecium: Carpels 2, bilocular, styles 2, distinct, filiform, ovary inferior one ovule in each cell (Locule).

Fruit: Drupe, laterally compressed, 0.8 cm diameter, endocarp crustaceous, albumen ruminate, seeds pendulous, embryo minute in dense fleshy albumen.

Floral formula: $\oplus$ ⚥ $K_{(5)}\ C_{(5)}\ A_5\ G_{(\bar{2})}$

Uses: *Heteropanax fragraus* serves as one of the important primary food plants of eri silkworm, *Philosamia ricini* throughout North- eastern region of India. It is considered as the second best food plant for eri silkworms next to Castor (FAO 1976, Sarkar 1988, Anonymous 1988, Seth 1995c, Sarmah 2000).

Wood used for turnery articles (Ambasta *et al.* 1986)

Cultivation: It is cultivated through seeds. In nature also it propagates through seeds. Sloppy well drained land and Feb-March is ideal season for nursery raising and Aug.-Sept. is ideal for Kesseru plantation. Overnight soaked fruits are rubbed with gunny clothes and seeds sinking in water are collected for sowing. Seeds are treated with endofil M-45 @ 2-3 gms/kg of seeds and 800 seeds are sown (2 seeds together) in each nursery bed maintaining a pit to pit distance of 15 cm and row to row distance of 20 cm. The nursery beds are made in 6x2 m size lifted up to 15 cm height by thoroughly mixing 6 ft FYM, sand and soil. After sowing seed bed is covered with layer of straw and irrigated regularly and a shed cover is also provided. Both covers are removed after 90% germination within 18-25 days. Seedlings are allowed to grow to a height of one foot by keeping one seedling/pit. The high and sloppy land is ploughed to a depth of 20-25 cm, leveled and pits of 30x30x30 cm size are made at 2x2 m spacing and FYM (2-4 kg/pit) is thoroughly mixed with the soil and subsequently one seedling is transplanted in each pit during Aug-Sept. (rainy season). Hoeing and weeding is carried out 3-4 times a year. FYM is applied @ 10 tonnes/ha. or 0.5 kg/plant once a year before monsoon and NPK is applied @ 125:75:25 kg/ha/yr in two split doses before and after monsoon after one year of plantation. Leaves can be harvested for silkworm rearing after two years and pruning may be attended after three years at 1.5 m height. (Anonymous 1998).

78. THEBOW

Botanical: ***Hodgosonia heteroclita*** H.f. & T. Syn. *Trichosanthes heteroclita* Roxb. *T. grandiflora* Wall. Cat.

Nepali: Darsani, Ghinphal

Assamese: Thebow — Bengali: Gulur — Lepcha: Kathior-pat

Lushai: Kha-um

SYSTEMATIC POSITION

Bentham & Hooker (1883)	**Engler & Prantl (1899)**	**Hutchinson (1959)**
Phanerogams	Phanerogams	Angiospermae
Dicotyledones	Dicotyledoneae	Dicotyledones
Polypetalae	Sympetalae	Lignosae
Thalmiflorae	Cucurbitales	Cucurbitales
Passiflorales	Cucurbitaceae	Cucubitaceae
Cucurbitaceae		

Geographical distribution: Sikkim, Assam, Meghalaya (Khasia hills), Bangla Desh (Chittagong) up to 3000 ft. ASL.

Habit: A large climber stem extending up to 100 ft, dioecious, (male and female flowers on separate plants).

Leaves: Simple, long and broad, 15-20 cm long, glabrous, deeply lobed, lobes entire or very sparingly denticulate, acute, lobes 3-5, petiole long, tendrils bi-trifid.

Flowers: Incomplete, large, dioecious, males in long racemes of 20 cm, bracts 1.2 cm, oblong, entire, deciduous, females solitary, epigynous, actinomorphic, white.

(A) Male:

Calyx: Sepals 5, gamosepalous, long, tubular, mouth shortly campanulate, teeth 5, rusty pubescent outside, tube 7-10 x 1.0 cm.

Corolla: Petals 5, connate at the base, very long, fibriate, corolla lobes 5 cm long, brown, pubescent, trinerved outside, white, tinged yellow inside.

Androecium: Stamens 3, filaments very short, anthers exserted, connate, linear, one monothecous, two dithecous.

Gynoecium: Nil.

(B) Female

Calyx: Sepals 5, gamosepalous, long, tubular, mouth shortly campanulate, teeth 5, rusty pubescent outside, tube 7-10 x 1.0 cm.

Corolla: Petals 5, connate at the base, very long, fibriate, corolla lobes 5 cm long, brown, pubescent, trinerved outside, white, tinged yellow inside.

Androecium: Absent.

Gynoecium: Carpels 3, syncarpous, ovary inferior, unilocular, style long, stigmas 3, oblong, bifid exserted, placentation parietal, placenta 3, ovules two on each side of placenta.

Fruit: Large, globular 10-25 cm wide, red-brown, densely tomentose, depressed, 12-grooved flesh hard.

Seeds: Perfect seeds usually 6, flat, ellipsoid, 5-7 x 2.5-4.0 cm.

Floral formulae: Male flower: $Br_r \oplus ♂ K_{(5)} C_{(5)} A_3 G_{(0)}$

Female flower: $Br_r \oplus ♀ K_{(5)} C_{(5)} A_0 G_{(3)}$

Uses: It serves as one of the important tertiary food plants of eri silkworms, *P. ricini* (Sarkar 1988, Seth 1995c).

Kernels eaten after roasting/baking, they yield a cooking oil (Ambasta *et al.* 1986).

Cultivation: It may be cultivated through seeds.

79. SAFED ARAND

Botanical: ***Jatropha curcas*** Linn. Syn. *J.* mollucana Herb. Russ.

Burmese: Thinbaw – Kyetsu Eng: Physic nut, Purging nut

Hindi: Jamalgota, Jangali arandi, Bagbherenda, Safed Arand

Assamese: Bhotera, Bongalibhotora Oriya: Jahazigaba

Bengali: Bag Verendi, Bagbherenda Garo: Borbondong

Gujarati: Jamalgota, Ratanjota Kannada: Kadu-haralu

Malayalam: Kattavanakka Marathi: Mogalierenda Manipuri: Awa – Kege

Sanskrit: Kanana eranda, Parvataranda Tamil: Kadalamanakku

Telugu: Adivi-Amidam

SYSTEMATIC POSITION

Bentham & Hooker (1883)	**Engler & Prantl (1899)**	**Hutchinson (1959)**
Phanerogams	Phanerogams	Angiospermae
Dicotyledones	Dicotyledoneae	Dicotyledones
Monochlamydeae	Archichlamydeae	Lignosae
(Incompletae)	Geraniales	Euphorbiales
Unisexuales	Euphorbiaceae	Euphorbiaceae
Euphorbiaceae		

Geographical distribution: Throughout India and Sri Lanka, common near villages, cultivated and naturalized. Indigenous in tropical America, distributed throughout tropics. Cultivated in Travancore, Punjab and Haryana.

Habit: A soft wooded shrub or evergreen tree of small size, makes excellent hedges, cut branches sprout profusely and readily.

Stem: Young shoots pubescent, older glabrous, soft wooded, branched, erect.

Leaves: Simple, alternate, angular or 3-5 lobed, petiole as long as blade, blade 10-15 cm long, lobes obtuse or acute, quite entire, stipules branched and hair like.

Inflorescence: Terminal cymes, glabrous or pubescent.

Flower: Unisexual, incomplete, actinomorphic, hypogynous, yellowish-green, monoecious, central flowers usually female.

Calyx: Sepals 5, imbricate, often petaloid.

Corolla: Petals 5, sympetalous, contorted, valvate, corolla tube of male villous inside.

Androecium: Stamens 10, arranged in two whorls of 5 each, 5 inner connate into a central coloumn, outer 5 free, anthers dithecous, present only in male flowers.

Gynoecium: Carpels 3, syncarpous, ovary superior styles connate below, stigma 2-fid, ovary and styles glabrous, trilocular, placentation axile, each locule with 1-2 pendulous ovules.

Fruit: Capsule of 2.5 – 4.0 cm.

Seed: Endospermic.

Pollination: Entomophillous.

Floral formulae: Male flower: ⊕ ♂ $K_5\, C_{(5)}\, A_{5+(5)}\, G_{(0)}$

Female flower: ⊕ ♀ $K_5\, C_{(5)}\, A_0\, G_{\underline{(3)}}$

Uses: It serves as one of the tertiary food plants of eri silkworm, *Philosamia ricini* (FAO 1976, Sarkar 1988, Seth1995c). Besides, in Travancore it is cultivated for seed oil used for manufacturing candles, soaps, varnishes and as a lubricant and for illumination (Pandey 1981). In Punjab and Haryana, it is largely cultivated for the seed oil, which is regarded as a bio-diesel and may subsititute petrolium in near future as a non-conventional source of energy. The cultivation of this bio-diesel, being pormoted by Sardar Vallabh Bhai Patel University, Meerut, is gaining popularity among the farmers of North India (Agarwal 2003).

Medicinally, seed oil is anthelmintic, purgative and externally applied in rheumatism and skin diseases. Seed oil is considered as hair stimulant, abortificient and taken in dropsy and applied in sciatica and paralysis. Bed bugs are removed by fumigation of leaves and bark extract in water is used as a remedy against rheumatism and leprosy. Leaf extract has been used for promoting lactation and treatment of toothache. Latex is applied to boils, skin sores and rheumatic joints as a remedy (Sinha 1996). Plant juice is used in scabies, eczema and ringworm. The oil yield from seeds is up to 50-60%. (Jain *et al*. 1990). Dried latex used as marking ink.

Bark yields a dark blue dye used for dyeing cloth, fishing nets and lines. Leaves also yield a dye (Ambasta *et al.* 1986).

Cultivation: It can be cultivated through seeds and seedlings.

80. BHOTERA

Botanical: *Jatropha multifida* Linn.

Eng: Coral plant Hindi: Bhotera

Kannada: Vilayati haralu Marathi: Chiniyerandi

Sanskrit: Bhadradanti, Brihaddanti, Jyotishka, Virechani

Tamil: Kattunervalam, Malaiyamanakku

SYSTEMATIC POSITION

Bentham & Hooker (1883)	Engler & Prantl (1899)	Hutchinson (1959)
Phanerogams	Phanerogams	Angiospermae
Dicotyledones	Dicotyledoneae	Dicotyledones
Monochlamydeae	Archichlamydeae	Lignosae
(Incompletae)	Geraniales	Euphorbiales

Unisexuales | Euphorbiaceae | Euphorbiaceae

Euphorbiaceae

Geographical distribution: It is a native of South America (Columbia) but cultivated and naturalized in various parts of India especially in Bengal, Assam and Peninsula.

Habit: A large shrub or small tree growing up to 1.8 – 2.4 m in height. It can be kept bushy by pruning branches periodically.

Stem: Soft woody, branched, glabresent.

Leaves: Simple, alternate, dark green, orbicular, palmately cut into numerous entire or subdivided caudate – acuminate segments, petiole as long as blade, blade glaucous below and 7-20 cm in diameter, stipulate, stipules capillary eglandular and multifid.

Inflorescence: Long and erect, peduncled, scarlet cymes. Inflorescence carries small coral red flowers on red branches.

Flower: Scarlet, petals free or nearly so, anthers linear, bracts and entire sepals glabrous. In other floral characteristics *J. multifida* Linn is similar to *J. curcas* Linn.

Floral formulae: Male flower: ⊕ ♂ K_5 C_5 $A_{5+(5)}$ $G_{(0)}$

Female flower: ⊕ ♀ K_5 C_5 A_0 $G_{\underline{(3)}}$

Uses: It serves as the tertiary food plant of eri silkworm, *Philosamia ricini* (Seth 1995c). Besides, it is also cultivated as a beautiful ornamental tree/shrub and serves as an excellent plant for rockeries and responses beautifully to pruning (Rao 1991).

Roasted tuberous roots eaten. Root decoction given in indigestion and colic. Seeds yield illuminating oil. Purgative leaves used in scabies. Roasted seeds used in fever and veneral diseases but used for criminal prisoning due to their dangerous cathartic nature (Ambasta *et al.* 1986).

Cultivation: It can be cultivated through seeds and seedlings.

81. SALAD

Botanical: ***Lactuca sativa*** Linn. syn. L. *scariola* Linn. var. *sativa* C.B. Clarke

English: Garden Lettuce, Lettuce | Hindi: Salad

Bengali: Salad | Manipuri: Salad | Tamil: Salathu

Telugu: Kavu

SYSTEMATIC POSITION

Bentham & Hooker (1883)	Engler & Prantl (1899)	Hutchinson (1959)
Phanerogams	Phanerogams	Angiospermae
Dicotyledones	Dicotyledoneae	Dicotyledones
Gamopetalae	Sympetalae	Herbaceae
Inferae	Campanulales	Asterales
Asterales	Compositae	Compositae
Compositae		(Asteraceae)

Geographical distribution: Native of South Europe and West Africa but now cultivated throughout India for its edible leaves.

Habit: Annual cultivated smooth (quite glabrous) herb of 2-4 ft. height. Latex present.

Root: Branched tap root.

Stem: Erect, cylinderical, hollow, herbaceous, round, corymbosely or paniculately branched.

Leaves: Suberect, sessile, auriculate, alternate, obovate, runcinate, pinnatifid or entire, very succulent, quite smooth, spreading, truncate - emarginate, blade 10-20 cm long, spinusely toothed margin, mid-rib distinct, buldging ventrally, radical when young, reticulate unicostate in venation, exstipulate.

Inflorescence: Heads of 1.5 cm long, erect, few flowered, panicled, glabrous, branches of panicle short, greenish white with numerous bracts, bracts appressed, green, cordate, involucre of bracts present, involucral bracts 3 or 4 seriate, outer involucral bracts ovate, triangular, inner linear, receptacle disk like, cylinderical, bearing many sessile flowers.

Flower: All flowers ligulate, zygomorphic, epigynous, homogamous, sessile, epigynous.

Fruit: A cypsela - achene, 0.6 cm including very slender beak, oblanceolate, many ribbed with persistent silvery pappus of many hairs, achenes black.

Seeds: Non-endospermic, embryo straight.

Flowering and fruiting: May - July.

Uses: It serves as alternative host plant for mulberry silkworms, *Bombyx mori* L. as a make shift only to keep the worms alive when due to some reasons mulberry is not obtained. Silkworms soon suffer in growth and vigour and diseases spread due to weakness. They spin filmsy cocoons if reared completely on lettuce (Ullal and Narasimhanna 1994). It is cultivated for its edible leaves for salad purposes (hence the name Salad) throughout the world (Pandey 1981). The plant is rich in vitamins and minerals (Sinha 1996).

Medicinally, plant is used in the treatment of bronchitis and asthma. Stem and root are used as poultice in burns and painful ulcers (Sinha 1996).

Cultivation: It can be easily cultivated in kitchen gardens through seeds/ seedlings.

82. SAONI

Botanical: *Lagerstroemia indica* Linn.

Eng: Common crepe Myrtle   Hindi: Saoni

Bengali: Pharash   Manipuri: Jarol

Punjabi: Saoni, Dhaura, Chinai mendhi

Tamil: Pavalak-kurinji, Sinappu   Telugu: Chinagortanta

SYSTEMATIC POSITION

Bentham & Hooker (1883)	Engler & Prantl (1899)	Hutchinson (1959)
Phanerogams	Phanerogams	Angiospermae
Dicotyledones	Dicotyledoneae	Dicotyledones
Polypetalae	Archichlamydae	Herbaceae
Calyciflorae	Myrtiflorae	Lythrales
Myrtales	Lythraceae	Lythraceae
Lythrarieae		

Geographical distribution: Indigenous to China and possibly to Shan hills in upper Myanmar. Cultivated in gardens throughout India due to its beautiful flowers.

Habit: *L. indica* var. *alba* is a deciduous handsome shrub with white flowers whereas *L. indica* var. *purpurea* bears light mauve coloured flower and *L. indica* var. *raiseda* is a shrub growing to the size of a small tree with large leaves and flowers of a light raised colour (Fig. 33).

Root: Branched tap root.

Stem: Erect, woody, aerial, smooth, branched, bark light gray, round, solid.

Leaves: Simple, alternate- sub opposite, oblong-oval, cauline and ramal, apex and base obtuse, lateral veins 4-6, glabrescent, upper surface dark green lower surface light green.

Inflorescence: Ample terminal panicles.

Flower: White, mauve or raised coloured, bisexual, regular, actimomorphic, hypogynous, hexamerous.

Calyx: Epicalyx 3, Sepals 6-7, gamosepalous, tubular, valvate, light green, calyx tube neither ribbed nor grooved.

Corolla: Petals 6-7, free, white, mauve or raised coloured, crumpled and wrinkled, alternating to sepals, imbricate spreading, ovate, clawed.

Androecium: Stamens numerous, inserted at the bottom of calyx tube, 6-7 often stouter and larger than the others, anthers versatile.

Gynoecium: Carpels 6-7, pistil syncarpous, ovary superior, 6-7 locular, 1-2 ovule in each cell, style long, bent, stigma capitate, placentation axile.

Fruit: Capsule dehiscing longitudinally.

Seeds: Non-endospermic, winged.

Flowering: April – May.

Fruiting: July - August.

Floral formula: ⊕ ⚥ $E_3\ K_{(6\text{-}7)}\ C_{6\text{-}7}\ A_{6\text{-}7+\alpha}\ G_{\underline{(6\text{-}7)}}$

Uses: It serves as one of the tertiary but recognised as primary/secondary food plant of tropical tasar silkworm, *Antheraea mylitta* D (FAO 1976, Seth 1995 a) and silkworm, Atlas *Attacus atlas* (Ambasta *et al.* 1986). Besides, it is cultivated as beautiful ornamental plant in the gardens and avenues throughout India. In addition, it is used medicinally also. Its bark is stimulant and febrifuge. Leaves and flowers are used as purgative and hydragogue, roots are astringent and used as gorgle while seeds are narcotic (Sinha 1996). In apththae of mouth fruits are locally applied (Ambasta *et al.* 1986).

Cultivation: It can be cultivated through seeds and seedlings like *L. speciosa.* Pers. Besides, it can also be propagated through hormonal treatment of cuttings. 1.0-1.5 cm thick and 18-20 cm long cuttings have been propagated by treating them with Serradix B_2/B_3 and inserting them in a soil or porous rooting media (FAO 1976).

83. SIDDHA

Botanical: ***Lagerstroemia parviflora*** Roxb.

Burmese: Zaung bale   Eng: Lendia

Hindi: Siddha, Sido, Seja, Lendi, Dhaura   Nepali: Borderi, Bhot-dunanyaro

Asamese: Sida, Dhauli   Bengali: Sida   Gujarati: Kakria

Kanada: Channangi   Lepcha: Kunhil-kung

Malayalam: Cimanii, Ventekku Marathi: Bonda, Bhorudar

Oriya: Salora Punjabi: Bakli, Dhaura

Tamil: Ventek, Benteak Telugu: Chennangi

SYSTEMATIC POSITION

Bentham & Hooker (1883)	**Engler & Prantl (1899)**	**Hutchinson (1959)**
Phanerogams	Phanerogams	Angiospermae
Dicotyledones	Dicotyledoneae	Dicotyledones
Polypetalae	Archichlamydae	Herbaceae
Calyciflorae	Myrtiflorae	Lythrales
Myrtales	Lythraceae	Lythraceae
Lythrarieae		

Geographical distribution: Assam, lower Bengal, Central India and the Western Peninsula in deciduous forests ascending up to 3000 ft. Also found in upper Myanmar.

Habit: A large tree, youngest shoots, leaves and inflorescence pubescent with very short and simple hairs.

Root: Branched tap root.

Stem: Erect, woody, aerial, smooth, branched, bark light gray, round, solid.

Leaves: Simple, ellptic, opposite, sessile or very shortly petiolate, 5-10 cm long.

Inflorescence: Axillary and terminal panicles.

Flower: White, complete, bisexual, regular, fragrant, 0.6 cm across, pedicels slender, hypogynous.

Calyx: Sepals 6, gamosepalous, calyx tube broadly turbinate or hemispherical cleft into 6 ovate lobes, neither ribbed nor grooved, hairy, 0.5-0.6 cm long from base to end of segments, valvate, segments erect and ultimately appressed to the capsule.

Corolla: Petals 6, free, clawed, imbricate or valvate, white, unguiculate, undulate or wavy.

Androecium: Stamens indefinite, free, filaments long, inserted at the bottom of calyx tube, anthers versatile.

Gynoecium: Carpels 6, syncarpous, polylocular, ovary superior, style long, stigma capitate, placentation axile and 1-2 ovule in each cell.

Fruit: Capsule, 1.8-2.5 cm long.

Flowering: Hot seasons.

Fruiting: Rainy seasons, remains longer on trees.

Leaf fall: March - April.

Sprouting: May.

Floral formula: $\oplus$ ⚥ $K_{(6)}\ C_6\ A_\alpha\ G_{(\underline{6})}$

Uses: It serves as one of the very important primary/secondary food plants of tropical tasar silkworms, *A. mylitta* D. especially for rearing of V stage silk worms and cocooning in Central India. Tira and Seoni races of *A. mylitta* D. are exclusively reared on *Lagerstroemia parviflora* in West Bengal (Bankuria and Purulia) and Chattisgarh (Seoni) (FAO 1976, Sengupta 1986, Thangavelu *et al.* 2000). Besides, it also serves as an important and valuable timber tree. Its wood is moderately hard and used for building, furniture, boxes and agricultural implements etc. (Singh 1990, Shukla and Mishra 1979). Wood is also used for boats, bridges carts, cooperages, bentwood furniture, fencing, golf stick shafts, oars, picker arms, poles, tool handles and railway sleepers (Ambasta *et al.* 1986).

It serves as a good fuel wood. It yields fodder, edible gum, inferior fibre and tannins (Ambasta *et al.* 1986).

Cultivation: It can be cultivated through seeds, seedlings as well as air layerings by use of even cheap rooting harmones like surrotex/serradix B. Besides, it can also be propagated vegetatively through stem cuttings by using root inducing IBA and NAA harmones.

For nursery raising, accoring to Regional Forest Research Centre, Rajahmundry, seeds should be soaked in cold water which gives 17% peak germination and 20% overall germination (Rao *et al.* 1998).

84. JARUL

Botanical: *Lagerstroemia speciosa* Pers. syn. *L. flos-reginae* Retz.

Eng: Queens crepe Myrtle, Pride of India, May/Crepe flower.

Hindi: Jarul Burmese: Eikmwe, Pyinma

Sinhalese: Murutagass

Assamese: Ajar, Ajhar, Thingdou thlado

Bengali: Jarool, Jarul, Ajar Kannada: Hole-dasvala, Challa, Holedachalla

Konkani: Tamonn

Malayalam: Manimaruthu, Katalpu, Atampu, Chemma-ruta

Manipuri: Jarol Marathi: Mota Bondara, Taman, Bondara-lendi

Oriya: Patoli, Ary	Punjabi: Jarul	Sanskrit: Arjuna
Santhali: Sekra	Tamil: Kadali	Telugu: Varagogu

SYSTEMATIC POSITION

Bentham & Hooker (1883)	Engler & Prantl (1899)	Hutchinson (1959)
Phanerogams	Phanerogams	Angiospermae
Dicotyledones	Dicotyledoneae	Dicotyledones
Polypetalae	Archichlamydae	Herbaceae
Calyciflorae	Myrtiflorae	Lythrales
Myrtales	Lythraceae	Lythraceae
Lythrarieae		

Lagerstroemia was named aftger Magnus V. Lagerstrom, a Swedish Botanist and *flos-reginae* means "Flower of the Queen". The plants are deciduous, shed their leaves before blooming especially in May and hence it is popularly known as "May Flower".

Geographical distribution: Lower Myanmar, Sri Lanka, Bangladesh (Chittagong) and moist lower country of Malay Peninsula. In India, it is found in Orissa, Bengal, South India and foot of Western Ghats ascending up to 2000 ft. ASL. Its true home is in the jungles of Assam, Myanmar, Sri Lanka and Kerala.

Habit: A deciduous tree, large in moist areas of eastern India, small (15-20 ft) in dry places of Northern India, crown rounded, bole usually short, leaf fall gradual, hence tree is rarely bare (Fig. 34).

Habitat: Common in swamps or banks of rivers. It grows well on rich deep alluvial soil in moderate light. Under heavey shades it is depressed.

Root: Branched tap root.

Stem: Erect, woody, aerial, smooth, branched, bark light gray, round, solid.

Leaves: Simple, oval (elliptic), glabrous, bright green above, pale green below, veined heavily on the under side, opposite - sub-opposite, spirally arranged, entire, apex acuminate, base obtuse, lateral nerves 10-15 pairs on both sides, turning red before falling in Feb.-March, blade 10-20 cm long, petiole stout, 0.6-1.2 cm long, cauline and ramal.

Inflorescence: Terminal branched panicles of 30-60 cm long.

Flower: Complete, bisexual, regular, hexamerous, hypogynous, pedicels long, purplish or pinkish, 10-12 cm across.

Calyx: Sepals 6, gamosepalous, tubular, strongly 12-14 ribbed, ribs rounded, broader than the furrows between, segments long - triangular, pinkish green, valvate, epicalyx-3.

Corolla: Petals 6, purplish or pinkish turning white before fall, crumpled and wrinkled in appearnce (hence crepe flower), free (polypetalous), alternating to sepals, round-ovate, clawed, spreading, imbricate.

Androecium: Stamens many, purple-yellow, inserted near the bottom of the calyx tube, filaments long and anthers versatile.

Gynoecium: Carpels 6, superior, sessile, syncarpous, hexalocular, ovules many, placentation axile, style long, bent, stigma capitate.

Fruit: Capsule broadly ovoid, green first but turn black, woody, opening by 5-6 longitudinal valves when ripe, 2-2.5 cm long, the lower third enclosed in persistent calyx, ripen during cold weather, dehisence during leaf fall.

Seeds: Many, pale brown with 1.2-1.8 cm long wing at apex, non-endospermic.

Flowering: April - May and July-Aug.

Fruiting: May – September.

Leaf fall: Feb. – March.

Sprouting: April - May.

Floral formula: $\oplus$ ⚥ $E_3\ K_{(6)}\ C_6\ A_{\alpha}\ G_{\underline{(6)}}$

Uses: It serves as tertiary but recognised as one of the primary/secondary food plant of tropical tasar silkworms, *Antheraea mylitta* D (FAO 1976, Seth 1995 a). Besides, it is well known ornamental tree and cultivated for the beauty of its flowers. It is also cultivated as an avenue tree. It is valued next to the teak as a timber tree and is used for building boats, carts etc (Randhawa 1983). Wood is used for bridges, dug outs and oars, cooperages, floor boards, motor lorry bodies, mortors, mine-props, rice pounders, railway carriages, turnery, well curbs, water tanks, furniture, wheels, boxes, poles, boot lasts, leather cutting blocks, match boxes, splints and after treatment railway sleepers (Ambasta *et al.* 1986).

Medicinaly, bark, leaves, flowers and fuirts are useful. The decoction of bark and leaves is used to promote urine flow and control of diabetes. The infusion of bark powder reduces or eliminates fever and diarrhoea. The fruit cures small ulcers in mouth while bark, leaves and flowers are used as purgative as well as laxative (Kurian 1995, Randhawa 1983, Sinha 1996).

Bark yields coarse fibre while leaves, fruit and bark yield tannins.

Cultivation: It is propagated through seeds. The slow growth of seedlings during first year gathers momentum during second year onwards and it starts flowering within 3-5 years. Transplantation of one-year-old seedlings is most suitable. Pruning of side branches is necessary to develop a well-shaped tree. It can also be cultured as Bonsai (Randhawa 1983).

85. SAHI

Botanical: ***Lithocarpus dealbata*** (Hk.f & Th. ex Miq.) Rheder. Syn. *Quercus dealbata* Hk. f. Th. ex Miq., *Pasania dealbata* (Hk.f. ex Th. ex Miq.) Kanji lal *et al.*

Vernaculars: Sahi, Dieng-sai, Fah, Keu-ko, Soh-rang, Khlaw, Rug-ying, To-ik

Manipuri: Sahi

SYSTEMATIC POSITION

Bentham & Hooker (1883)	Engler & Prantl (1899)	Hutchinson (1959)
Phanerogams	Phanerogams	Angiospermae
Dicotyledones	Dicotyledoneae	Dicotyledones
Monochlamydeae	Archichlamydeae	Lignosae
(or incompletae)	Fagales	Fagales
Unisexuales	Fagaceae	Fagaceae
Cupuliferae		
Quercineae		

Geographical distribution: Arunachal Pradesh, Assam, Meghalaya (Khasi hills), Nagaland, Mizoram and Manipur at an elevation between 2700-6000 ft. ASL. Also found in Bhutan.

Habit: A monoecious shrub or small tree forming a scrub forest in hills where fire is a annual feature. Young shoots softly pubescent, occasionally furfurious (Fig. 35).

Root: Branched tap root.

Stem: Aerial, erect, woody, hard, branched, bark ashy grey, splitting in squarish flakes, 7-12 mm. thick, coarsely granular, light brown within.

Leaves: Simple, lanceolate, entire, apex usually acuminate, 8-18 x 2.5-7.0 cm, glabrous above on maturity, more or less ashy-grey tomentose beneath, base acute or acuminate usually oblique, secondary nerves 8-14 pairs, tertiaries parallel and transverse, very fine, petiole 5-17 mm long.

Inflorescence: Terminal panicled spikes, usually androgynous.

Flowers: Unisexual, actinomorphic, incomplete, epigynous, bracteolate, bracteoles unequal.

Male flowers: Sessile, in clusters of 3, perianth segments 6, stamens about 10-12.

Female flowers: In clusters of 3, confluent at base, styles 3, linear, ovary tricarpellary, inferior.

Fruit: Acorns (Nuts), glan turbinate or pyriform, 12-18 mm across, fasicled on a stout long rachis on maturity or 1-3 together. Cupules sessile, woody, completely enclosing the glans, scales pubescent, connate with a free apex forming tubercles.

Flowering: Jan - Feb.

Fruiting: Next cold season

Floral formulae: Male flower: Brl. ⊕ ♂ $P_{(6)}\ A_{10-12}\ G_0$

Female flower: Brl. ⊕ ♀ $P_{(6)}\ A_0\ G_{(\bar{3})}$

Uses: It serves as one of the important secondary food plants next to *Q. acutissima* and *Q. grifithii* for temperate tasar viz., *Anthereae proylei, A. pernyi* and *A. roylei.* For *A. frithii* it serves as primary food plant because *A. frithii* is multivoltine and *L. dealbata* is an evergreen tree as compared to other oak tasar silkworms which are uni/bivoltine and food plants which are deciduous in nature (FAO 1976, Mishra *et al.* 2000, Srivastav and Singh 2002). The wood is used as a fuel and for making agricultural implements. Bark consists of 10-13% tannins (Negi and Naithani 1995). Fruits may be eaten in very limited quantity.

Cultivation: It may be propagated through either direct sowing or transplantation of seedlings raised in nursery.

86. KUHI

Botanical: *Lithocarpus fenestrata* (Roxb.) Rheder syn. *Quercus fenestrata* Roxb. *Pasania fenestrata* (Roxb.)

Vernaculars: Arkaula, Katy-chakma, Ka-diengjing, Dieng-sai, Kashiengdung, Dieng-sa-sei-saw

Nepali: Palte katus, Arkaula

Assam: Kuhi, Ka-diengjing   Bengali: Kala-chakma   Lepcha: Kashiendung

Manipuri: Kuhi

SYSTEMATIC POSITION

Bentham & Hooker (1883)	**Engler & Prantl (1899)**	**Hutchinson (1959)**
Phanerogams	Phanerogams	Angiospermae
Dicotyledones	Dicotyledoneae	Dicotyledones

Monochlamydeae	Archichlamydeae	Lignosae
(or incompletae)	Fagales	Fagales
Unisexuales	Fagaceae	Fagaceae
Cupuliferae		
Quercineae		

Geographical distribution: Sikkim to Arunachal Pradesh, Assam, Meghalaya (Khasi hills), Nagaland, Mizoram and Manipur between 5000-8000 ft. ASL. Also found in Bangladesh and upper Myanmar between 4500-7500 ft. ASL.

Habit: A medium sized or large evergreen monoecious tree, young shoots deciduously tawny-pubescent.

Root: Branched tap root.

Stem: Aerial, erect, woody, branched, bark grey, rough with cracks exfoliating into irregularly shaped woody flakes 1.25-1.5 cm thick, reddish-brown, coarsely fibrous and granular inside.

Leaves: Simple, oblong - lanceolate or elliptic - lanceolate, entire, coriaceous, glabrous above, 15-25 x 8-9 cm, underside clothed with dense felt of very minute stellate hairs, secondary nerves 11-13 pairs, slender but prominent beneath, apex acuminate, base cuneate often oblique, petiole 7-15 mm long, stipules pilose, linear - lanceolate.

Inflorescence: Panicled or fasicled spikes at apex of branchlets, tomentose and bearing both male as well as female flowers or unisexual flowers only. Male spike terminal panicles, grey pubescent with long acuminate deciduous bracts. Female spikes generally long and branching.

Flower: Unisexual, actinomorphic, incomplete, epigynous, bracteate, crowded but not confluent.

Male flowers: in glomerules of 2-3, 3-bracteate, perianth segments 5-6, tomentose outside, stamens about 12.

Female flowers: in glomerules of 3, single bracteate, styles 3, ovary inferior.

Fruit: Acorns (Nut). Glans 1.5-2.0 cm across, smooth, globose, apiculate, matured fruits in 1-2 or rarely 1-3 together. Cupules completely enclosing the glans, scales ovate-acute, connate at the base, pubescent, apex acute, incurved.

Flowering: Sept. - Oct.

Fruiting: April - May.

Floral formulae: Male flower: Br. ⊕ ♂ $P_{(5-6)}$ A_{12} G_0

Female flower: Br. ⊕ ♀ $P_{(5-6)}$ A_{12} $G_{(\bar{3})}$

Uses: It serves as secondary food plant of temperate tasar silkworms viz., *Antheraea proylei, A. pernyi, A. roylei* and *A. frithii.*

Used for fuel and timber in tea estates and also for building purposes and agricultural implements. Bark and acorns are used for dyeing in Manipur. Bark contains 10-16% tannins and can be used as a transtuff to give leather of light colour (Negi and Naithani 1995).

Cultivation: Either it may be cultivated through direct sowing of seeds *in situ* or transplantation of seedlings raised in nursery.

87. *LITHOCARPUS XYLOCARPA*

Botanical: ***Lithocarpus xylocarpa*** (Kurz.) Markgr. Syn. *Quercus xylocarpa* Kurz., *Pasania xylocarpa* (Kurz.) Schottky.

SYSTEMATIC POSITION

Bentham & Hooker (1883)	Engler & Prantl (1899)	Hutchinson (1959)
Phanerogams	Phanerogams	Angiospermae
Dicotyledones	Dicotyledoneae	Dicotyledones
Monochlamydeae	Archichlamydeae	Lignosae
(or incompletae)	Fagales	Fagales
Unisexuales	Fagaceae	Fagaceae
Cupuliferae		
Quercineae		

Geographical distribution: Arunachal Pradesh, Meghalaya, Nagaland, and Manipur at 4000-6000 ft. ASL. Also found in Myanmar at 3600-4500 ft. ASL.

Habit: A large, 30-40 ft. high, evergreen, monoecious tree with thinly hairy youngest shoots.

Root: Branched tap root system.

Stem: Aerial, erect, woody, hard, bark whitish, plain with vertical fissure and faint horizontal wrinkles, inside pale-brown, fibrous, turning darker brown, soft 1.5 cm thick.

Leaves: Simple, elliptic-oblong-lanceolate, thinly coriaceous, apex, caudate-acuminate, base acute or obtuse, entire, opaque and glabrous above with a slender pubescent mid-rib subsilvery beneath, lateral nerves 10-12 pairs, arched, ascending, conspicuous beneath, petiole 7-10 mm long.

Inflorescence: Unisexual spikes.

Flower: Unisexual, incomplete, epigynous, actinomorphic.

Fruit: Acorns (nut), 4-10 cm long, 1.8 cm across, nut globose, the umbo only exserted, adnate to the involucre except glabrous convex top, the upper fourth polished, the lower three fourth rough. Involucres confluent in sessile clusters of three, globose, clothed with soft spinous pubescent bracts or tubercles enclosing entirely the glans, woody 1.8-3.2 cm across, walls 0.3 cm thick.

Fruiting: December.

Uses: It serves as one of the secondary food plants of temperate tasar silkworms viz., *Antheraea proylei, A. roylei, A. pernyi* and *A. frithii.*

The nuts are roasted or eaten raw (Negi and Naithani 1995).

Cultivation: It can be cultivated through seeds either directly or through transplantation of seedlings raised in nursery.

88. MEJANKARI

Botanical: ***Litsaea citrata*** Blume. Syn. *L. cubeba* Pers., *Tetranthera polyantha* Wall.

Nepal: Siltimber

Assamese: Mejankari, Adakori, Edunkeri — Garo: Zeng-Jil

Khasi: Dieng-si-ing — Lepcha: Terhilsok, Tanghaercherking

Lushai: Sernam — Mizo: Sernam

SYSTEMATIC POSITION

Bentham & Hooker (1883)	Engler & Prantl (1899)	Hutchinson (1959)
Phanerogams	Phanerogams	Angiospermae
Dicotyledones	Dicotyledoneae	Dicotyledones
Polypetalae	Archichlamydeae	Lignosae
Thalmiflorae	Ranales	Laurales
Laurales	Lauraceae	Lauraceae
Lauraceae		

Geographical distribution: Eastern Himalayas covering Sikkim to Mishmy hills in Arunachal Pradesh and Assam, Manipur, Garo, Khasi and Naga hills up to an altitude of 6000 ft. ASL. Sub-himalayan tracts and outer ranges from

Nepal eastwards ascending to 9000 ft. ASL, upper Myanmar and Java at high elevations.

Habit: A deciduous bush or small tree of dioecious and aromatic nature, smells like oranges.

Habitat: Humid and warm climatic conditions with high rainfall along the sub-Himalayan hill ranges and Brahmputra valley in North-eastern India consisting of sandy loam soil with acidic pH between 4.5-5.0 are their natural habitat. High and sloppy land with good drainage system is suitable for their plantation.

Root: Branched tap root.

Stem: Aerial, erect, solid, branched, branches slender and glabrous but youngest shoots pubescent (hairy), branches black when dry, cylinderical, bark green, warty, thin, blaze yellowish and mucilaginous when young but turning brownish when mature.

Leaves: Simple, alternate, deciduous, penninerved, petiole slender 1.2-2.2 cm long, exstipulate, membranous, lanceolate or ovate–lanceolate, bright green above and dull glaucous beneath, black when dry, youngest hairy, apex acute – subacuminate, margin entire, venation reticulate, midrib purplish beneath, lateral nerves 7-11 on both sides, slender, base acute or somewhat oblique, blade 7-12 cm long and 2-3.5 cm broad.

Inflorescence: Solitary or in sessile or shortly stalked corymbs, 4-6 flowers in each head. Heads on slender peduncles.

Flower: Bracteate and pedicellate, enclosed by 4 membranous glabrous bracts which are glabrous or sparsely silky and ciliate at the edges, pedicels hairy and white.

Perianth: Tepals 6, polytepalous, membranous, obovate, sub-equal and valvate.

Androecium: Found only in male flowers, stamens 9-10, polyandrous, filaments usually silky, anthers 4-celled, adnate, introse, opening by a valve from below to upwards, stamens imperfect in female flowers.

Gynoecium: Found only in female flowers, monocarpellary, unilocular with single ovule, stigma disciform, ovary superior, placentation basal.

Fruit: Fleshy, berry, small, globose, 0.6-0.8 cm long.

Seeds: Exalbuminous.

Floral formulae: Male flower: $Br._4 \oplus ♂ P_6 A_{9\text{-}10} G_0$

Female flower: $Br._4 \oplus ♀ P_6 A_0 G_{\underline{1}}$

Uses: In Assam valley chiefly in Dibrugarh and Sibsagar districts, it is used to feed silkworms which yield the valuable Mezenkuri silk which was the exclusive wear of the Ahom kings and elites as revealed by historical records. The Mejenkuri silk is creamy white, very tough but soft in comparison to conventional muga silk worms reared on primary food plants. Mejankari is categorized as an endangered species and located in the neighbouring hill states of Assam.

The size of Mejankari cocoons is much smaller than that of Som cocoons but SR% of Mejankari cocoons (15.23%) are much higher than the Som cocoons (9.64%) because pupa of Mejankari cocoons is smaller than that of Som cocoons. The filament of Mejankari (479-500 mtrs) is comparatively shorter than that of Som cocoons (536-550 mtrs.). But Mejankari silk is marketed at very high price, which compensates smaller size of its cocoons (Saikia and Goswami 1997).

Fruits are edible, carminative, used for dizziness, hysteria, paralysis and loss of memory (Sinha 1996). Fruits are substituted for *Piper cubeba* Linn.

Aromatic floral oil used in perfurmary is identical to Rose and Coriander while that of fruit is reminescent of Verbena oil. Seeds yield an illuminating fat (Ambasta *et al.* 1986).

Cultivation: *L. citrata* may be propagated through seeds and airlayerings like *L. polyantha.*

89. KATHALUA

Botanical: *Litsaea nitida* Roxb. Syn. *Tetranthera nitida*

Nepal: Lhopre

Assamese: Kathalua, Kotalhua, Loban, Supin-urn-rong.

SYSTEMATIC POSITION

Bentham & Hooker (1883)	Engler & Prantl (1899)	Hutchinson (1959)
Phanerogams	Phanerogams	Angiospermae
Dicotyledones	Dicotyledoneae	Dicotyledones
Polypetalae	Archichlamydae	Lignosae
Thalmiflorae	Ranales	Laurales
Laurales	Lauraceae	Lauraceae
Lauraceae		

Geographical distribution: Eastern Himalayas, Bhutan, Bangladesh (Sylhet), Jharkhand (Singhbum) and North-eastern India viz., Assam.

Habit: Middle sized, sometimes a large dioecious tree.

Habitat: Evergreen forests having warm climate with high humidity and high rainfall.

Root: Branched tap root.

Stem: Branched, branchlets stout, shining, angular, aerial, erect, woody, solid.

Leaves: Simple, dark green, shining, rigidly coriaceous, oblanceolate or elliptic, mostly obtuse, petiole 1.7 to 2.5 cm, blade 20-30 cm long. Secondary nerves faint, extipulate, entire.

Inflorescence: Axillary racemes, male raceme 2-4, female raceme 1-2 inch long.

Flowers: Bracteate, regular, pedicellate, unisexual, hypogynous.

Perianth: Tepals 6, polytepalous, small, thick, valvate.

Androecium: 9-12 stamens present in male flowers only, polyandrous, arranged in 3-4 whorls, anthers adnate opening by a valve from below upward, imperfect stamens in female flowers.

Gynoecium: Monocarpellary, unilocular, single ovuled, ovary superior, placentation basal, in male flowers abortive.

Fruit: Fleshy berry, ellipsoid, purple when ripe, 0.8 cm long, when unripe enclosed entirely in the indistinctly lobed perianth.

Seeds: Exalbuminous.

Foral formulae: Male flowers: $Br._4 \oplus ♂ P_6 A_{9-12} G_0$

Female flowers: $Br._4 \oplus ♀ P_6 A_0 G_{\underline{1}}$

Uses: *L. nitida* serves as one of the secondary food plants of muga silkworm (Thangavelu *et al.* 1988, Das *et al.* 2000, Seth 1995d).

Wood used for poles (Ambasta *et al.* 1986).

Cultivation: *Litsaea nitida* propagates through seeds in nature but it may be raised through air layerings also.

90. SOALU

Botanical: *Litsaea polyantha* Blume Syn. *L. monopetala* (Roxb.) Pers., *Tetranthera monopetala* Roxb., *L. polyantha* Juss

Burmese: Ondon

Hindi: Meda, Soalu, Katmarra, Singran

Nepali: Kadmero, Ratamanti

Assemese: Munga-songbiphang, Kakuri, Sualu, Ketmara, Patoia, Muga, Hoanlu

Bengali: Bara-Kukurchita Gond : Leja

Lepcha: Sunyok-kung, Sapot-king Manipuri: Tumitla

Marathi: Ranamba, Rapamba Oriya: Balghoari, Kulya, Bastura

Punjabi: Rian, Gwa, Karkawa Tamil: Maidalagadil, Muchaippeyetti, Picinbattaw

Telugu: Naramamidi

SYSTEMATIC POSITION

Bentham & Hooker (1883)	Engler & Prantl (1899)	Hutchinson (1959)
Phanerogams	Phanerogams	Angiospermae
Dicotyledones	Dicotyledoneae	Dicotyledones
Polypetalae	Archichlamydeae	Lignosae
Thalmiflorae	Ranales	Laurales
Laurales	Lauraceae	Lauraceae
Lauraceae		

Geographical distribution: Extending northward from Punjab up to salt range along the foothills of Himalayas ascending up to 3000 ft. ASL eastwards to North-eastern India and southwards to Satpura range. Prevalent throughout Assam (Cachar), Megahalaya, (Khasi hills) Mizoram, Nagaland, Manipur and Bengal plains. Also found in Chittagong (Bangla Desh) and Myanmar in the plains and valleys of the principal rivers.

Habit: Evergreen, small to large fast growing tree, attaining a height of more than 20 m and a girth of 2 m with large spreading crown (Fig. 36)

Habitat: Humid and warm climatic conditions with high rainfall along the sub-himalayan hill ranges and Brahmputra valley in North-eastern India consisting of sandy loam soil with acidic P^H between 4.5-5.0 are their natural habitat. High and sloppy land with good drainage system is suitable for their plantation.

Root: Branched tap root.

Stem: Aerial, erect, woody, branched, branches tomentose or pubescent, bark rough, dark grey or pale brown, exfoliating in corky scales.

Leaves: Simple, alternate, rarely opposite or sub-opposite, obovate-oblong, oblanceolate or elliptic – oblong, obtuse or acute, entire, coriaceous, petiolate, petiole hairy and 1- 2.5 cm long, dark green and glabrescent above, glaucous and rusty pubescent beneath, 8-30 cm long and 5-11 cm broad, exstipulate, base rounded or cordate, Nerves or veins strong beneath, lateral nerves 15-18.

Inflorescence: Sessile or pedicellate umbels in clusters, 1.2-2.6 cm in diameter, axillary, 4-6 flowered, flower buds enclosed by 4-5 concave, overlapping, tomentose bracts in two whorls forming involucre

Flower: Unisexual, dioecious, sometimes monoecious, bisexual flowers exist when one sex is abortive, pedicellate, pedicel villous, about 0.2 cm long, incomplete stalked, white or pale greenish yellow, bracts 5, persistent in flower, imbricate in bud, concave, deciduous, orbicular, membranous, hypogynous, actinomorphic.

Perianth: Tepals 4-6, small, thin, hairy, almost free, tubular, campanulate, persistent.

Androecium: Stamens usually 9-12 in male flowers only in 3-4 whorls, rarely more, filaments villous, not longer than perianth, filaments of first and second whorl eglandular while filaments in third and fourth whorls bear a pair of lateral gland, glands stipitate, anthers adnate, 4-celled, introse, opening by a valve, in female flowers 9-12 staminodes present and arranged as in male flowers.

Gynoecium: Monocarpellary, unilocular with single ovule in female flowers only, ovary superior, ovule solitary and pendulous, placentation marginal/basal, style thick, stigma irregularly lobed. In male flowers ovary absent or with a slender style and stigma.

Fruit: Fleshy berry, ovoid, oblong, globose or ellipsoid, black when ripe, 0.5-1.1 cm long, seated on the persistent bases of perianth segments.

Seed: Exalbuminous (Non-endospermic), pendulous.

Flowering: February-May. Flowers appear chiefly on the upper leafless branches while lower branches still bear leaves.

Fruiting: May - August

Floral formulae: Male flower: $Br._5 \oplus ♂ P_{(4\text{-}6)} A_{3+3+3+3} G_0$

Female flower: $Br._5 \oplus ♀ P_{(4\text{-}6)} A_{Std\ 9\text{-}12} G_{\underline{1}}$

Varieties: In Soalu two varieties have been identified on the basis of leaf shape, size and texture. The oblong variety bears oblong narrow and thin leaves while ovate variety bears oval, broad and thick leaves. The silkworms prefer the ovate variety. Recently, 10 morphotypes of Ssoalu have been isolated (Benchamin 2000). F5, F1, M1 and F6 exihibited qualitative superiority through biochemical assay (Srivastava *et al.* 1998).

Uses: Besides being used as primary food plants in sericulture for rearing of Muga silkworms, *Antheraea assama* Ww., *Litsaea monopetala* (Roxb). Pers leaves are also applied in muscle pains of legs and arms. Its bark is astringent and used in diarrhoea. Seeds yield a fat useful in candle making and rheumatism.

Powdered bark is applied for relieving pains caused due to blows, bruises or hardwork. It is also applied to fractured bones in animals (Sinha 1996). The Muga silkworms reared on Soalu produce relatively low quality yarn than those reared on Som (FAO 1976, Thangavelu *et al.* 1988, Seth 1995 d, Das *et al.* 2000).

Wood used for agricultural implements, furniture, house-building and oars, suitable for internal construction, plywood and tea-boxes (Ambasta *et al.* 1986).

Cultivation: *L. polyantha* (*L. monopetala*) may be cultivated through seeds, air layerings, stem cuttings and leaf bud cuttings. Air layers may be made by using NAA 400 and IBA 500 ppm root inducing harmones through which 70% to 77.77% rooting has been achieved. (Thangavelu *et al.* 1988, Sengupta *et al.* 1991). The stem cuttings exhibited 53.33% rooting and 95.83% of post transplantation survival, whereas leaf bud cuttings rooted to an extent of 66.6% at Regional Muga Research Station, Boko, Assam (Sengupta *et al.* 1991).

For cultivation through seeds, Soalu seeds are collected during May and June. The pulp is washed off by kneeding in water and seed is dried in shade. Seeds remain viable for very less period hence they should be either sown soon after collection or be stored in moist seed beds for six to eight weeks under low temperature to prolong their viability. The seed viability is tested through floating test and seeds sinking in water are selected for nursery raising. The seed beds are prepared by ploughing, leveling and raising the soil upto 20 cm to avoid water logging and FYM @ 30 tons/ha is thoroughly mixed. Seeds are sown 2 cm deep at 15 x 15 cm spacing from each other and irrigated. These seed beds are made under tree shades and covered with moist gunny bags followed by sprinkling of water for maintaining the moisture. The germination starts after four weeks. Germinated seeds are dibbled in polythene tubes of 9x6 inch size of 150-220 gauze filled with soil, sand and FYM in 1:1:1 ratio and kept under shades. (Isa *et al.* 1989, Paliwal *et al.* 1991). The seedlings are transplanted to the pits of 1x1x1 feet size manured with 0.5 cft FYM after becoming three months old from July-September and irrigated regularly. A spacing of 5x5 or 4x4 meter is suitable accommodating 400 and 625 plants/ha respectively. Training, pruning/ pollarding/clipping and application of fertilizer/ organic manure scheules may appropriately be adopted in accordance with the practices followed for Som (Thangavelu *et al.* 1988).

91. DIGLOTI

Botanical: ***Litsaea salicifolia*** Roxb. Syn. *Tetranthera lancifolia* Kurz., *T. glauca* Wall.

Nepali: Sampat, Sanu-pahenle

Assam: Digloti Khasi: Dieng-lati Lepcha: Senashelkung

SYSTEMATIC POSITION

Bentham & Hooker (1883)	Engler & Prantl (1899)	Hutchinson (1959)
Phanerogams	Phanerogams	Angiospermae
Dicotyledones	Dicotyledoneae	Dicotyledones
Polypetalae	Archichlamydeae	Lignosae
Thalmiflorae	Ranales	Laurales
Laurales	Lauraceae	Lauraceae
Lauraceae		

Geographical distribution: Found in sub-Himalayan tract, often associated with Sal on damp ground and outer ranges from Nepal eastwards ascending up to 6000 ft. ASL viz., Assam, Garo, Khasi and Naga hills, Bengal plains, Sunderban Chittagong (Bangladesh) and lower and upper Burma (Myanmar).

Habit: Small, evergreen, dioecious tree or shrub (Fig. 37).

Habitat: Like *Litsaea polyantha* and *L. citrata* humid and warm climatic conditions with high rainfall.

Root: Branched tap root.

Stem: Aerial, erect, branched, branches, minutely hairy, slender, horizontal, bark greyish or chocolate brown and 0.3 cm thick, blaze yellowish which turns to dark brown.

Leaves: Simple, exstipulate, petiolate, petiole 0.7-2.5 cm long, very variable in shape from elliptic to linear-lanceolate, acute or acuminate, glabrescent above, glaucous and pubescent beneath, charataceous, reticulate, lateral nerves 10-15 pairs, margin entire, length 8-25 cm, breadth 3.0-5.0 cm.

Inflorescence: 4-6 flowered umbel (axillary clusters) in male and panicled receme in female.

Flower: Small, short pedicelled, bracts-4, imbricate, deciduous, hypogynous, actinomorphic, unisexual and pale white.

Perianth: Tepals 6, petaloid, linear – oblong, campanulate, 0.2 cm long, valvate aestivation.

Androecium: Stamens 6-9, absent in female flowers, polyandrous, filaments short and villous, anthers 4-celled, adnate, opening by a valve from below upwards.

Gynoecium: Abortive in male and present in female flowers, monocarpellary, unilocular with single ovule, ovary superior, style short, villous, stigma capitate, placentation basal/marginal.

Fruit: Fleshy berry, dark purple, narrowly elliptic, 0.8 cm long seated on the funnel or club shaped perianth, which is narrower than the fruit.

Seed: Exalbuminous.

Floral formulae: Male flower: $Br._4 \oplus ♂ P_6 A_{6\text{-}9} G_0$

Female flower: $Br._4 \oplus ♀ P_6 A_0 G_{\underline{1}}$

Uses: It is one of the very important secondary food plants of muga silkworms. The muga silkworms reared on Digloti produce relatively low quality silk yarn than those reared on Som (FAO 1976, Thangavelu *et al.* 1988, Seth 1995 d, Das *et al.* 2000)

Cultivation: *L. salicifolia* may be cultivated through seeds (Choudhury 1981) and air layerings, stem cuttings and leaf node cuttings (Raja Ram *et al.* 1993b, 1995). Stem cuttings, leaf node cuttings and apical bud cuttings have responded 89%, 83% and 75% rooting respectively by using IBA 200 ppm dose for 10 minutes by treating basal regions of cuttings. The aforesaid cuttings exhibited 89.80%, 85.66% and 85.33% final survival. The sand has been found as the best rooting media for cuttings whereas moist saw dust served as the best rooting media for raising air layers at Regional Muga Research Station, Boko. Seeds exhibit 94% germination and 90% survivality whereas airlayers prepared with the help of IBA 200 ppm dose exhibited 91% rooting and 54.50% survivality after transplantation (Raja Ram *et al.* 1995).

The Dighloti seeds are collected during last week of May to first week of July from the mature plants in polythene bags. After two days of storage, the seeds are washed in running water 2-3 times to remove the pulp and clean seeds thus obtained are dried for 2-3 hrs under shade. The seeds should be sown in the nursery bed containing soil, sand and FYM in equal proportion at the depth of 0.5-1.0 cm. The seeds germinate after 17 days.

The air layerings may be prepared during the month of April for which 1.5-2.5 cm thick branches should be selected. 30-40 cm long shoot from the top should be ringed by removal of bark with the help of a sharp knife. IBA 200 ppm should be applied on the ring with the help of cotton and it should be air-layered with moist sawdust as rooting media and wrapped with polythene sheet. After two months, rooted air layerings may be separated from mother plants and planted either in nursery or at the transplantation site. Air layers root within 25-30 days.

Stem cuttings of 20-23 cm long with 2.5-3.5 cm diameter and leaf node cuttings of 7-8 cm length with 1.2-1.5 cm diameter should be prepared from basal portion of juvenile shoot whereas apical bud cuttings of 4-5 cm length may be prepared from apical portion and such apical bud cuttings should have two leaves intact. The basal portion of all the cuttings should be treated with IBA 200 ppm for 10 minutes and inserted in the sand media filled in polythene bags kept in underground pit. The pit is subsequently covered with polythene sheet and irrigated regularly. All such cuttings may be prepared during January. Rooting occurs within 25, 20 and 22 days in stem cuttings, leaf node cuttings and apical bud cuttings respectively.

92. SOM

Botanical: ***Machilus bombycina***. King ex. Hook. f. Syn. *Persea bombycina* Kaestern

Hindi: Som, Soom

Assamese: Som, Artucheknam Manipuri: U-laphoi

SYSTEMATIC POSITION

Bentham & Hooker (1883)	Engler & Prantl (1899)	Hutchinson (1959)
Phanerogams	Phanerogams	Angiospermae
Dicotyledones	Dicotyledoneae	Dicotyledones
Polypetalae	Archichlamydeae	Lignosae
Thalmiflorae	Ranales	Laurales
Laurales	Lauraceae	Lauraceae
Lauraceae		

Geographical distribution: India, Nepal, Myanmar (Burma), Cambodia, Malaysia and Indonesia. In India these plants commonly grow in Assam, Meghalaya, Arunachal Pradesh, Tripura, Nagaland, Manipur and West Bengal upto an altitude of 1500 ft. ASL.

Habit: Evergreen, terrestrial, monoecious and middle-sized tree with spreading branches (Fig. 38).

Habitat: Humid and warm climatic conditions with high rainfall along the sub-Himalayan hill ranges and Brahmaputra valley in North- eastern India having alluvial to sandy loamy soil with acidic pH (4.5-5.0) are their natural habitat. High and sloppy lands with good draining system are suitable for their plantation.

Root: Branching tap root system.

Stem: Aerial, erect, woody, solid, cylinderical, bark light or dark grey, rather rough, profusely branched, young shoots silky.

Leaves: Simple, alternate, exstipulate, elliptic – lanceolate or obovate – lanceolate, acute or sub-acuminate or rarely obtuse, glabrescent above, pubescent beneath, coriaceous, minutely pubescent when young, margin entire or semi wavy, base cuneate or acute, venation reticulate, 6-8 lateral nerves on either side, length x breadth varies from 2.5-5.0 x 0.8-2.0 inch.

Inflorescence: Axillary or terminal panicles.

Flower: Bisexual, hypogynous, small, actinomorphic, 0.5-0.8 cm x 0.6 cm in size, yellow, regular, pedicellate, pedicels up to 0.3 cm long, ebracteate, fragrant.

Perianth: Tepals-6, persistent, polytepalous, oblong or oblong-oblanceolate, obtuse arranged in two whorls of three each, pubescent on both surfaces, imbricate in bud and valvate when mature.

Androecium: Stamens 12, perigynous, cordate, arranged in 4 whorls, polyandrous, exserted, perfect stamens 9, innermost whorl of 3 stamens reduced to staminodes, a pair of lateral glandular outgrowth found in the stamens of third whorl. Anthers usually 4-celled, adnate, opening by a valve from below upward. Filaments short and hairy at the base.

Gynoecium: Carpel-1 (monocarpellary), unilocular with single, solitary, pendulous, anatropous ovule, ovary superior, placentation marginal, style terminal, stigma 3-lobed.

Fruit: Fleshy, berry (drupe), globose or oblong, 6-8 mm, dark purple when mature.

Seeds: Pendulous, exalbuminous.

Flowering: December - March.

Fruiting: March - May.

Floral formula: $\oplus$ ⚥ $P_{3+3}\ A_{3+3+3+3\,(std.)}\ G_{\underline{1}}$

Varieties: Depending upon leaf shape and size four varieties have been identified which are Naharpatia (leaves resemble with that of Nahar tree), Ampatia (leaves like Mango tree), Jampatia (leaves like Jamun tree) and Kathalpatia (leaves like Jack fruit tree). Naharpatia has been considered as the best variety for muga silkworm rearing. On the basis of leaf taste two varieties viz., sweet and bitter or sour have been recognised out of which sweet variety has been considered to be the best for muga silkworm rearing by rearers (Thangavelu *et al.* 1988).

Uses: It is used as primary food plants of muga silkworm, Antheraea assama Ww. in Assam (FAO 1976, Thangavelu *et al.* 1988, Seth 1995d, Das *et al.* 2000). Besides, extract of Som leaves is given in Mouth ulcers and bark extract is useful in asthma (Sinha 1996). Muga silk reared in Som produce the best cocoons (FAO 1976).

Weed used for tea boxes (Ambasta *et al.* 1986).

Cultivation: *M. bombycina* may be cultivated through seeds (Isa *et al.* 1989, Paliwal *et al.* 1991), air layerings, stem cuttings, single leaf and shoot bud cuttings and juvenile shoot cuttings. Air layers are made by using NAA 400 ppm or IBA 500 ppm and upto 70% rooting has been achieved (Thangavelu *et al.* 1988). Further modifications in technologies yielded maximum rooting percentage of 27.8, 81.0, 60.0, 90.0 and maximum survival percentage of 53.3, 88.7, 93.0 and 94.86 in stem cuttings, air layerings, single leaf and shoot bud cuttings and juvenile shoot cuttings (Choudhury 1965, Yadav and Goswami 1992, Siddiqui *et al.* 1996, Devnath *et al.* 1992) respectively at Regional Muga Research Station (RMRS), Boko, Assam. While spring is best season for air layerings, June-Nov. is best period for leaf node cuttings. According to Yadav and Goswami (1992) ringing increases rooting %, root vigour and post transplantation survival % of foliated cuttings. The effect is further increased when growth regulators like NAA and IAA are used. NAA 100 ppm and 50 ppm recorded better for all the parameters studied.

Devnath *et al.* (1992) studied potentials of a leaf bud cutting consisting of a leaf blade, petiole and a short piece of the stem with the attached axillary bud and found that about 53.75 days on an average are taken for root initiation. The maximum rooting, number of primary roots and survival % of such cuttings to the tune of 51%, 6, 60% and 93.5%, 4.3 and 87.5% were recorded during I and IV seasons (Sept-Nov. and June – Aug) respectively. June – Oct. is the best period for propagation of *Machilus bombycina* King by this method.

Vegetative propagation through juvenile shoot cuttings

The simple operational and cost effective technology for vegetative multiplication of Som has been described by Siddiqui *et al.* (1996) as here under.

1. A pit of 4 x 1 x 0.5 meter size in length, breadth and depth should be prepared in high lands to avoid inundation during rainy season and drainages are prepared to prevent entry of water.
2. A slanting overhead shed, slightly bigger than the pit size is erected with bamboo of 20 x 15 cm size.
3. June-Aug. and Sept.-Nov. seasons are better than March-May and Dec.-Feb for raising juvenile shoot cuttings because atmospheric temperature ranges between minimum of 19.8-27.81°C and maximum of 26.68-33.04°C, while atmospheric humidity ranges between 80-92% and rainfall ranges from 110-355 mm in different months in Boko, Assam.
4. Juvenile shoot cuttings are prepared with the help of sharp knife from 30-45 days sprouted shoots from basal portion along with some trunk bark tissues of the pollarded/pruned Som plants and collected in a bucket half filled with water

during morning. Such cuttings are treated with 0.1% Bavistin solution for 2 minutes.

5. The juvenile cuttings are inserted for about 3-4 cm length into the rooting medium of clean river sand in polythene tubes kept in the pit.
6. High amount of atmospheric moisture is maintained in the pit by watering twice a day during summer and once daily during rainy and winter seasons.
7. Rooting initiates within 30-40 days during June to Nov. with 92% rooting and 94.86% survival during June to Aug. and 67.33% rooting and 92.10% survival in Sept. to Nov. period.
8. Sufficient rooting and sprouting takes place within 60 days and during this period 1/5th rooting media of polythene tube is replaced by decomposed FYM and soil mixed with 2-3 gm of 10% BHC powder for faster growth.
9. Before transplantation pits are filled with 0.5 cft FYM, soil and 5-10 gm BHC (10%) or aldrin (5%) and mixed thoroughly.
10. Saplings are transplanted when they are 90-105 days old during evening hours by making two longitudinal slits in polythene for easy proliferation of roots.
11. The cost of each saplings raised through juvenile shoot cuttings comes out to be Rs. 0.95 as compared to Rs. 4-5 per air layer charged in the open market.
12. While saplings raised through juvenile shoot cutting require 3 months, those raised through stem cuttings, air layerings, single leaf and bud cuttings require 6, 4-5 and 6 months period for transplantation.

The technology for nursery raising and plantation has been developed at Regional Muga Research Station (Anonymous 1998). For raising economic plantations for muga silkworm rearing, seeds of Som are collected during April-May and viable seeds are screened by floating test in water. Sunken viable seeds are selected and dried in shade. Seeds above 0.3 gm in weight and 0.7 mm in diameter are ideal for sowing. (Anonymous 1998).

According to Raja Ram *et al.* (1992) studies on the germination behaviour and longevity on Som (*M. bombycina*) seeds revealed that the germination % is much higher (84.75%) in depulped seeds against 53.5% germination of pulped seeds at the initial stage. Pulped seeds take a longer period for germination and seed longevity of depulped seeds reduced drastically from 84.75% initially to 40.5% after one month and 5% only after a three months period of preservation Yadav (1999) analytically reviewed research on seeds of *M. bombycina* Syn. *Persea bombycina*.

The well-drained, high land in shady place is selected for seed bed preparation, ploughed up to 30 cm depth and leveled. The size of the beds may be 5.0 x 1.5 meter, which should be, raised up to 15-20 cm. A gap of 30 cm should be maintained between two beds for drainage as well as cultural operations. 5 cft of well decomposed FYM and 200 gms of Forate may be mixed thoroughly in each bed regularly (Anonymous 1998).

The selected seeds are soaked in water for 24 hrs. before sowing. Approximately 8000 seeds are sown in each bed at a depth of 1 cm and 15 cm distance from each other. The seed beds are mulched with thatch grass and irrigated regularly (Anonymous 1998).

Seeds may also be sown in polythene tubes containing FYM, sand and soil rooting media in equal ratio. 2-3 seeds should be sown in each polythene tube (15 x 20 cm) and after germination only one seedling is kept in each polythene tube and others are transferred (Anonymous 1998).

Seedlings with some soil should be transferred from seed beds to polythene bags containing above rooting media (soil, sand, FYM) after 2-3 months of sowing for providing better spacing and facilitating transportation to transplantation sites safely besides ensuring their survival after transplantation.

9-12 months old seedlings are transplanted in pits of size 1'x1'x1 at 2x2, 3x3 or 5x5 meter spacing between plants as well as rows. Before transplantation each pit should be filled with app. 0.5 cft FYM mixed with 5 gm forate and polythene bags are removed gently without disturbing the root system. Regular cultural operations are carried out by weeding and loosening of soil around the plants (Isa *et al.* 1989, Paliwal *et al.* 1991, Thangavelu *et al.* 1988, Anonymous 1998).

Well-decomposed FYM should be applied by making 6-8 inch deep ring around the plants @ 0.5 cft/plant at the distance of 2 feet from the plants during 1st – 4th year and @ 1.0 cft/plant from 5th year onwards.

The recommended NPK doses for 2x2 meter spacing are 100:50:50 kg/ha/yr for which 87 gm urea, 125 gm single super phosphate and 34.5 gm murate of potash may be applied in two split doses during April-May and September from 5th year onwards. This NPK dose should be reduced by 50% in young plantations of 1st- 4th year age.

When plants are raised at 3x3 or 5x5 meter spacing the suitable doses of NPK for such plantations will be N175, P75, K50 and N220, P300, K80 kg/ha respectively in two split doses during Feb-March and Sept.-Oct. and March and August respectively (Anonymous 1998, Thangavelu *et al.* 1988).

Hence, from economics point of view either 2x2 or 3x3 meter spacing is most suitable. The NPK may be applied by making rings as in the case of FYM or through 5-6 holes in the ground around the plants at a distance of 2 ft. After NPK application plot should be irrigated if there are no rains. To save expenditure on mandays FYM and NPK applications may be attended simultaneously. More than 30,000 kg of leaves/ha/year may be obtained by such fertilizations.

The plants should be trained by cutting side branches after 3 years of plantation by keeping the main trunk without any branch upto 3 feet. After attaining five years of age, light pruning should be resorted annually and heavy pollarding/pruning may be resorted after every three years subsequently. Pollarding may be done at 5-6 feet height from ground level after 6-7 years of plantation and branches should be clipped off after raising every Muga silkworm crops. The pruning/defoliation of the plants should be done 4 months and 5 months prior to summer and winter rearings respectively to obtain uniform mature leaves (Anonymous 1998).

According to one rough estimate a full grown tree of 12-20 years can support the rearing of 5-10 disease free layings of Muga silkworms and may yield 500 cocoons in one season and one tree can be utilized for two rearings in a year alternately during spring and autumn seasons. Thus five trees can produce 5000 cocoons thereby yielding 1 kg of Muga silk. However, in economic plantations one plant can sustain hardly one disease free laying in every crop and 2 dfls every year.

93. KAULA

Botanical: ***Machilus odoratissima*** Nees.

Hindi: Kaula, Kaunla, Kawala Nepali: Lali Kawla

Khasi: Dieng-la-ngiar-iong Lepcha: Rohun-kung

Punjabi: Kaula, Mithapatta

SYSTEMATIC POSITION

Bentham & Hooker (1883)	**Engler & Prantl (1899)**	**Hutchinson (1959)**
Phanerogams	Phanerogams	Angiospermae
Dicotyledones	Dicotyledoneae	Dicotyledones
Polypetalae	Archichlamydeae	Lignosae
Thalmiflorae	Ranales	Laurales
Laurales	Lauraceae	Lauraceae
Lauraceae		

Geographical distribution: Outer Himalayan ranges and valleys from the Sutlej eastwards up to 2000-7000 ft. ASL on Khasi hills (Meghalaya)

Habit: Moderate sized, evergreen, monoecious tree.

Root: Branched tap root.

Stem: Erect, woody, cylinderical, branched, solid.

Leaves: Simple, exstipulate, entire, glabrous, alternate, petiolate, petiole 1.3 to 2.5 cm long, dark green, shining above, pale beneath, blade 10-15 cm long, secondary nerves slender, minute, raised, reticulate veins prominent on both sides.

Inflorescence: Shortly pedunculate, axillary or pseudo terminal panicles of 7-8 cm length.

Flower: Bisexual, complete, pedicellate, actinomorphic, regular, hypogynous, small.

Perianth: Tepals 6, persistent, glabrous, polytepalous, valvate, arranged in two whorls.

Androecium: Stamens 12, polyandrous, filaments short, arranged in 4 whorls, inner most reduced to staminodes, anthers 4 celled, adnate, basifixed.

Gynoecium: Monocarpellary, unilocular, ovule solitary and pendulous, ovary superior, placentation basal.

Fruit: Purple, 1.5 cm long, oblong or ellipsoid berry, supported by the reflexed tepals.

Seeds: Exalbuminous, pendulous, non-endospermic.

Uses: It serves as secondary food plant of Muga silkworms in Assam (Seth 1995d).

Seeds yield a semi-drying oil. Wood used for beams, frames, doors, planking, rafters, window sashes, furniture and plywood (Ambasta *et al.* 1986).

Cultivation: *M. odoritissima* may be grown through seeds as well as through air layerings.

94. MAHWAH

Botanical: ***Madhuka indica*** J.f. Gmel. syn. *M. latifolia* Macbr., *Bassia latifolia* Roxb. *B. villosa* Wall.

Eng: Indian Butter tree — Hindi: Mahwah, Mahua, Mohwa

Burmese: Kansan — Singhalese: Mi

Bengali: Mohwa, Mahua, Banmahuva, Maul

Gujarati: Mahuda — Kannada: Doddippe, Hippe

Konkani: Mohwa — Marathi: Mohwa, Mahuva, Maha, Mohura

Malayalam: Illupa, Irippapu — Oriya: Mohuka, Moha, Madgi, Mahula

Tamil: Illupai, Kat Illupi — Telugu: Ippi, Sanaippa, Adaviyappa

Sanskrit: Atavimadhuka — Santhali: Matkom

SYSTEMATIC POSITION

Bentham & Hooker (1883)	**Engler & Prantl (1899)**	**Hutchinson (1959)**
Phanerogams	Phanerogams	Angiospermae

Dicotyledones	Dicotyledoneae	Dicotyledones
Gamopetalae	Sympetalae	Lignosae
Heteromerae	Ebenales	Ebenales
Ebenales	Sapotaceae	Sapotaceae
Sapotaceae		

It is named after Fernando *Bassia,* curator of the Botanical Gardens at Bologna in Italy and *latifolia* due to its broad leaves.

Geographical distribution: Throughout Central India, wild in Kumaon Terai, indigenous in the sub-Himalayan tract from Ravi to Gandak, in the Satpura range and the Peninsula, from Bengal to Western Ghats. It is found along with mango in entire countryside of Uttar Pradesh.

Habit: A large deciduous tree usually with short trunk and a rounded crown, height up to 50 ft., all the young parts densely rusty woolly.

Habitat: Sub-tropical to tropical climatic conditions with high temperature, moderate rainfall and humidity.

Root: Branched tap root

Stem: Erect, aerial, woody, branched, bark wrinkled, grey or brown, young shoots tauny tomentose.

Leaves: Clustered at the end of branches, firm, leathery, elliptic, oblong-elliptic or obovate to broadly, lanceolate, apex acute or acuminate, base acute or cuneate, coppery red and densely woolly beneath when young, glabrescent when old, primary parallel nerves 12 on each side, petiole 2.5-3.0 cm long, stipules linear and caducous, blade 13-20 cm long, L/B ratio approximately 5/3.

Flowers: Axillary, fascicled in dense clusters at among terminal tufts of leaves, pedicels woolly, 2.5-4.0 cm long, complete, hermaphrodite, actinomorphic, epigynous.

Calyx: Sepals 4, rarely 5, ovate-subacute, coriaceous, densely rusty tomentose outside, imbricate,·polysepalous, plum coloured.

Corolla: Petals 7-14 (Usually 8-9), yellowish white (cream) coloured, tube fleshy, short erect, ovoid or campanulate.

Androecium: Stamens usually 24-26, anthers hairy, sub- sessile hairy, usually twice the number of petals, anthers lanceolate, acute.

Gynoecium: Ovary 4-12 (usually 6-8) celled, villous, style long exserted, inferior.

Fruit: Berry, green, fleshy, juicy 2.5-4.0 cm long ovoid, 1-4 seed, seeds 1.3-2.5 cm long, ripening during June-July, edible.

Flowering: Feb-April when tree is almost leafless. The tree blooms at night and flowers fall to the ground at dawn. The fresh leaves sprout when corolla has fallen.

Floral formula: $\oplus$ ⚥ $K_{4\text{-}5}\ C_{(7\text{-}14)}\ A_{24\text{-}26}\ G_{(\overline{4-12})}$

Uses: It serves as the tertiary but recognised as secondary food plant for tropical tasar silkworm, *Antheraea mylitta* D (Jolly *et al.* 1974, Seth 1995a). Other uses of the tree are of immense value. The ripe fruits are eaten and cover of the fruits and seeds are used as vegetables. Flowers form staple food for the farmers of Uttar Pradesh and forest dwellers of Central India. They are eaten raw or roasted or cooked or made into sweet meats. Spirit is made after their fermentation and distillation. Seeds yield (35-40% oil) a substitute for ghee used for cooking that is why it has been named as butter tree. This oil is also used for making margarine or soap. While the bark is used as a dye, Mahua oil is also used for illumination and Mahua cake is used as manure. (Sagreiya 1967, Randhawa 1983, Singh 1990, Jain *et al.* 1990). Mahua cake has insecticidal and piscidal properties (Ambasta *et al.* 1986). Wood is of good quality and is used for carriages, beams, furniture, turnery, musical instruments, oil and sugar cane presses, sports goods, house and ship building and also for agricultural implements after well seasoning (Ambasta *et al.* 1986). It is considered as a bad fuelwood (Sagreiya 1967).

Medicinally, the milk sap from cut flowers, stem and branches/branchlets is used as remedy for rheumatism.

The decoction of bark expels worms from body and cures ulcers, leprosy, fractures and biliousness or congestion of the liver. The decoction of flowers helps in curing heart diseases, ear complaints, removal of mucous secretions from bronchial tubes, soothing of alimentary tract, relieving inflammation and cough. Decoction of fruits cures blood disease, bronchitis and tuberculosis and helps in curing heart diseases and ear complaints. Decoction of seeds helps in increasing secretion of milk in lactating mothers. Dried flowers are used as fomentation as a remedy for ortitis or inflammation of the testes. Oral consumption of flowers fried in ghee treats piles and skin diseases (Randhawa 1983, Kurian 1995).

Cultivation: Planted in most parts of India through seedlings which are transplanted when they are big enough otherwise they die. It propagates itself through self-sown seeds. Protection from frost is required when seedlings are young (Randhawa 1983).

95. PANCHAPA

Botanical: ***Magnolia pterocarpa*** Roxb. Syn. *M. sphenocarpa* Wall.

Nepali: Pat pate

Assamese: Panchapa, Thouthua, Baramp thuri sopa

Bengali: Anda champa, Duli champa

SYSTEMATIC POSITION

Bentham & Hooker (1883)	**Engler & Prantl (1899)**	**Hutchinson (1959)**
Phanerogams	Phanerogams	Angiospermae
Dicotyledones	Dicotyledoneae	Dicotyledones
Polypetalae	Archichlamydeae	Lignosae
Thalmiflorae	Ranales	Magnoliales
Ranales	Magnoliaceae	Magnoliaceae
Magnoliaceae		

Geographical distribution: Sub-Himalayan belt from Nepal eastwards, Assam, Meghalaya (Khasi hills) and Chittagong hills (Bangladesh).

Habit: A large evergreen tree, youngest parts tawny pilose.

Root: Branched tap root.

Stem: Erect, aerial, woody, young branches hairy, old sclid.

Leaves: Ramal, alternate, simple, stipulate, leaf buds enveloped in the convolute stipules, petiolate, leathery (coriaceous), quite glabrous when full grown, younger pilose, 20-40 cm long, secondary nerves 12-18 pairs, entire, venation unicostate reticulate.

Inflorescence: Solitary, terminal flowers.

Flower: Complete, bisexual, pedicellate, bracteate, ebracteolate, regular, actinomorphic, hypogynous, thalmus elongated and core like, fragrant, campanulate, 5-8 cm broad, on a stout 3.0-5.0 cm long pedicel, white.

Calyx: Sepals 3, coriaceous, green outside, free, valvate.

Corolla: Petals 6-12 in whorls of 3, fleshy, white, free, valvate.

Androecium: Stamens numerous, polyandrous (free), filament short, anthers dithecous, without a connective, introse, dehiscence longitudinal.

Gynoecium: Carpels indefinite (numerous), apocarpous, ovary superior, spirally arranged, each carpel with a basal swollen ovary, a long style and coriaceous stigma beak shaped or simple, unilocular with two collateral ovules.

Fruit: Dense spike of 12-17 cm x 3.8-6.3 cm size (Etaric of follicles)

Seeds: Endospermic, orange, embryo small, erect.

Floral formula: ⊕ ⚥ $K_3\ C_{3+3+3+3}\ A_{\alpha}G_{\alpha}$

Uses: It serves as one of the tertiary food plants of Muga silkworms, *Antheraea assama* Ww. but recognized as secondary host plant (Seth 1995d, Thangavelu *et al.* 1988, Das *et al.* 2000).

Wood suitable for tea boxes and plankings but does not tolerate exposure to moisture (Ambasta *et al.* 1986).

Cultivation: It can be cultivated through seedlings and air layerings.

96. SEB

Botanical: ***Malus pumila*** Mill. Syn. *Malus sylvestris* (L.) Mill. Syn. *Pyrus malus* Linn., *M. domestica* Borkh.

English: The Apple Tree
Hindi: Seb, Sew, Seo
Bengali: Seb, Sev
Kannada: Sebu, Sevu
Laddakhi: Kushu
Punjabi: Seo
Sanskrit: Seb
Vernacular: Chur, Pangi

SYSTEMATIC POSITION

Bentham & Hooker (1883)	Engler & Prantl (1899)	Hutchinson (1959)
Phanerogams	Phanerogams	Angiospermae
Dicotyledones	Dicotyledoneae	Dicotyledones
Polypetalae	Archichlamydeae	Lignosae
Calyciflorae	Rosales	Rosales
Rosales	Rosaceae	Rosaceae
Rosaceae		

Geographical distribution: Indigenous in Europe and Western Asia, supposed also to be wild in the North-west Himalaya (Brandis 1906). It is believed to have originated in the Caucasus Mountains of Western Asia where wild apple trees exist even today. From its homeland, nearly 5000 years ago, apples were carried to Israel, Turkey and many parts of Europe and therefrom to all other continents. France, USA, Italy, Germany, Spain, Japan, Australia (especially Tasmania), Turkey, Canada, and South Africa are major apple producing countries. Large scale introduction of 100 apple varieties was made by Alexander Coutts in India in 1887 in his orchard near Shimla. Therefrom, commercial cultivation of apple spread to Kashmir, Jammu, Uttaranchal, Karnataka, Tamil Nadu, Sikkim and Arunachal Pradesh.

Habit: Low spreading, round crowned, long lived, unarmed tree, young shoots, underside of leaves and inflorescence clothed with thin, white, silky tomentum. Tree is seldom allowed to exceed a height of 12.5 meter under cultivation (Fig. 39).

Habitat: Hills which have 1500-2500 m ASL altitude, deep, fairly fertile, well aerated, moist and drained soil containing a small amount of lime combined with annual well distributed rainfall of 60-75 cm and winter conditioning or chilling of between –2 and 5°C for 2-3 months to provide rest to the trees and force them to flower. It thrives best in medium loam, but it has been successfully grown in deep fertile loams of Kashmir to light loams of Kullu valley and brown or reddish brown sandy loams of Kumaon.

Root: Tap, branched.

Stem: Aerial, erect, woody, branched, cylinderical.

Leaves: Simple, ovate, acuminate, obtusely serrate, petiole usually half the length of leaf or shorter, dull green, very downy underneath, mostly clustered on short shoots or spurs, stipules caducous.

Inflorescence: Umbellate cyme, terminal, usually borne on two or more years old spurs, intermixed with leaf buds, covered with white thin tomentum especially when young.

Flowers: White tinged with red, 1.2-1.5 cm in diameter, hermaphrodite, regular, epigynous.

Calyx: Sepals 5, gamosepalous, reflexed, bell shaped, adnate to the ovary, persistent, becoming enlarged and fleshy in fruits.

Corolla: Petals 5, orbicular, white or pink, shortly clawed, alternating with sepals on the epigynous cup, polypetalous.

Androecium: Indefinite, 20 or more, free, outer whorl of 10 in 5 anti-sepalous pairs, followed by a whorl of 5 anti-petalous and third whorl of 5 anti-sepalous stamens, anthers in young flowers incurved.

Gynoecium: Tetra-pentacarpellary, syncarpous, tetra-pentalocular, enclosed within and adnate to the calyx tube, ovary inferior, styles 4-5 free to the base, stigma terminal, small, ovules 2 in each carpel.

Fruit: Pome, on a short pedicel, large, globose, intruded at both ends, crowned by the persistent calyx-lobes, mature ovaries surrounded by enlarged and fleshy receptacle and parchment like, each containing 1-2 seeds, seeds non-endospermic. The fruit skin green, yellow, red or combination of two or all three colours.

Flowering: March – May.

Fruiting: July – Sept.

Floral formula: $\oplus$ ⚥ $K_{(5)}\ C_5\ A_{\alpha}\ G\ (\overline{4-12})$

Varieties: Nearly 5600 horticultural forms of *M. pumila* are known today, out of which, nearly 1000 diploid (2n=34), triploid (2n=51) and tetraploid (2n=68) varieties are under cultivation. The few best known cultivars are "Red Delicious", "Royal Delicious", "Golden Delicious", "York Imperial," "McIntosh", Baldwin", "Beverly Hills", "Tropical Beauty", "Jonathan", "Lady Sudelay" "Beauty of Bath", "Prince Albert", "Charles Raised", "Newton Wonder", "Laxton," "Superb", "Red Soldier", "Orange pispin", "American Mother" and "Blanheim orange".

Uses: Out of total world production of 21427000 tonnes, Indian production was only 270000 tonnes annually during 1974. "An apple a day keeps doctor away" and "To eat an apple before going to bed will make the doctor beat his heart" are the two old sayings. Apple is very useful in child dysentery. It is mild laxative, best antibiotic for stomach and elementary canal and brain tonic. It is also used in jaundice, kidney stones, liver diseases, arthrities, joint pains, headache, acidity, indigestion, asthma, bile imbalance and dysentery/ diarrhoea (Gala *et al.* 1997).

Nearly half of the apple production is consumed as fresh fruit and reminder is processed into butter, jelly, canned applesauce and juice. Apple juice (Sweet cider) may be fermented to an alcoholic product (hard cider) or turned into acetic acid (Vinegar) through controlled bacterial inoculation. Dehydrated apples, apple flour and canned apples are commercially important products. Apple pomace, the refuse from the cider press is used in the preparation of jellies, jam and marmalades. The peeled fruits cut into pieces are cooked with water until soft and then sugar is added to prepare apple murraba, a preserve popularly regarded as a stimulant for heart. The hard cider may be used to produce calvados or apple jerk, an excellent brandy through distillation (Kocchar 1981).

A ripe apple has 84.1% water, 11% sugar, 1% fibre and small quantities of proteins, fats, minerals and fruit acids. Malic acid constitutes 90-95% of the total acid content. The characteristic flavour and aroma of the apple is due to the presence of esters and essential oils viz., amyle esters of formic, acetic and caproic acids.

The apple trees also serve as host plants of wild silkworms viz., *Antheraea roylei, A. pernyi* Linn. *Attacus atlas* Linn. in North-west as well as North-East sub-Himalayan belt of India.

Wood used for knobs, mallet heads, rulers, tool handles and turnery (Ambasta *et al.* 1986).

Cultivation: Through grafting selected cultivars on to seedling root stocks or clonally propagated Malling or East Malling dwarfing root stocks of a nearly related form or species. The dwarf root stocks retard the vegetative growth of the scion due to limited absorbing capacity without altering the size and quality of the fruits. The apple trees start fruiting within 6-8 years and continue to do so for approximately 100 years. Yet, most orchards are renewed with better cultivars within 40-50 years.

97. MANGO

Botanical: *Mangifera indica* Linn.

English: Mango tree	Hindi: Am, Amba, Amri	
Burmese: Thayet	Assamese: Am	Bengali: Am
Gujarati: Amri	Kannada: Mavu	Kol: Uli
Malayalam: Amram, Cutam	Manipuri: Heinou	Marathi: Ambi, Amba
Oriya: Amba	Santhali: Ul	Sanskrit: Amra chuta
Tamil: Maambazham, Mau	Telugu: Mamidi, Mavi	

SYSTEMATIC POSITION

Bentham & Hooker (1883)	Engler & Prantl (1899)	Hutchinson (1959)
Phanerogams	Phanerogams	Angiospermae
Dicotyledones	Dicotyledoneae	Dicotyledones
Polypetalae	Archichlamydeae	Lignosae
Disciflorae	Sapindales	Sapindales
Sapindales	Anacardiaceae	Anacardiaceae
Anacardiaceae		

A full description of Mango was given in Sabhaparva of Mahabharatha by sage Vyasa. Though a native of South-East Asia, it has been cultivated in India for more than 4000 years ago, which is evident from representation of mango trees on the Buddhist stupas of 110 B.C. including the famous Barhut Stupa. The old Sanskrit and Urdu poets have written many verses on mango. It holds an important place in Hindu mythology, religion, ritual and customs. In all Hindu observations and festivals, the festoons of mango leaves are used as decoration and symbol of piousness. A mango grove was presented by "Amradharika" to Lord Buddha so that he could relax in the shade of trees. The famous "Lakh Bagh" an orchard of 100000 mango trees was planted by Akbar, the great Moghul Emperor near Dharbhanga (Bihar). The most beautiful lady of ancient "Vaishali" kingdom was found as baby by her father in mango orchard and hence she was named as Amrapali. After abjuring violence and conversion to Buddhism, the great emperor Asoka planted mango trees along the roadsides to give shade and shelter.

Geographical distribution: Indigenous to Assam (Nambar forest), Sikkim, Meghalaya (Khasi hills), parts of sub-Himalayan tract, Satpura range and Western Ghats up to 1200 m ASL. It is cultivated in all parts of India, chiefly Uttar Pradesh, Andhra Pradesh and West Bengal. Although 87 other countries are also cultivating it, India accounts for about 65% of total world output. Yet it ranks only fourth among the exporters of fresh mango fruits due to heavy domestic consumption.

The other producers of mango in order of importance are Pakistan, Brazil, Mexico, Bangladesh, the Phillippines, Tanzania, Dominican Republic, Colombia, Haiti, Venezuela, Egypt, Zaire, Peru, Hawaii, South Florida, California and Tropical Australia (Brandis 1906, Sagreiya 1967, Kocchar 1981).

Habit: A large spreading, magnificently crowned, evergreen tree attaining upto 10-25 m height (Fig. 40).

Habitat: Mango trees prefer well-drained soils with a p^H between 5.5-6.0. The optimum temperature is 24-27°C and annual rainfall is 75-190 cm. The tree is fairly sensitive to frost and heavy rains during flowering and fruiting period. Dry weather and clear sky are essential for a good crop.

Root: Tap root, branched, deep.

Stem: Erect, branched, woody, solid, cylinderical, bark resinous, rough, transversely cracked, gray to black and thick.

Leaves: Single, narrowly elliptic or lanceolate, glossy, leathery, 8-40 x 2-10 cm, dense, pointed, yellowish green to purple when young and dark green on maturity, petiolate, exstipulate, alternate, spiral, petiole 3/4 - 1-½ inch, secondary nerves arching, many.

Inflorescence: Terminal large panicles, erect, pubescent, sometimes bearing more than 4000 flowers.

Flower: Hermaphrodite and staminate, staminate more than perfect ones, actinomorphic, pentamerous, with a glandular disc located outside the stamens, hypogynous, small, pinkish-white, nearly sessile, fragrant.

Calyx: Sepals 5, polysepalous, aestivation imbricate.

Corolla: Petals 10, arranged in two whorls, twice the length of calyx lobes, polypetalous outer whorl imbricate, inner gamopetalous whorl valvate, pinkish white.

Androecium: Stamens 5, one fertile, 4 reduced to staminodes, anther oval, purple, anti sepalous.

Gynoecium: Carpel 1 with sessile ovary, obliquely placed, greenish yellow, unilocular, surmounting a white disc, style lateral, simple, ovule 1, pendulous, attached to the side of the cavity above its base.

Fruit: A large fleshy drupe, ovoid-oblong, asymmetrical, 3-30 cm long, with a thick or thin skin (exocarp), 0.06-1.1 kg sometimes 1.8-2.3 kg in some varieties, exocarp green, yellowish green, yellow or orange, edible mesocarp yellow, orange or red, fibrous or fibreless, endocarp, thick, hard and fibrous enclosing a single large flattered, exalbuminous seed, seeds with mono-embryonic true zygote or polyembryony with several adventitious embryos arising from nucellus in addition to the zygotic embryo.

Flowering: Dec. – March, sometimes 2-3 times during the season.

Fruiting: May - July.

Floral formula: ⊕ ⚥ $K_5\ C_{5+(5)}\ A_{1+4}\ G_{\underline{1}}$

Varieties: About 1000 varieties of mango are known in India, of which only 20 are grown on a commercial scale. Each variety has its peculiar taste, flavour and consistency of the pulp. Many varieties taste like something between a peach and a pineapple, other poor varieties are stringy with turpentine like taste. In poor varieties, the flesh is more or less fibrous while in selected varieties it is fibreless. The important varieties and their hybrids are detailed below:

1. **Popular delicious varieties:** Alphonso, Bangalora, Banganapalli, Bombai, Chausa, Cherukuresam, Daphool, Dasheri, Fajli, Fajri, Gulab khas, Himasagar, Jehangir, Krishnabhog, Langra, Malda, Mulgoa, Mundappa, Neelam, Raspuri, Ranee Pasand, Rumani, Safdar Pasand, Samar, Swarnarekha, Vanyaraja and Willard.
2. **Superior famous varieties:** Alphonso, Banganpalli, Dusheri, Chausa, Humayuddin, Langra and Mulgoa.
3. **Hybrids:**
    1. Hy-10 (Banganapalli x Alphonso): Good yield, large sized delicious fruit.
    2. Hy-13 and Hy-15: Quality fruits and long shelf life.
    3. Amrapalli and Mallika (Dusheri x Neelum): Early bearing, high yielding, dwarf varieties.
    4. Neeleshan (Neelum x Baneshan); Neeluddin (Neelum x Himayudin), Neelgoa (Neelum x Mulgoa), Swaranajahangir (Swarnarekha x Jahangir), Amrapalli (Dashehari x Neelum), Hy-51 (Alphonsa x Bangalora): High degree of productivity.

Many varieties are recognized by different names in different regions/states as here under.

i. **Aphonso:** Apus, Badami, Gundu, Khader.

ii. **Bangalora:** Collector, Kili Mukku, Totapuri, Sundersha.

iii. **Banganpalli:** Chatta, Chaptai, Safeda.

iv. **Bobai:** Bombai, Malda.

v. **Himayauddin:** Himampasand, Himayat, Imampasand.

vi. **Neelum:** Kasaladdu.

Uses: Mango is one of the most popular, choicest and highly priced dessert fruit of India. The young unripe fruits are astringent (acidic and rich in starch) and used in pickles, chutneys and other culinary preparations like amchur powder, which is used as a souring agent for soups, chutneys and vegetables. Ripe fruits are mostly eaten fresh as a dessert or made into jams, jellies, squash, preserves, canned or its juice is dried in the sun to form solid sheets of mango leather (Am-papad), which is pliable and sour-sweet product. Mango shake, a refreshing drink is made by mixing mango flesh with milk, sugar and ice. Mangoes are also fermented to form vinegar, wine and brandy. The kernel inside the stone is roasted powdered and cooked for use in several parts of India as a human food in times of scarcity. The ripe mango fruits contain 84% water, 10-20% sugar and small amounts of proteins, minerals, fats, fruit acids (citric, malic, oxalic, succinic and two other acids), carotene, riboflavin, potassium and vitamins A, B and C. The kernels contain 70% carbohydrates (sucrose, fructose, glucose and maltose), 10% fats and 6% proteins (Kocchar 1981, Pandey 1981).

The mango fruit is digestive, tonic, nutritive, invigorating, refreshing, fattening, laxative and diuretic. They are given in haemorrhoids of uterus, lungs and intestine. Unripe fruits are used in opthalmia, heat apoplexy, skin diseases, stomach disability, diarrhoea, dysentery and constipation. Bark is used in diptheria and rheumatism. Gum is tonic (Sinha 1996). Oil from kernel is used in arthritis, while kernel is used for diarrhoea, vomiting and burning sensations in the heart. The soft leaves of mango are used as appetizer and expectorant and removes bile. The kernel is use as anthelmintic and it increase blood in the body. The infusion of bark is used in dysentery, stimulates blood synthesis and stops bleeding from lungs, intestines and uterus.

It is a very good shade tree for high ways but not suitable for town avenues as it harbors mosquitoes. Its timber is durable and used for tea chests, packing cases, planking, canoes, doors and for building construction (Sagreiya 1967).

Wood also used for floor and ceiling boards, window frames, packing boxes, brush-backs, agricultural implements, inferior furniture, plywood, shoe heels, beams, rafters, trusses and window shutters (Ambasta *et al.* 1986).

Non-mulberry wild silkworms *Cricula trifenestrata* help, which yields an elegant golden silk, are commercially reared on mango plantations in

Indonesia. It is believed that mango trees infested with these silkworms produced more fruits than non-infested ones. In India, this silk is known as Amkoni or Amresham and found in Bihar, Assam, Arunachal Pradesh and few other states but their commercial exploitation is yet to be achieved (Mishra 2002, 2003).

Cultivation: The seeds of mango (stored) are used for raising seedlings. The seeds are collected during May-June and seedlings are raised during June-July. The seedlings are transplanted during July (Chundawat and Gautam 1993). Inarching and grafting of selected varieties has been practiced in India for long time. However, in Myanmar, the tree is generally not grafted due to the belief that the seeds of a good kind as a rule produce good fruit of the same description. The mango fruits start fruiting after 4-6 years. Heavy flowering and fruiting considerably depletes the nitrogen reserve of the tree resulting into a light crop during the next year. However, adequate fertilization with nitrogen, potassium and phosphorus can correct such erratic production.

98. SAKARKAND (TAPIOCA)

Botanical: ***Manihot utilitissima*** Pohl Syn. *M. esculenta* Crantz.

Eng: Tapioca, Cassara, Manioc Hindi: Sakarkand, Maravvli

Burmese: Pooloopinanmyouk French: Mandioc Singhalese: Manyokka

Assamese: Simul Alu Kannada: Maragenasu Malayalam: Marachini

Manipuri: U-mangra Tamil: Maravalli kijahangu

Telugu: Karraapendalamu

SYSTEMATIC POSITION

Bentham & Hooker (1883)	Engler & Prantl (1899)	Hutchinson (1959)
Phanerogams	Phanerogams	Angiospermae
Dicotyledones	Dicotyledoneae	Dicotyledones
Monochlamydeae	Archichlamydeae	Lignosae
(Incompletae)	Geraniales	Euphorbiales
Unisexuales	Euphorbiaceae	Euphorbiaceae
Euphorbiaceae		

Geographical distribution: Cultivated in West coast Kerala, Tamil Nadu, Andhra Pradesh and Karnataka in South India, Meghalaya, Manipur and Assam in North-east India. It is also popular in Myanmar and Malawi and other countries of Africa. Being a native of South America, it seems to have originated from coastal peru.

Habit: Herbaceous perennial plant growing to a height of 0.9-2.7 meters (Fig. 41).

Root: Branched tap root, tuberous and rich in starch.

Stem: Erect, herbaceous, unarmed, glabrous, cylinderical, contain milky latex (Euphorbiaceae character), stem marked by prominent scars.

Leaves: Alternate, rarely opposite, 3-7 deeply parted palmate, clustered leaves at top add beauty to the plant, lower leaves shedded, stipules absent, petioles long, foliage dark green usually but light green, red, yellow and other shades of purple pigmentation in centre in ornamental species.

Inflorescence: Racemes.

Flower: Unisexual, incomplete, regular, actinomorphic and hypogynous, pedicellate, monoecious, male flowers towards lower portion, female flowers towards apical portion of inflorescence, white or cream coloured.

Gynoecium: Carpels 3, syncarpous, ovary trilocular, superior, each locule contains one pendulous, anatropous ovule, placentation axile.

Fruit: Winged capsule.

Seeds: Endospermic.

Pollination: Usually entomophillous (through insects), rarely anemophillous (through wind).

Uses: It serves as one of the best secondary food plants (foliage) of eri silkworm *Philosamia ricini* (FAO1976, sarkar 1988, Seth 1995c, Sarmah 2000). Besides, the root is largely used in adhesives, cosmetics, puddings, buiscuits and confectionary items due to high quantity of starch. The tubers are also used for making starch and sago and are rich in minerals viz., calcium and phosphorus. Starch used for sizing and finishing of textiles, laundering, paper making etc. Glue made from the starch used to prepare syrup, acetone etc. Plant also yields an alkohlic beverage (Ambasta *et al.* 1986). Tuber flour is given in fever and stomach disorders, while juice is aperient, leaves are purgative and vermifuge and used for sores, ascites, eczema, cariasis, scabies and sycosis. (Kurian 1995, Sinha 1996).

Cultivation: It can be raised in kitchen garden as it needs little care. It is propagated by mature stem cuttings. The underground tubers mature within 6-9 months and are dug out individually and periodically whenever required as they remain in good condition for long period in ground.

99. PHUTKI

Botanical: ***Melastoma malabathricum*** Linn. syn. *M polyanthum* Blume

Burmese: Myetpye, Shame Eng: Indian Rhododendron Hindi: Phutki

Nepal: Tulasi, Choulisy

Assamese: Futki	Kannada: Ankerki	Lepcha: Tungbram
Malayalam: Kalampatti	Manipuri: Bobuchunmei (Rongmei)	Marathi: Palore
Oriya: Gongai, Koroti	Tamil: Nakkukaruppan	Telugu: Pattuda

SYSTEMATIC POSITION

Bentham & Hooker (1883)	**Engler & Prantl (1899)**	**Hutchinson (1959)**
Phanerogams	Phanerogams	Angiospermae
Dicotyledones	Dicotyledoneae	Dicotyledones
Polypetalae	Archichlamydeae	Lignosae
Calyciflorae	Myrtiflorae	Myrtales
Myrtales	Melastomaceae	Melastomaceae
Melastomaceae		

Geographical distribution: Found throughout the country upto 6000 ft. ASL in sub-Himalayan tract and outer hills from Nepal eastwards Sikkim, Assam, Meghalaya (Khasi hills), Andaman and Western Peninsula. Also found in Bangla Desh (Chittagong), Myanmar, Sri Lanka, Seychelles, Malay Peninsula and Archipelago, China, Australia and New Caledonia.

Habit: An evergreen small tree or shrub, densely clothed with acute, or acuminate, often serrulate scales.

Root: Branched tap root.

Stem: Erect, branched, bushy, branchlets clothed with serrulate scales.

Leaves: Simple, broadly lanceolate, 7-10 cm long, 3-5 nerved, hairy ventrally, petiole 0.6-0.8 cm long, exstipulate.

Inflorescence: 1-5 flowered cymes.

Flower: Bisexual, purple, complete, regular, actinomorphic, cyclic, pentamerous, epigynous, 2.5-4.0 cm across.

Calyx: Sepals 5, densely clothed with pale linear, acuminate, entire or serrate scales, gamosepalous, calyx teeth long, ovate-lanceolate - narrow lanceolate, glabrous inside the tube or few short hairs present near the tip, deciduous, teeth 1.2-1.8 cm equally long to calyx tube, aestivation valvate.

Corolla: Petals 5, free, purple.

Androecium: Stamens usually 10, inserted with petals on the calyx rim, unequal considerably, 5 short and 5 long, longer ones with bithecous, purple anthers and long connective while the shorter ones alternating to longer ones with yellow anthers without connective.

Gynoecium: Carpels-5, adnate to base of calyx tube, syncarpous, polylocular, apex of ovary bristled, placentation axile, placentae large, ovules numerous.

Fruit: Pulpy, 0.6-0.8 cm across, shortly ovoid, truncate.

Seeds: Minute, non-endospermic.

Flowering: Throughout the year especially during March - July.

Floral formula: ⊕ ⚥ $K_{(5)}\ C_5\ A_{5+5}\ G_{(\bar{5})}$

Uses: It serves as one of the tertiary but recognised as one of the secondary food plants of tropical tasar silkworm, *Antheraea mylitta* D. in forests (Seth 1995 a) and *Attacus atlas* silkworms (Ambasta *et al.* 1986).

Leaves, flowers and fruits eaten. Fruit yields a purple or black dye while leaves and roots yield a pink dye. Ashes serve as a dye mordant (Ambasta *et al.* 1986).

Bark and leaves are used for skin troubles. In diarrhoea, dysentery and leucorrhoea plant is very useful. The root and bark extract are used as gorgle and considered as antiseptic (Sinha 1996).

Cultivation: It can be propagated through seeds or seedlings.

100. BAKAIN

Botanical: ***Melia azedarach*** Linn.

Burmese: Jamarhka English: Persian Lilac, Beed tree

German: Glatter Zedrach French: Acacie d'Egupte Nepali: Bakaina

Hindi: Bakain, Drek, Dake Persian: Tak

Assamese: Jhamaga, Thamaga Bengali: Mahanim, Ghora nim

Gujarati: Bakam limbodo Kannada: Arebevu, Hutchu Bevu

Khasi: Dieng-jah-rasang Konkani: Phirangi nimb

Malayalam: Karin Vembu, Vembu, Vayambu

Manipuri: Seizrak Marathi: Pejri, Padrai

Punjabi: Drek Sanskrit: Mahanimba

Tamil: Malai Vembu, Vembu, Mallay, Vembu, Puvempu

Telugu: Turaka Vepa, Yerri Vepa

Urdu: Drek

Uttaranchal: Dek

SYSTEMATIC POSITION

Bentham & Hooker (1883)	Engler & Prantl (1899)	Hutchinson (1959)
Phanerogams	Phanerogams	Angiospermae
Dicotyledones	Dicotyledoneae	Dicotyledones
Polypetalae	Archichlamydeae	Lignosae
Disciflorae	Geraniales	Meliales
Geraniales	Meliaceae	Meliaceae
Meliaceae		

Geographical distribution: Native of South-western Asia (sub-Himalayan and Shivalik tracts). However, cultivated in other parts of world also. Cultivated and naturalized throughout India, Persia and Myanmar. Due to wild nature, it is supposed to be indigenous in Baluchistan and the Jhelum valley in Kashmir, common in Punjab. Stands more cold than the neem and grows at 9000 ft. ASL in Bashahr (Chini).

Habit: A middle sized deciduous tree, 8-12 meters in height. Young shoots and inflorescence sparsely clothed with deciduous stellate hairs. Very fast growing with spreading crown (Fig. 42).

Habitat: Sub-temperate – temperate climate with humid conditions and various types of soil.

Root: Branched tap root.

Stem: Erect, branched, solid, woody with characteristic smell, stout, spreading, bark smooth, dark gray or brown, cracked vertically in long fissures, heartwood light red, annual rings marked by a belt of large vessels.

Leaves: Bipinnately compound, occasionally pinnae 3-4 pairs, more or less opposite, leaflets 3-12, ovate-lanceolate, more or less deeply serrated, occasionally lobed, green turning yellow.

Inflorescence: Cymose.

Flower: Hermaphrodite, actinomorphic, regular, complete, pedicellate, hypogynous, pentamerous, lavender (purplish), lilac with a strong honey scent / fragrance.

Calyx: Sepals 5, gamosepalous, companulate, green, aestivation valvate.

Corolla: Petals 5, polypetalous, twisted, purplish, fragrant.

Androecium: Stamens usually 8-12, monoadelphous (united), staminal tube purple, 1/3 inch long, teeth 20-30, linear, anthers glabrous, shorter than or as long as teeth.

Gynoecium: Carpels 5, syncarpous, stigma clavate, 5 toothed, ovary superior.

Fruit: Drupe, 3/4 inch long, yellow when ripe, round, smooth but turning to wrinkled afterwards, five seeded, poisonous.

Seed: Each seed has a natural perforation through the center, which makes them ideal for being used as beads.

Flowering: March – April (Spring) or hot season (May-June)

Fruiting: Cold season.

Leaf fall: Dec. – Jan.

Sprouting: March.

Floral formula: $\oplus$ ⚥ $K_{(5)}\ C_5\ A_{(8\text{-}10)}\ G_{\underline{(5)}}$

Uses: It is cultivated mainly for it's ornamental and shade value. Due to quick growth, it is ideal for places where shade is required but not suitable for avenues being shallow rooted and easily blown off characteristics by strong winds (Randhawa 1983).

The fruits are used in Manipur for starching the warps and wefts during weaving of fabrics in looms. The pounded leaves mature tumors when applied as a poultice. The bark and roots give tone to the body as a tonic and expel intestinal worms. The poultice of the bark and roots together applied externally cures enlarged lymph nodes and glands. 2-3 berries (fruits are eaten to treat leprosy and tuberculosis (Kurian 1995). The juice of the leaves is used as an anthelmintic and the seeds oil in rheumatism (Pandey 1981). Boiled extract of leaves is used for washing sores and skin diseases. Leaves and flowers cure nervous headache when used as poultice. Leaf juice is diuretic and emmenagogue. Leaves, fruits and bark are insect repellant. Gum is used in spleen enlargement. Bark infusion is used in ascariasis. Wood extract is used to cure asthama (Sinha 1996).

Wood used for cigar and ammunition boxes, toys packing and museum cases, agricultural implements, turnery, musical instruments and ornamental plywood. Seeds yield drying oil for soap and hair oils (Ambasta *et al.* 1986).

Cultivation: Through seeds or seedlings in rainy season. The seeds have poor viability, so they should be sown as soon as they ripe. It attains a height of 8-10 ft. within a year or 15 months due to fast growth. Saplings are tender to frost, but are more resistant than the neem. It is not attacked by rats or white ants due to its bitter bark (Randhawa 1983).

The seeds are collected during November-December and sown during February-April for raising nursery. The seedlings are transplanted during July. The seeds may also directly be sown in July. The plants also propagate through cuttings and suckers (Chunadawat and Gautam 1993).

101. CHAMPA

Botanical: ***Michelia champaca*** Linn. Syn. *M. aurantiaca* Wall.

Burmese: Saga

Eng: Champak

Hindi: Champa, Champ, Champaca

Nepali: Ouliachamp

Assamese: Titasopa

Bengali: Champa, Champaca

Kannada: Sampige

Gujarati: Champ

Malayalam: Champakam

Manipuri: Leihao

Marathi: Pivala-champa

Oriya: Chompa, Kanchana

Sanskrit: Champaka

Tamil: Shembuga

Telugu: Champakam

SYSTEMATIC POSITION

Bentham & Hooker (1883)	Engler & Prantl (1899)	Hutchinson (1959)
Phanerogams	Phanerogams	Angiospermae
Dicotyledones	Dicotyledoneae	Dicotyledones
Polypetalae	Archichlamydeae	Lignosae
Thalmiflorae	Ranales	Magnoliales
Ranales	Magnoliaceae	Magnoliaceae
Magnoliaceae		

Full description of ornamental Champak (*M. champaka*) was given by sage Vyasa in Sabhaparva of Mahabharatha (Rao 1991).

Geographical distribution: Wild on the Western Ghats in the Southern portion of Peninsula, Sikkim (up to 3000 ft. ASL), lower Burma (Myanmar). It appears to have originated from Java and India (Himalaya). Cultivated in gardens throughout India.

Habit: Tall, evergreen, graceful tree with dark grey bark, cultivated (Fig. 43).

Habitat: Cultivated in moister parts of India and Burma.

Root: Branched tap root.

Stem: Aerial, erect, branched, woody, solid.

Leaves: Ramal, alternate, simple, stipulate, stipules convolute, petiolate, ovate or oblong- lanceolate, entire, apex acute, coriaceous, glabrous, dark green, venation unicostate reticulate.

Inflorescence: Cluster of axillary/solitary axillary flowers.

Flower: Complete, regular, actinomorphic, bracteate, ebracteolate, bisexual, pedicellate, pedicel short, hypogynous, pale yellow, very fragrant, large.

Calyx: Sepals 3, free (polysepalous), greenish, inferior, valvate.

Corolla: Petals 9, (sometimes 15-20) in three whorls of 3 each, free (polypetalous), pale yellow, fragrant, inferior, outer whorl valvate, inner whorl narrow and twisted.

Androecium: Stamens indefinite (numerous), free (polyandrous), filaments distinguished, spirally arranged, anthers dithecous, basifixed, dehiscence longitudinal, extrose.

Gynoecium: Carpels numerous, free (apocarpous), superior, arranged spirally on cone like thalmus, each carpel bears a swollen ovary, style curved, stigma beak shaped, unilocular, placentation marginal, ovules many in every carpel, ovaries pubescent and separated from stamens by a distinct node.

Pollination: Entomophillous (through insect agencies).

Fruit: Etario of follicles (developing from many pistils of the same apocarpous single flower), dark coloured with white specks.

Seeds: Large, endospermic (albuminous), endosperm copious, embryo small.

Floral formula: $\oplus$ ⚥ $K_3\ C_{3+3+3}\ A_{\alpha}\ G_{\alpha}$

Uses: Besides serving as one of the tertiary but regarded as one of the secondary food plants of *Antheraea assama* Ww. (Muga silkworms) (Seth 1995d), this taxa also yields timber in South India, Western Ghats, Assam and Bengal (Shukla and Mishra 1979). The wood is used in millwork and making of boxes, musical instruments, boats, toys, drums etc. Wood is also used for posts, boards, veneers, furniture, decorative fittings, carriage, bobbins, general joining works, battery separators and tea chest plywoods (Ambasta *et al.* 1986). It is grown in gardens for large and sweet smelling flowers. The oil obtained from the flowers of *M. champaca* is used in perfumery and flowers yield a dye also (Shukla and Mishra 1979). *Michelia champaca* is of considerable medicinal value. Flowers are considered as tonic, diuretic, stomachic, carminative and used in dyspepsia, nausea, fever, renal and veneral diseases like gonorrhoea and gout. Dried root and root bark are used as purgative and emmenagogue. Roots are also used for dyspepsia, gonorrhoea and in stomach complaints also. Bark is stimulant, diuretic and febrifuge. Flower oil is useful in cephalagia (Sinha 1996, Shukla and Mishra

1979). Flowers are insect repellents and extract of leaves and flowers eradicates lice when applied to head (Sinha 1996).

The seeds yield 32.2% edible oil hence; it may also be used commercially for production of oil (Jain *et al.* 1990).

Cultivation: Cultivated through seedlings as well as air layers. Air layers may be made by application of IBA 200 ppm doses. The plants with flowers of excellent fragrance and colour may be grafted easily. The grafted plants usually begin flowering within 16-18 months with large size reddish-yellow flowers bearing very pleasant fragrance and require very less space.

102. BOR SOPPA

Botanical: ***Michelia oblonga*** Wall.

Assamese: Bor soppa, Phul-sopa

Garo: Chambisersang, Bewachh amphe Khasi: Dienga-ta-roi

SYSTEMATIC POSITION

Bentham & Hooker (1883)	**Engler & Prantl (1899)**	**Hutchinson (1959)**
Phanerogams	Phanerogams	Angiospermae
Dicotyledones	Dicotyledoneae	Dicotyledones
Polypetalae	Archichlamydeae	Lignosae
Thalmiflorae	Ranales	Magnoliales
Ranales	Magnoliaceae	Magnoliaceae
Magnoliaceae		

Geographical distribution: Assam (Sylhet) and Khasi hills (Meghalaya) forest.

Habit: A very tall, entirely glabrous tree.

Habitat: Localities with moist climatic conditions.

Root: Branched tap root.

Stem: Aerial, erect, woody, branched, solid, glabrous.

Leaves: Simple, thinly coriaceous, petiole 1.2-2.5 cm, blade 10-15 cm long, 5-7 cm broad, obovate to oblong, glabrous above, glaucous benath, secondary nerves 7-9 pairs, somewhat prominent beneath, obtusely acuminate.

Inflorescence: Cluster of axillary flowers.

Flower: Yellowish white, 7-10 cm wide, flower buds 2.5 cm, glabrous, hypogynous.

Calyx: Sepals 3, free, greenish, valvate, 3.5 cm long, narrow-obovate.

Corolla: Petals 9, arranged in three whorls, lanceolate, free.

Androecium: Stamens indefinite, free, filaments flat, anthers introse, dithecous, basifixed.

Gynoecium: Carpels numerous, free, sessile, woody, lenticellate, blunt, ovary superior.

Pollination: Entomophillous.

Fruit: Lax.

Seeds: Endospermic.

Floral formula : ⊕ ⚥ $K_3\ C_{3+3+3}\ A_\alpha\ G_\alpha$

Uses: It serves as one of the tertiary but regarded as secondary food plant of muga silkworms, *Antheraea assama* Ww. in Assam (FAO 1976, Seth 1995d).

Wood used for cabinet work, canoes, planking, rough furniture and tea chests (Ambasta *et al.* 1986).

Cultivation: It can be cultivated through seedlings, airlayerings and graftings.

103. SAINJNA

Botanical: ***Moringa oleifera*** Lam. Syn. *M. pterygosperma,* Gaertn.

Burmese: Dunt-dalon, Danthalun English: Drumstick

Hindi: Sainjna, Soanjana, Saonjna, Mungna, Sahjan, Shajna, Subunjna, Sohanjna

French: Moringa Sinhalese: Murunga

Assemese: Sajana, Saijana Bengali: Sajina, Shajna Gujarati: Saragavo

Kannada: Nugge-Mara, Nugge Konkani: Mashingajhad

Malayalam: Sakhta, Shakta, Muringa Manipuri: Shajna

Marathi: Shevaga Oriya: Sajana Punjabi: Saonjna

Sanskrit: Shobhanjana Santhali: Munga arak Sindhi: Singuin

Tamil: Moringa, Monaga, Murungai, Morunga

Telugu: Munga, Munaga Urdu: Saonjna

SYSTEMATIC POSITION

Bentham & Hooker (1883)	Engler & Prantl (1899)	Hutchinson (1959)
Phanerogams	Phanerogams	Angiospermae
Dicotyledones	Dicotyledoneae	Dicotyledones
Polypetalae	Archichlamydeae	Lignosae
Disciflorae	Rhoedales	Capparidales
Sapindales	Moringaceae	Moringaceae
Moringeae		

Geographical distribution: Indigenous to the Western sub-Himalayan tract and Oudh. Commonly cultivated throughout India and Myanmar.

Habit: A fair sized elegant looking deciduous tree growing up to nine meters height with soft white, rough, corky trunk and gummy bark (Fig. 44).

Habitat: Sub-temperate to sub-tropical agro-climatic conditions in more or less all types of soils.

Root: Tap and branched.

Stem: Erect woody, cylinderical, solid, branched, trunk straight, attaining a girth of 120-150 cm, bark thick, soft white, rough, corky, gum passages present.

Leaves: Compound, 30-75 cm long, opposite, divided and sub-divided, usually tripinnate, leaflets on short petiolules, ovate or obovate, opposite, entire, caducous, pale beneath, secondary nerves indistinct.

Inflorescence: Axillary panicles.

Flowers: Hermaphrodite, actinomorphic, complete, hypogynous, creamy white, strongly honey scented, borne in large loose clusters alongwith the leaves, 2.5 cm across, bracteate, bracts linear, shorter than pedicels.

Calyx: Sepals 5, gamosepalous, cup-shaped, segments petaloid, deciduous, somewhat unequal.

Corolla: Petals 5, linear-spatulate, with yellow dots at base, polypetalous (free), showy, upper smaller.

Androecium: Stamens 5, perfect, opposite to petals, alternating with 5 filaments without anthers, anthers dorsifixed, oblong, single celled (monothecous), base of filaments hairy.

Gynoecium: Carpel 1 (Monocarpellary), ovary superior, unilocular, style long, stigma simple, ovary hairy, placentation parietal.

Fruit: Pods, 22-50 cm long, pendulous, 9-ribbed but triangular, containing many seeds, seeds rounded, trigonous with broad wings at the angles.

Flowering: Jan.-April, with or before the fresh leaves.

Fruiting: May- July.

Leaf fall: Dec. – Jan.

Sprouting: January.

Floral formula: ⊕ ⚥ $K_{(5)}$ C_5 $A_{5+5 \text{ staminode}}$ $G_{\underline{1}}$

Uses: Leaves, flowers and fruits are eaten as vegetables. Oil, commonly known as Ben oil and fibre are also extracted from seeds and fruits respectively. The root has strong pungent taste and used as a vesicant. The tree yields a gum similar to Tragacanth. Young roots serve as a substitute for horse-radish (Brandis 1906, Randhawa 1965).

The bark of root promotes urine flow and its juice is used in earache and toothache. The bark of stem is used as a cardiac stimulant in asthma, cough and other respiratory disorders. Fomentation with hot decoction of the bark relieves inflammation. Cooked young leaves are eaten to fight scurvy and their poultice is applied in wounds and swellings. The juice of leaves mixed with honey is used in eye diseases while fresh root of young tree is good for intermittent fever, epilepsy, hysteria, palsy, chronic rheumatism, gout, dropsy, dyspepsia and enlargement of liver and spleen. The paste of root mixed with salt is applied to inflammatory swellings, rheumatic joints and paralytic parts while decoction of root is used for gorgling in hoarse throat. The tender leaves are useful in liver and spleen problems, tetanus, articular pains and paralysis. The seed oil is used in painful gout and rheumatic joints (Kurian 1995). The fruit is a good nutritive, digestive and stomachic vegetable. The leaves are given in catarrhal affections, emetic and antiseptic. The seeds are antipyretic and roots are antibiotic (Sinha 1996). The fruits and leaves are especially useful in diabetes (Dash 1993).

Twigs lopped for fodder. Pods rich in leucine, leaves rich in vitamins A and C. Flowers tonic, diuretic and cholagogue. Oil used in foods, cosmetics and illumination. Cake used for fertilization and gum in calice printing (Ambasta *et al.* 1986).

Cultivation: Through seeds, seedlings and cuttings. Seeds are collected during March-April and nursery is raised during April-June. The seeds may be directly sown in July. The seedlings grow fastly and young plants are hardy and easy to handle (Randhawa 1965). The seedlings and stem cuttings are also transplanted during July (Chundawat and Gautam 1993).

104. MULBERRY

Botanical: ***Morus alba*** L., *M. indica* L., *M. serrata Roxb.*, *M. laevigata Wall.* ex. Brand and *M. nigra* L.

Burmese: Posa, Poosarpin
Chinese: Sang
English: Mulberry
Hindi: Tut, Tutri, Toot, Shahatoot
French: Amourie, Muries
German: Maulbeer
Italian: Gelso
Japanese: Lewwa
Persian: Tut
Sinhalese: Mulberry
Assamese: Bola, Nooni, Nuni
Bengali: Tuth, Toonth
Gujarati: Shetur, Tutri
Kannada: Hipnerie
Konkani: Amor
Malayalam: Tulam
Manipur: Kobangchak
Marathi: Ambat, Tuti, Tut
Mizo: Thingdamdawi, Hmubelbing
Oriya: Tuto
Punjabi: Turi, Hmubelbing
Sanskrit: Brahmdaru, Tuta, Brahmakashtha, Kramuka, Mridupippali, Palashika, Madhupippali, Pusa, Tula, Shalmali, Surupa, Tuda
Sindhi: Tut
Tamil - Musukottai
Telugu - Reshme chattu
Urdu - Shahetuta

SYSTEMATIC POSITION

Bentham & Hooker (1883)	Engler & Prantl (1899)	Benson (1957)	Hutchinson (1959)
Phanerogams	Phaenrogams	Spermatophyta	Angisospermeae
Dicotyledones	Dicotyledoneae	Angiospermae	Dicotyledones
Monochlamydeae	Archichlamydeae	Dicotyledonae	Lignosae
Unisexuales	Urticales	Urticales	Urticales
Urticaceae	Moraceae	Moraceae	Moraceae

The sacred, largest and oldest mulberry tree of the world belonging to *M. serrata* having nearly 6.5 metre girth and estimated to be more than 1200 years old has been reported from Joshimath, Uttarakhand (Tikader *et al.* 1999).

Geographical Distribution: Chinese and Indo-China centre have been considered as the natural home of several crop plants including mulberry. Mulberry has been found distributed in all countries except Brazil, which are located in north of the equator. In Asia, 24 species are found in China, 19 species in Japan, 6 in Korea, 4 each in India and Taiwan, 3 each in Burma and

Indonesia, 2 each in Thailand, Vietnam and Afganistan and 1 each in Arabia, Oman and Muscat whereas in America, 14 species are found in North America and the remaining 7 species are found in Central and South America. China has 17 endemic species out of 24, Japan has 14 out of 19 and North America has 9 endemic species out of 14 species. Distribution pattern of mulberry shows that great species diversity occurs in Sino-Japanese region of the old world and the Rocky Mountains southwords to Andes of Continental Americas in the New World. Hence, primary gene centre of *Morus* appears to be in Sino-Japanese region and/or North America (Sanjappa 1989) and, mulberry may have mono as well as polyphyletic origin. India and other regions might have been secondary centre of species diversity for mulberry (Sarkar 1998). However, according to another school of thought, the mulberry is believed to be a native either of India or China and it is believed to have originated on the lower slopes of the Himalayas. Chin-Nong, one of the successors of Empheros Fo-Hi taught cultivation of mulberry in China during 2800 BC. Silk industry took its origin in the province of Chang Tong and spread to India from China through Kotan (Tibet) by about 140 BC (FAO 1976). The distribution of various mulberry species in India is as here under (FAO 1976).

1. ***Morus alba* L.**: These are cultivated in Punjab and North West Himalayas ascending up to 3500 m ASL
2. ***Morus indica* L.**: They are distributed in the lower Himalayan and sub-Himalayan tracts from Kashmir to Sikkim, ascending up to an altitude of 2500 m ASL. Also found in Bengal and Assam in the north-east and in southern plateau in Karnataka and Tamil Nadu ascending up to 1500 m ASL.
3. ***Morus serrata* Roxb.**: Found in temperate Himalayas from Kumaon hills westward up to an altitude of 3000 m ASL.
4. ***Morus laevigata* Wall ex Brand.** : Distributed in tropical and sub-tropical regions from Indus valley to Assam and cultivated ascending up to 1500 m ASL.

Besides above four species, *M. nigra, M. multicaulis, M. sinensis* and *M. phillipinensis* also occur as exotic species in this country.

Habit: Trees up to 20--25 m height with 8 m girth of the trunk or shrubs, perennial, cultivated or wild, profusely branching (Fig. 45).

Habitat: An atmospheric temperature ranging from 24-28°C is found to be optimum for mulberry growth. The growth and sprouting of buds of mulberry does not occur at temperature below 13°C. Places with annual rainfall ranging from 600 mm to 2500 mm are suitable for mulberry. On an average 50 mm rainfall once in 10 days and 65-80% relative humidity has been considered as ideal for mulberry. It grows well in our country with a sunshine range of 9-13 hrs a day at altitudes between 300-800 m ASL but ± 700 m ASL has been considered desirable for luxuriant growth of mulberry. Deep, fertile, well drained, clayey loam to loam, slightly acidic (6.2-6.8 pH), friable and porous soil with good moisture holding capacity is most suitable but in India mulberry is cultivated on a variety of soil like sandy loam to loam in West Bengal and Kashmir, red loam in Karnataka, lateritic red loam in Bihar/ Jharkhand and Black soil in Maharashtra.

Stem: Stem colour varies based on species, climate and origin. The Mysore local, Kanva-2 and Berhampore varieties of *M. indica* are white to grayish white in colour.

Leaves: The leaves are simple, alternate, stipulate, petiolate. Petiole 2.5-7.5 cm with a groove on the dorsal side. The size of leaf varies in different species and varieties. While Kanva-2 bears large leaves, Mysore local have small leaves. Kanva-2 and Kosen have only unlobed leaves while Mysore and Berhampore locals have mostly leaves with 1-7 lobes. Lamina is usually glabrous, ovate or ovate-cordate. The leaf base may be curved (cordate), truncate or ordinary. The leaf tip may be acute or acuminate. The leaf margin may be serrate, dentate, dentate-serrate or serrate-dentate while venation is multicostate reticulate type, usually 3 nerved in *M. alba*. Stipules 1.5-2.0 cm in *Morus nigra* and chinar variety of Kashmir. Phyllotaxy is 1/2, 1/5 or 2/5 depending upon varieties.

Inflorescence: Pendent or drooping peduncle of Catkin inflorescence in mulberry bear unisexual flowers. Generally, male and female catkins are found in separate plants (dioecious) but both may be found on the same plants also (monoecious). Development of pistillodes and staminodes may also rarely cause occurrence of bisexual flowers.

Male catkins: Axillary, rarely branched, usually clustered and 2.5-5.0 cm long with loosely attached flowers to peduncle. They become pink, dry and droop off after dehiscence of pollens.

Female catkin: Axillary, rarely branched and clustered usually 1.5 - 3.5 cm but 10-12 cm in *M. laevigata*. Attachment of flowers to peduncle is stiff and entire inflorescence turns into a multiple fruit of deep pink to black colour.

Flowers: Small, usually sessile or rarely short pedicellate, regular, hypogynous and unisexual.

Male flowers: Perianth leaves (tepals) four, the two inner flat, the two outer more or less keeled, arranged in two whorls with imbricate aestivation, green homo-chlamydous, sometimes with hairy margins and outer surfaces. Stamens four, arranged opposite to perianth leaves. Filaments broad, bent inward in bud stage. Anthers versatile, dithecous, dehiscing longitudinally. Pollen grains round, dry, light and dust like with smooth exine.

Female flowers: Tepals four, similar to male, persistent, become thick, fleshy and juicy in fruit after fertilization. Ovary bicarpellary, syncarpous, superior and unilocular. Ovules pendulous attached to the margin of the ovary. Style single, longer (in Dolichostylae) or shorter (in Macromorus), united near the base (connate). Stigma bifid, rarely trifid, papillose or pubescent (hairy).

Flowering period: Feb-March in plains and April-May in Himalayas.

Pollination: Mostly anemophillous (through wind).

Fruit: Cylinderical and short or long depending upon species/varieties, sweet, mucilaginous, purple, red, reddish-purple or black when mature, edible with pubescent (hairy) and stout stalk. Fruits ripen during March-April in *M. alba* and during June-July in *M. serrata* in Himalayas.

Seed: Oval, flat at the micropylar region. An elongated streak of Hilum is found on one edge. Micropyle (small pore through which seeds absorb water) is found at one end of Hilum. Seed coat consists of outer hard and brittle layer of Testa and the inner thin, papery and slightly brownish layer of Tegmen. Kernel found inside seed coat consists of endosperm with stored material and a curved embryo embedded in endosperm. Embryo consists of plumule, radicle and two cotyledons. Plumule gives rise to shoot system while radicle gives rise to root system on germination.

Floral formulae: Male flower: ⊕ ♂ $P_{2+2} A_4 G_0$

Female flower: ⊕ ♀ $P_{2+2} A_0 G_{(\underline{2})}$

Distinguishing features of important species: (after Jain and Kumar 1989 and Sarkar 1998)

1. ***M. indica***: Leaves ovate or broadly ovate, caudate, sharply serrated and lobed, both leaf surfaces scabrous, styles short, papillose and nearly connate to tips. Fruit medium sized, 1.5" - 2" long, ovoid or subglobose and turns black after ripening.
2. ***M. alba***: Leaves much longer than *M. indica,* ovate, obtuse, acute or sharply acuminate, serrate or crenate serrate with three basal veins, styles short, papillose and free to base. Fruit medium and less than 2" long, dark red or blackish when ripe.
3. ***M. nigra***: Leaves wider than that of *M.alba,* more or less coriaceous (leathery), thick, 5-nerved, styles densely pubescent (hairy), fruit purple.
4. ***M. serrata***: Large tree, young parts softly hairy, dioecious, leaves pubescent and serrated, styles pubescent, fruit 1"-2" long.
5. ***M. laevigata***: Large tree, young parts softly hairy, leaves serrated and pubescent. Spikes drooping, ripe fruits yellowish to dark red, very sweet and 3"-6" long.

Recommended varieties as bush plantation

Following varieties have been recommended for sericulture in various agro-climatic regions.

1. **Rainfed regions:** S799, S1 (Mandalaya), K2 (M5), Sujanpur-5, S14, S13, S34, RFS-135, RFS-175 (indigenous). Kosen, KNG, Ichinose (exotic).
2. **Hilly regions:** Kosen, MR2, K2, S54, BC259.
3. **Alkaline regions:** G9, TR10.
4. **North-East India:** S1635, BC259, TR8, TR10, S1.
5. **South India** (Karnataka, Tamil Nadu and Andhra Pradesh): K2, S30, S54, V1, S36, S13, RFS175, RFS135, Viswa, Vishala and TG-Thalaghattapura.
6. **West Bengal** (Malda, Murshidabad, Birbhum): S1, S799.
7. **Darjeeling Dist:** Kosen (Japanese variety)
8. **North-Western Himalayan region:** Goshoerami, Kokuso-27, Kairyoroso, K2, S146, S1635.
9. **Chotanagpur** (Jharkhand): S1, C763, K2, S54.

Varities suitable as tree plantation:

S1301, S36, S523, TR4, C776, S1, K2 (M5), S41, Sujanpur-5, S30, TR8, RFS-135, S146 and *M. laevigata* (Indigenous). Thailand var., Aftrican var. and Shrim-2 (Exotic).

Varieties with fungal resistance: Kosen, MR2.

Varieties with bigger fruits: *M. leavigata, M. latifolia, M. serrata* and *M. nigra*.

Uses: Mulberry serves as the food plants of mulberry silkworms, *Bombyx mori* L. and accounts for 95% of the world silk production (Ullal and Narsimhanna 1994) and 87% of our country's silk production. Besides providing foliage to silkworms for feeding, the *Morus* species are extensively exploited in sports, food processing, timber, tannin, dairy, poultry and pharmaceutical industries and hence they are capable of transforming economic scenario of any state/ country (Singhvi and Sinha 1991, Chinnaswamy *et al.* 1993). Nearly 70% of the silk proteins produced by a silkworm is directly derived from the proteins of the mulberry leaves (FAO 1976).

The wood of *M. alba* (white mulberry) is very much valued for house building, agricultural implements, sports goods and furniture etc. It is considered as the medium grade fuel wood and their shoots can be used in paper industry when grown as bush. The wood contains tannins for tanning and colouring purposes (Singhvi and Sinha 1991).

The leaves of *M. alba* (white mulberry) and *M. serrata,* owing to their high carotene content, can form a valuable source of vitamin-A for sheeps and poultry birds. The mulberry leaf stalk remnants left after feeding silkworms can also be used for feeding cattles without any adverse effect on their health and performance (Anonymous 1962, Singh 1982, Dwivedi 1988, Singhvi and Sinha 1991).

The growing mulberry leaves contain 6.0-6.8% crude protein, 4% soluble carbohydrate and 0.6% crude fat. Vit-A, B, C, D and several compounds of Ca, P, Si, Mn, Mg, Fe, Cu, Zn etc. are also found in mulberry leaves. Hence, they also satisfy the dietary requirements of human beings as vegetables like Amaranthus, Pea, Cucumber and Drumstick leaves (Sreekumar *et al.* 1994).

The fruits of *M. laevigata*/*M. macrouro*/*M. nigra* reach a length of 8 cm and are very sweet and soft and are eaten fresh or made into a jam, jelly and syrup. In Eruope, a wine is also prepared by fermentation of the fruit (Phillip 1989).

The seeds of *M. alba* yield as much as 25-35% oil content of edible nature and hence may be exploited commercially (Jain *et al* 1990).

The medicinal value of mulberry is mentioned in scriptures. The mulberry leaves are considered diaphoretic and emollient. They are used as gorgle in the inflammation of throat. The aqueous and alkali extract of the leaves and stem are active against Gram +ve bacteria and yeasts. An infusion of leaves causes drooping of blood sugar, sometimes diuresis and reduction in arterial pressure. The stem bark is used as purgative and vermifuge. The roots posses anthelmintic and astringent properties. (Chinnaswamy *et al.* 1993). From the roots of black mulberry (*M. nigra*) an alkaloid **moranoline** or **deoxyjirimycin (DNJ)** is extracted which inhibits the glucosidases and hence reported to be promising potential medicine against AIDS. The other alternative from of this drug **homonojirimycin** has been found to effectively control diabetes by checking rise in blood sugar (Ray 1990). The ripe fruits of black mulberry are nutritive, laxative, cooling and their juice is a grateful drink for convalescent and for anemics (Kurian 1995).

In general, the mulberry fruits are acrid, sweetish, sour convalescent, laxative and used against diarrhoea, ulcerated intestines, biliousness, burning sensation and 'Vata' and 'Kapha' doshas in Ayuyved. Fruits improve appetite, enrich blood against anemia and are considered good for brain, heart and spleen.

In Combodia while leaves of *M. alba* are used for treatment of conjunctivitis, in China its root bark is considered as restorative, tonic and astringent remedy against nervous disorder (Sarkar 1998).

In Manipur, fruit juice of *M. australis* is used in gargling for throat troubles whereas plant juice of *M. laevigata* is considered as antiseptic and applied as plaster on sores (Sinha 1996). In Mizoram, bark of *M.alba* is regarded as purgative, root and bark as anthelmintic and leaf as diaphoretic and used as remedy against toothache and dyspepsia (Pradhan 1998).

Cultivation: Mulberry can be cultivated through seeds, stems, cuttings, saplings, air layings and tissue culture.

Fertile, deep, porous, slightly acidic (pH 6-7), clayey loam to loam soils with good water holding capacity are ideal for raising mulberry plantation. Areas with atmospheric temperature ranging from 20-30°C, rainfall of 50 mm fortnightly and 9-13 hours of sunshine are ideal for good growth of mulberry. While mulberry should be planted after onset of monsoon in rainfed areas, in irrigated areas it can be planted at any time excepting very hot or cold seasons.

Cuttings of suitable varieties are prepared by slantly cutting 6-8 months old 10-12 mm thick hardwood branches. The size of cuttings should be 20 cm having 3 or more live buds. 3-4 feet wide nursery bed is prepared by digging the soil well up to 30-40 cm depth and raising the same upto 7-8 inches above the ground level. FYM is thoroughly mixed and cuttings are planted 4-5 inch deep in the soil at a spacing of 10-15 cm keeping atleast one bud above the ground. Cuttings are irrigated immediately and afterward periodical irrigation should be made at the interval of 4-5 days during dry seasons. NPK fertilisers @ 25:25:25 kg/ha may be applied after 5-6 weeks when cuttings sprout well and the root system is well established. Vigorous saplings of 3-4 ft. height become ready for transplantation within 4 months (Anonymous 1998).

The land is prepared by deep ploughing 2-3 times during April-May for loosening and making fine tilth of the soil. Pits of 1x1x1 ft. size (LxWxD) at 3x3 ft spacing be prepared and basal dose of 1 kg. FYM or compost should be applied in each pit in rainfed areas. Irrigated lands however, may be applied 20-25 tons FYM/ha by broadcasting evenly in the field. Every sapling should be transplanted in these pits during June-July after on set of monsoons.

When cuttings, instead of saplings, are directly used for plantation purposes, three cuttings are planted in triangular fashion at a distance of 15 cm in each pit in slanting fashion.

In South India, chemical fertilisers @ 100:50:50 kg. NPK/ha may be applied in two split doses in plough furrows and FYM may be applied @ 10 metric tons/ha annually in rainfed areas. On the contrary, in irrigated areas 300: 120:120 kg. NPK/ha in four-five split doses in deep furrows and FYM @ 20-30 metric tons/ha doses are applied to mulberry orchards for harvesting enough foliage for silkworm rearing. The fertilizer application is attended 2-4 weeks after pruning (Ullal and Narasimhanna 1994).

Mulberry garden is pruned at 10-15 cm from ground (bottom pruning) during July-August and Dec-Jan at 45-60 cm above ground to improve the foliage quality in rainfed areas whereas in irrigated areas pruning may be resorted four to five times approximately, two and half months before initiation of silkworm rearing at different heights during different seasons combined with split fertilisation and irrigation.

Doses of chemical nitrogen and phosphorus may be reduced by 50% by use of biofertilizers viz., *Azotobacter* and/or *Azospirillum* (for nitrogen) and vesicular arbuscular mycorrhizae and/or phosphatica (for phosphorus). Use of Mussorie-phos/ rock phosphate is more advisable than single super phosphate due to lesser cost as well as reduction of acidity in soil by M-phos due to inherent calcium carbonate in it. Application of M-phos reduced lime application generally resorted as a soil corrective measure especially when it is very acidic for neutralising acidity and improving the physical properties of soil which enhances terminal growth of shoots and roots (Srivastav and Singh 1996b).

The fertilizer requirement in West Bengal (336 N: 180 P : 112 K in 3-4 split doses by deep furrows) and Kashmir (300 N: 150 P : 200 K in two split doses by application in root zone) is entirely different from that of south India (Karnataka, Tamil Nadu and Andhra Pradesh). In Kashmir organic manures are applied during winter @ 7.5-10.0 metric tons/ha and mulberry is grown as bushes and bottom pruning is practiced during spring and summer in alternate years (Ullal and Narasimhanna 1994).

Thus, the fertilizer, organic manure and pruning requirements of mulberry greatly vary according to climate, geographical conditions and methods of silkworm rearing.

105. KADIPATTI

Botanical: *Murraya koenigii* Spreng syn. *Bergera koenigii* Linn.

Burmese: Pidosin English: Curry Leaf Tree Sinhalese: Karapincha

Hindi: Kadipatti, Gandhle, Gandhela, Kathnim, Mithnim

Assmese: Bishahari Bengali: Barsanga, Kariaphulli

Gujarati: Goranimb Kannada: Karibevu, Karibevana Mara

Malayalam: Kariyapala Marathi: Karinimb, Karepacha

Oriya: Basango Punjabi: Bovala

Sanskrit: Mahanimb, Surabhininiba

Tamil: Karuvembu Telugu: Karepeku

The Indian curry leaf tree is very popular tree in South Indians as it is found in the backyard of almost every home due to strong smelling leaves used regularly to flavour curries.

SYSTEMATIC POSITION

Bentham & Hooker (1883)	**Engler & Prantl (1899)**	**Hutchinson (1959)**
Phanerogams	Phanerogams	Angiospermae
Dicotyledones	Dicotyledoneae	Dicotyledones
Polypetalae	Archichlamydeae	Lignosae
Disciflorae	Geraniales	Rutales
Geraniales	Rutaceae	Rutaceae
Rutaceae		

Geographical distribution: Outer Himalaya, from the Ravi eastward to Sikkim ascending up to 5000 ft. ASL in Uttarakhand, Assam, West Bengal, Bihar, Jharkhand, Chittagong (Bangladesh), Upper and Lower Myanmar. Evergreen and deciduous forests of the Peninsula and often as underwood in Kerala and Tamil Nadu.

Habit: A small, pubescent tree with a short trunk and a close shady crown. All parts emit a powerful peculiar smell (Fig. 46).

Habitat: Tropical to temperate humid habitat with varying types of soil and climatic conditions.

Root: Tap and branched.

Stem: Erect, branched, cylinderical, woody, green, solid, bark dark grey.

Leaves: Compound, exstipulate, petiolate, alternate, imparipinnate, green, 6-12 inches long, leaflets 11-25, mildly serrate or entire, ovate-lanceolate, oblique, 1 inch long, shining, smooth, coriaceous, venation unicostate reticulate, aromatic.

Inflorescence: Terminal corymbose panicles.

Flower: Pedicellate, bracteate, actinomorphic, complete, hermaphrodite, pentamerous, hypogynous.

Calyx: Sepals 5, gamosepalous, oblong, green, qunicuncial.

Corolla: Petals 5, polypetalous, white, gland dotted, alternating with sepals, imbricate.

Androecium: Stamens 10 in two whorls, polyandrous.

Gynoecium: Carpels 2, style short, cylinderical, syncarpous, ovary superior, placentation axile.

Fruit: An ovoid, dark red or purplish black, rugose, berry, ½ inch in diameter.

Seeds: Endospermic.

Flowering: Hot season.

Leaf shedding: Hot season for a short time.

Leaf sprouting: Hot season.

Floral formula: Br $\oplus$ ⚥ $K_{(5)}\ C_5\ A_{5+5}\ G_{\underline{(2)}}$

Uses: Leaves are used as condiments for flavouring curries and chutneys. The bark root and leaves are crushed and applied externally to cure skin eruptions. The green leaves cure dysentery when eaten raw and cure bruishes when applied externally after crushing. An infusion of roasted leaves stops vomiting. The decoction of root is mild purgative. The decoction of leaves is tonic, carminative and antipyretic. The decoction of leaves is used for strengthening and toning of the stomach, purification of the blood, elimination of worms, curing of tuberculosis, burning pains, haemorrhoids, acute abodominal pain and swelling various parts of the body (Kurian 1995).

Leaves yields an essential oil for soap perfume. Fruits edible, yield an essential oil. Wood used for agricultural implements (Ambasta *et al.* 1986).

Cultivation: Through seeds and seedlings.

The foliar spray of 0.25% DAP spray has been found to be good for influencing all the attributes of seedlings of curry leaf (Masilamani *et al.* 2000).

106. YONGCHAK

Botanical: *Parkia javanica* Merr. Syn. *P. roxburghii* G. Don

Assamese: Khorial, Zongto Bengali: Supota Manipuri: Yongchak

SYSTEMATIC POSITION

Bentham & Hooker (1883)	**Engler & Prantl (1899)**	**Hutchinson (1959)**
Phanerogams	Phanerogams	Angiospermae
Dicotyledones	Dicotyledoneae	Dicotyledones
Polypetalae	Archichlamydeae	Lignosae

Calyciflorae	Rosales	Leguminales
Rosales	Leguminosae	Mimosaceae
Leguminosae	Mimosoideae	
Mimosoideae		

Geographical distribution: Assam, Manipur, Nagaland, Mizoram and Bangla Desh (Chittagong). Mostly cultivated, occasionally wild.

Habit: A moderate sized unarmed tree attaining 20-meter height and 0.6 metre diameter (Fig. 47).

Habitat: Moist evergreen forests.

Root: Tap root, branched, deep.

Stem: Aerial, erect, branched, branches spreading, bark rough, gray.

Leaves: Bipinnate, pinnae 20-30 pairs, leaflets 40-80 pairs, linear, 0.8 cm long, opposite, green.

Inflorescence: Globose cymose heads pendant, supported by a coriaceous bract, the upper flower bisexual, the lower male or sterile, greenish white-creamy white.

Flower: Complete or incomplete, bisexual, male or sterile, actinomorphic, hypogynous, white.

Calyx: Sepals 5, gamosepalous, unequal, imbricate, odd sepal posterior.

Corolla: Petals 5, polypetalous, creamy white.

Androecium: Stamens 10, monoadelphous, filaments longer than petals, cannote at base and adnate to calyx tube.

Gynoecium: Monocarpellary, ovary superior, unilocular, stigma simple, style long, filiform, placentation marginal.

Fruit: Pod flat, lomentous, coriaceous, 38-50 cm x 2.5-4.0 cm smooth, green-dark brown.

Flowering: Oct. - Nov.

Fruiting: Dec. – Jan.

Uses: Fruits are edible. Vegetables, salad, chutney and iromba etc. are prepared from it in Manipur and other North-eastern states. Pods are used in bleeding piles while bark extract is given in diarrhoea and dysentery. Pods and seeds are used in stomach disorders. Bark and leaves are applied as lotion to sores and skin affections (Sinha 1996). Woods are used as firewoods.

Cultivation: Through seeds and seedlings.

107. COPPER POD

Botanical: *Peltophorum inerme* Lanos. Syn. *P. ferrugineum* Benth., *P.roxburghii.* (G.Dou) Degeher, *Caesalpinia inermis*

English: Copper pod Rusty shield-bearer

Tamil: Ivalvagai, Iyalvgai, Iya-vakai, Perungondarai

Telugu: Kondachinta

SYSTEMATIC POSITION

Bentham & Hooker (1883)	Engler & Prantl (1899)	Hutchinson (1959)
Phanerogams	Phanerogams	Angiospermae
Dicotyledones	Dicotyledoneae	Dicotyledones
Polypetalae	Archichlamydeae	Lignosae
Calyciflorae	Rosales	Leguminales
Rosales	Leguminosae	Caesalpiniaceae
Leguminosae	Caesalpinioideae	
Caesalpinioideae		

Geographical distribution: A native of Sri Lanka and Andamans, it is commonly found in Bihar, West Bengal and Western Ghats. It is found in North Australia, Malay Peninsula and Archipelago.

Habit: A partially deciduous tree with dark handsome foliage, fairly spreading crown and smooth or slightly rough gray bark.

Habitat: Sub-tropical to tropical places with hot and humid climate and almost all types of soil.

Root: Tap, branched, deep.

Stem: Aerial, erect, woody, cylinderical, solid, branched, branches spreading, bark smooth, gray, branchlets ferrugineous pubescent.

Leaves: Compound, bipinnate, large, feathery or fern like, yellowish green, rachis and mid-rib ferrugineous pubescent, leaflets opposite, obtuse, 1.2-1.8 cm long, obliquely-oblong.

Inflorescence: Panicles, erect, rusty-tomentose, large, terminal.

Flower: Bisexual, zygomorphic, large, yellow, pedicellate, complete, hypogynous.

Calyx: Sepals 5, more or less connate (gamosepalous), segments imbricate.

Corolla: Petals 5, polypetalous, with long ferrugineous hairs on back, bright golden yellow.

Androecium: Stamens usually 10, free, filaments with tufts of silky hairs of base.

Gynoecium: Carpel 1, ovary superior, unilocular, placentation marginal, style long, stigma peltate.

Fruit: Large copper pods, conspicuously hanging on the tree, flat, indehiscent, 5.0-10.0 cm long, with a firm broad wing along each suture.

Flowering: March – May.

Leaf fall: January.

Leaf sprouting: February.

Floral formula: $\bullet | \bullet$ ⚥ $K_{(5)}\ C_5\ A_{10}\ G_{\underline{1}}$

Uses: It is highly ornamental tree for planting in parks, garden and avenues. The bright yellow crowns provide a contrasting colour with the scarlet heads of the gulmohur when planted together. The timber is used for making cabinets (Randhawa 1983). It is mostly planted in coffee and tea plantations and also along the coasts. It also provides good fuel and fodder and constitutes an effective multipurpose forest tree production system (Chandawat and Gautam 1993).

Bark, wood and leaves contain tannins. Bark used in dysentery, forms an ingradient of gargles, tooth powders and lotions used in eye troubles, sores and musculer pains. Wood suitable for cabinet works, coach building and planks (Ambasta *et al.* 1986).

Cultivation: Easily propagated through seeds, seedlings and cuttings (Randhawa 1983).

108. MAKHANPHAL

Botanical: *Persea americana* Mill.

Burmese: Htaw-bat-thee	English: Avocado	French: Avocat
German: Avogado	Hindi: Makhanphal	
Bengali: Kulunashpati	Kannada: Benney-hannu	Urdu: Makhan phaul

SYSTEMATIC POSITION

Bentham & Hooker (1883)	**Engler & Prantl (1899)**	**Hutchinson (1959)**
Phanerogams	Phanerogams	Angiospermae

Dicotyledones	Dicotyledoneae	Dicotyledones
Polypetalae	Archichlamydeae	Lignosae
Thalmiflorae	Ranales	Laurales
Laurales	Lauraceae	Lauraceae
Lauraceae		

Geographical distribution: Indigenous to Mexico or South America, it is cultivated in tropical and sub-tropical countries. It has been introduced in India and growing in Bangalore, Pune and few other places (Kurian 1995, Kocchar 1981).

Habit: A moderate sized tree growing up to 15 meters height but under cultivation, the height is kept below 10 meter.

Habitat: Tropical-sub-tropical, moist hot.

Root: Tap root, branched.

Stem: Aerial, erect, woody, cylinderical, branched, bark grey, slightly rough-smooth, aromatic.

Leaves: Simple, ovate, glabrous, coriaceous, 10-20 x 5-8 cm, petiole 1.5-2.5 cm long, green, exstipulate, entire, apex acute, veins buldging below, venation unicostate reticulate.

Inflorescence: Large panicles arising from the axils of upper most leaves.

Flower: Bisexual, small, regular, actinomorphic, hypogynous.

Perianth: Small, grey, undifferentiated, biseriate, 3 in each whorl, free, valvate, persistent.

Androecium: Stamens 12, arranged in 4 whorls of 3 each, the innermost reduced to staminodes, anthers 4 celled, basifixed.

Gynoecium: Monocarpellary, ovary superior, unilocular, ovules solitary, pendulous, anatropous.

Fruit: Berry, large, pear shaped, purple – green with thick skin.

Floral formula: $\oplus$ ⚥ P_{3+3} $A_{3+3+3+3std.}$ $G_{\underline{1}}$

Uses: Fruit is edible and grown in India as a commercially less important fruit (Kocchar 1981). Fruits eaten as salad flavoured with peppers, salt and sugar and used in ice-cream also. Bark and root yield essential oil. Roots yield an antibiotic for food preservation (Ambasta *et al.* 1986). Wood provides valuable timber used for various purposes (Shukla and Mishra 1979). The oil from fruit is rubbed on the painful parts in rheumatism and inflammed joints. This oil is rubbed for removing dandruff, stopping fall of hairs and promoting growth of hairs also. The infusion of the fresh leaves is diuretic, checks flatulence, expels gases, strengthen the gums and cures pain of stomach and bowels. Hot compression of decoction of leaves is used in headaches and pain of facial nerves. The tea made from flowers promotes menstrual flow. Two teaspoonfuls of roasted and ground seed in a cup of lukewarm water is used in diarrhoea and dysentery. Applying this powder as cataplasm may treat inflammations of the finger and toes accompanied with pus formation. It also relieves pain from dental caries. Decoction of this powder is used in gorgles for toothache (Kurrian 1995).

Avocado plantations are extensively utilized for commercial production of wild silk from *Attacus atlas* L. silkworms in Indonesia (Mishra 2002).

Cultivation: Through seeds and seedlings.

109. KHAJUR

Botanical: ***Phoenix dactylifera*** Linn.

Burmese: Sun-balun	English: Date Palm	Hindi: Khajur, Khaji
Bengali: Khajur	Gujarati: Khajur	Kannada: Khajura
Malayalam: Ittappazham	Oriya: Khorjjuri	Tamil: Karchuram
Telugu: Kharjuramu		

SYSTEMATIC POSITION

Bentham & Hooker (1883)	Engler & Prantl (1899)	Hutchinson (1959)
Phanerogams	Phanerogams	Angiospermae
Monocolyledones	Monocotyledoneae	Monocotyledones
Calycinae	Principes	Corolliferae
Palmae	Palmae	Palmales
		Palmae or Palmaceae

One of the oldest known fruit, date palm is a fruit symbolic of desert climate. It is believed to have originated around the Persian gulf (Euphrates, tigris doab and oasis of Sahara desert) and become naturalized in Arabia, North Africa and Northern India. It was probably first raised in Iraq, the erstwhile Mesopotamia as early as 5000 B.C (Brandis 1906, Kocchar 1981).

Geographical distribution: The major date producing countries are Iraq, Iran, Egypt, Saudi Arabia, Algeria, Pakistan, Sudan and Moroco. In India, it is growing wild or cultivated mostly in the dry regions of Gujarat, Rajasthan, Punjab, Uttar Pradesh, Madhya Pradesh, Andhra Pradesh and Karnataka.

Habit: A tall, unbranched, robust, dioecious tree growing upto 25-40 meters height with a crown of feathery leaves and root – suckers, if not cut off, developing into subsidiary trunks, giving the plant a clumpy habit.

Habitat: Around water holes or oasis and water courses (in Africa and the Middle East) or warm and humid regions where adequate irrigation is available.

Root: Adventitious roots, sufficiently long, growing mostly in the upper strata of the soil and foot surrounded by a dense mass of root-suckers.

Stem: Trunk more or less straight, cormous or bulbous and shaggy, roughened and armoured by persistent rhomboidal leaf bases of old fallen leaves.

Leaves: Crowny, 100-120, pinnate, feathery, 4-8 meters long, leaflets making acute angle with common petiole, stiff, glaucous, grayish green, 20-40 cm long, mostly two ranked, basal four ranked and modified into spines.

Inflorescence: Panicles, axillary, branched, enclosed within a deciduous fibrous spathe during initial development. Male plant producing 30-50 inflorescences, each with 100-150 branches of 12-15 cm length. Female plant producing 8-12 bunches of inflorescences, each bearing 10-30 branches of 30-50 cm length.

Flowers: Unisexual, incomplete, actinomorphic, hypogynous, male and female on separate plants.

Male flowers: Smaller, borne on slender, erect branches, waxy white, perianths 3+3, stamens 6, ovary rudimentary and centrally placed.

Female flowers: Bigger than male flowers, borne on stouter rachises which become pendulous on fruiting, greenish yellow, perianth 3+3, stamens nil, carpels 3, apocarpous, stigma short and hooked.

Fruit: One seeded oblong berry, 2.5-7.5 cm long, a thick sweet flesh surrounding the deeply grooved, hard, cylinderical seed with a brown papery taste. Fruits are hard and green first but later turn soft and yellow or red in colour.

Flowering: Feb- June.

Fruiting: After wind pollination, one of the three carpels enlarges and the remaining two fall off. Fruit ripening occurs from June to December. The average fruit yield is 45-68 kg/tree annually. The tree starts fruiting at 5-7 years of age, comes to full bearing by about 15 years of age and continues production upto 80 years of age.

Varieties: There are several varieties of date palm. One kind of date palm bears a seedless fruit.

Floral formulae: Male flower: $\oplus$ ♂ $P_{3+3}\, A_6\, G_0$

Female flower: $\oplus$ ♀ $P_{3+3}\, A_0\, G_{\underline{3}}$

Uses: Dates contain 70.6% carbohydrates, 1.9% protein, 2.5% fat, 13.8% water, 10.0% fibre and 1.2% ash (minerals). Besides, dates contain vitamins A, B_1, B_2 and nicotinic acid. In Arab, dates are eaten after replacing stone with a lump of butter. They are often eaten fresh or mixed with milk or curd. A paste of raosted locusts or grasshoppers is pounded and spread on the bread before eating. Jellies and jams are also prepared from ripe dates.

Fermentation and distillation of inferiror grade, cull dates yields a strongly alcoholic liquor, which becomes "Arrak" the strongest and most dreadful drink after flavouring with licorice. Date milk, a mild drink is obtained by tapping the stem near the base of the young terminal bud or the flowering spathe. The sap can be fermented or boiled down to provide sugar. The terminal bud of a young tree may also be used as a vegetable.

Nearly 800 uses of various parts of date palm have been recorded. The trunks are used to build doors, fence posts, rafters, shutters and walls. The leaflets are made into baskets, fans and mats besides using them for packing dates. The tender leaflets are used for making finest matting. The fibres are extracted from the leaflets to make ropes (Kocchar 1981). Walking sticks are prepared from the leaf stalks (Pandey 1981). This is also one of the principal tannin yielding plants (Koccher 1981).

Sweat nutritive sap is laxative and used for Jaggery and sugar. Gum from tree used in diarrhoea. Terminal leaf bud used as vegetable. Dates are demulcent, expectorant, laxative and used in respiratory diseases and fever. Ground or water soaked seeds used for feeding camels, goats, horses and poultry (Ambasta *et al.* 1986).

Cultivation: They can be propagated from seeds or offshoots that spring up around the base of the trunk.

The date palm needs its feet in water and its head in the fire of the sky for the production of good quality dates. Hence, long intensely hot summer with hot nights as well as hot days, an average temperature of 30°C, a low humid environment without rains and adequate supplies of underground water or irrigation are essentially required for cultivation of date palms. Recently, micropropagation of *P. dactylifera* has been achieved through somatic embryogenesis (Rao *et al.* 2001).

110. KHASI PINE

Botanical: ***Pinus insularis*** Endl. Syn. *P. khasya* Hook.f., *P. kesiya* Royle

Burmese: Tinyu

English: Khasi Pine

Bengali: Saral

Khasia: Dingsa

Lushai: Far

Manipuri: Uchan

Mizo: Far

SYSTEMATIC POSITION

Bentham & Hooker (1883)	Engler & Prantl (1899)	Hutchinson (1959)
Phanerogams	Phanerogams	Gyminospermae
Gymnospermae	Gymnospermae	Coniferales
Coniferae	Coniferae	Pinaceae

Geographical distribution: Uttarakhand, Mizorom (Khasi hills), Nagaland (Naga hills) and Manipur (Ukhrul and other hill areas) in India. Also found in Myanmar (Shan hills, upper chindwin, Sitang and Salween) between 2500-7000 ft ASL.

Habit: A tall evergreen, pyramidal or conical tree attaining 100-150 ft. height. When young they are conical but become irregular and extremely picturesque at maturity (Fig. 48).

Habitat: Open wind swept, sunny locations and well drained shallow soils present on rocks. Flourish well on lands too lean in nutrients and too stony and steep for agriculture.

Root: Tap, branched, shallow in the soil.

Stem: Aerial erect, woody, solid, thick, columnar, cylinderical, branched, branches more or less whorled, excurrent, confined to upper part, lower columnar stem naked, bark thick deeply cracked, resin duct numerous in the outer and middle belt of each annual ring, wood without vessels. The branches are of two types, long shoots or branches of unlimited growth present on main stem bear terminal bud hence they grow indefinitely and dwarf shoots or branches of limited growth develp on long shoot in the axil of scaly leaf without apical bud hence their growth is limited. Dwarf shoots bear foliage leaves surrounded by a scaly membranous sheath at the base, which combine to form spur.

Leaves: Foliage leaves 15-25 cm long, slender, needle like (acicular) in clusters of 3, green with convex back, sheath persistent, grey, lacerated at apex.

Male cone: Below apical bud on long shoot, clustered.

Microsporophyll: 60-125 / cone spirally arranged around central axis; microsprorphyll apex sterile.

Microsporangia: 2, ventrally present on each microsporophyll, contain numerous pollen grains.

Pollengrains: Spherical, triple layered, laterally double winged.

Female cone: In the axil of scale leaf on young branch near its apex, 1-4 clustered together; first year small, erect, ovoid, pale green, 2.5 cm long, comprising compactly arranged mega sporophylls; second year large, woody, reddish, 10-15 cm long.

Megasporophylls: Spirally arranged around central axis; lower scale thin, membranous and small called as bract scale; upper scale large, stout, woody and triangular in shape called as ovuliferous scale; each ovuliferous scale bearing two naked ovules on upper surface near central axis.

Seed: Approximately 12-15 months required between pollination and fertilization. Cone formation to seed dispersal needs approximately two years.

Uses: Timber is used for building, construction purposes, furniture, agricultural implements tool handles, matches, crates, boxes and rough carpentary works. It is a source of oleoresin, which yields oil of turpentine. Plant decoction is used along with honey for cough. Oleoresin is applied to skin in meningitis. Paste of wood is applied to forehead in headache (Sinha 1996).

Wood is suitable for planking and packing cases. It also yields pulp suitable for bag and wrapping papers (Ambasta *et al.* 1986).

Cultivation: Through seeds and seedlings. Seeds remain dormant for sometime or may germinate without undergoing resting period.

111. JANGAL JALEBI

Botanical: *Pithecolobium dulce* Benth. Syn. *Mimosa dulcis* Roxb., *Inga dulcis* Willd.

English: Manila Tamarind   Hindi: Jangal Jalebi, Vilayati Imli

SYSTEMATIC POSITION

Bentham & Hooker (1883)	**Engler & Prantl (1899)**	**Hutchinson (1959)**
Phanerogams	Phanerogams	Angiospermae
Dicotyledones	Dicotyledoneae	Dicotyledones
Polypetalae	Archichlamydeae	Lignosae
Calyciflorae	Rosales	Leguminales
Rosales	Leguminosae	Mimosaceae
Leguminosae	Mimosoideae	
Mimosoideae		

Geographical distribution: Indigenous to America, naturalized in the western Peninsula. Cultivated throughout India.

Habit: A middle sized, glabrous, evergreen tree, armed with straight stipulary spines.

Habitat: Sub-tropical to tropical moist places in more or less all types of soils.

Root: Tap, branched, deep.

Stem: Erect, aerial, branched, bark rough, dark grey.

Leaves: Compound, pinnate, pinnae 2, each with one pair of pale green, coriaceous, unequal sided, usually obtuse leaflets, 2.5-5.0 cm long, common and partial rachis slender terminating into short bristles.

Inflorescence: Globose, sessile or short pedunculate, heads; arranged in long panicled racemes.

Flower: Bisexual, actinomorphic, hypogynous, complete, regular, white.

Calyx: Sepals 5, gamosepalous, campanulate, 5 toothed, valvate.

Corolla: Petals 5, gamopetalous, white, valvate

Androecium: Stamens indefinite, without glands, monoadelphous, much exserted.

Gynoecium: Monocarpellary, ovary superior, unilocular.

Fruit: Lomentous pod, curved, turgid, twisted, 10-13 cm long.

Flowering: Jan.- March.

Uses: It serves as an excellent hedge plant. Besides, its fruits are eaten due to which it was cultivated throughout India in the past but now they occur only sporadically.

Cultivation: Through seeds and seedlings. The seeds are collected during April – June and nursery is raised immediately. The seeds may directly be sown during July or else seedlings raised in nursery be transplanted during July-Aug. It may also be propagated through cuttings (Chundawat and Gautam 1993).

112. GOLAINCHI

Botanical: ***Plumeria acutifolia*** Poir, *P. acuminata* Roxb., *P. acuminata* Ait, *P. rubra* Lin., Forma *acutifolia* Poir.

Eng: Temple, Frangipani or Pagoda Tree

Hindi: Golainchi, Chameli, Gul-e-chin, Gobur champa,

Burmese: Tayok – saga

Sinhalese: Araliya

Assamese: Gulanchi, Goalanchi, Galauchi

Bengali: Dalan phaul, Dalama phula

Gujarati: Dholo champo, Rhada champo

Kannada: Kadusampige, Deva ganigile, Ganagala

Malayalam: Arali, Vella Champakan

Marathi: Khair champa, Son Champa

Oriya: Golochi

Sanskrit: Khira champa

Santhali: Champa pungar

Tamil: Ilattalari, Perungali

Telugu: Arhataganneru, Vuru Varahaalu

SYSTEMATIC POSITION

Bentham & Hooker (1883)	**Engler & Prantl (1899)**	**Hutchinson (1959)**
Phanerogams	Phanerogams	Angiospermae
Dicotyledones	Dicotyledoneae	Dicotyledones
Gamopetalae	Sympetaleae	Lignosae
Bicarpellatae	Contortae	Apocyanales
Gentianales	Apocynaceae	Apocynaceae
Apocyanaceae		

It is named after Charles Plumier; a French Botanist while acutifolia or acuminata refers to the tapering or sharp pointed leaves.

Geographical distribution: A native of Mexico and Guatemala. Introduced and cultivated all over India from time immemorial.

Habit: A small deciduous tree (Fig. 49).

Root: Branched tap root.

Stem: Erect, branched, aerial, solid, cylinderical, crooked, branches thick, full of tenacious white milk which oozes out when bark is broken, stem and branches appearing lime washed in a variety usually found in Kerala (Trivendrum).

Leaves: Simple, broadly lanceolate, or oblanceolate, entire, tapering at both ends, alternate, quite distinct due to prominent parallel veins, borne in crowded spirals at the end of long branches, secondary nerves numerous, straight, transverse, joined by straight intramarginal veins, blade 15-35 cm long, petiole 2.5-5.0 cm long. Leaf fall by December and sprout during rainy season but young trees do not shed leaves.

Inflorescence: Compound penduncculate axillary cymes.

Flower: Pedicellate, bracteate, regular, complete, actinomorphic, pentamerous, hypogynous, highly fragrant, waxy white with yellow centres, borne in upright clusters, bisexual.

Calyx: Sepals 5, gamosepalous.

Corolla: Petals 5, gamopetalous forming a campanulate corolla tube, aestivation contorted (twisted), white with yellow centres, sometimes tinged with pink below.

Androecium: Stamens 5, alternating with petals, epipetalous, filaments short, anthers introse, polyandrous, opening by longitudinal slits.

Gynoecium: Carpels 2, semi-inferior ovary.

Fruit: A paired follicle or two horn like pads on a long pendent stalk, rarely produces fruits or seeds in India.

Seeds: Winged, with tassel of silk.

Flowering: March – April and again from July – Oct.

Floral formula: $\oplus \, ⚥ \, K_5 \, \overbrace{C_{(5)} \, A_5} \, G_{(\bar{2})}$

Uses: Besides serving as one of the tertiary food plants of eri silkworm, *Samia Cynthia ricini* (FAO 1976, Sarkar 1988, Seth 1995c, Sarmah 2000). "Pagoda tree" has also got several other uses. It is highly valued for ornamental planting and often associated with temples, hence it has also been named as "Temple tree". The flowers are used in wreath making and often offered as an offering to the deities in the temples. It also serves as a medicinal plant. Its decoction of leaves or bark brings down fever and prevents or cures asthma while decoction of root is used as remedy against sores. The massage of joints with milky latex extracted from plant mixed with coconut oil relieves rheumatic pains and heated leaves are applied as a poultice over swellings as a remedy (Kurian 1995, Randhawa 1983).

Bark is stimulant, purgative, febrifuge, emmenagogue and anti herpatic. Latex is rubefacient and purgative and root cathartic. Wood is used for making drums and other musical instruments (Ambasta *et al.* 1986).

Cultivation: It can easily be propagated through cuttings, which should be allowed to wilt before planting in July-August. Sandy and stony soils with rainfall of 70 cm or above are suitable for its survival. It does not wilt even when its roots are exposed and replanted in a new location and continues growing and blooming hence it has also been venerated as the "Tree of life" (Randhawa 1983).

113. KARANJ

Botanical: ***Pongamia pinnata*** Pierre. Syn. *P. glabra* Vent.

Burmese: Simizu, Thinwin
English: Indian Beech, Pongam
French: Pongami
Hindi: Karanj, Papar, Kanji, Papri
Persian: Khaiulmalisa
Sinhalese: Magulkaranda
Assamese: Karchaw
Bengali: Karanj
Gujarati: Karanj
Kannada: Honge
Konkani: Kerongi
Malayalam: Minnari, Unne
Marathi: Karanj
Oriya: Koronjo
Punjabi: Sukhchein
Santhali: Kuruinj
Sanskrit: Karanja
Tamil: Ponga, Pangam, Kanuga
Telugu: Pungu, Gaanuga
Urdu: Karanjwah Punku

SYSTEMATIC POSITION

Bentham & Hooker (1883)	**Engler & Prantl (1899)**	**Hutchinson (1959)**
Phanerogams	Phanerogams	Angiospermae
Dicotyledones	Dicotyledoneae	Dicotyledones
Polypetalae	Archichlamydeae	Lignosae
Calyciflorae	Rosales	Leguminales
Rosales	Leguminosae	Papilionaceae
Leguminosae	Papilionaceae	
Papilionatae		

Geographical distribution: Wild in both the Peninsulas and Andamans. Very common in the forests of Central India, West Bengal and Travancore. In Travancore, it ascends up to 800 ft. ASL while in Sub-Himalayan tract it ascends up to 2000 ft. ASL. Also common in tidal and beach forests of Sri Lanka, Myanmar, Malay Archipelago, South China, the Figi islands and tropical Australia (Brandis 1906, Randhawa 1983).

Habit: A moderate sized deciduous tree with short bole and spreading shady crown having smooth grey bark.

Habitat: Sandy, Banks of streams, water courses and the sea coasts.

Root: Tap, branched.

Stem: Erect, aerial, woody, cylinderical, branched, crown spreading, bark grey, smooth, soft, wood yellowish white, without heartwood.

Leaves: Compound, imparipinnate, glabrous, bright green, alternate, trifoliate, glossy, leaflets opposite, without stipules, 2-4 pair, ovate-oblong, shortly acuminate, 5-12 cm long, often marked by the presence of ugly patches of insects under the surface tissue.

Inflorescence: Axillary racemes, shorter than leaves.

Flower: Hermaphrodite, complete, zygomorphic, hypogynous, pedicellate, pedicels slender pubescent, 1.0-1.5 cm long, white or purple, papilionaceous, flowering in leafless condition, worth seeing when laden with flowers.

Calyx: Sepals 5, gamosepalous, below the disc in a tubular calyx, 5 toothed, aestivation ascending imbricate.

Corolla: Petals 5, unequal, upper most and largest petal known as standard or vexilum, the two free lateral petals known as wings or alae, the anterior united pair of petals known as keel or carina encloses stamens and pistil, white or purplish in colour.

Androecium: Stamens 10, inserted on a disc below the calyx, in two bundles (diadelphous) of 9+1, the tenth filament free at the base, in the middle connate with the rest, anthers bithecous, dorsifixed.

Gynoecium: Carpel one, free, ovary superior, unilocular, placentation marginal, style bent, short and hairy, stigma simple.

Fruit: Pods, woody, yellowish – grey, turning to dark-grey, indehiscent, turgid, more or less falcate, 1-seeded, 3.5-5.0 cm long, obliquely oblong, bilaterally compressed, seeds oily, single, kidney shaped.

Leaf shedding: March – April.

Leaf sprouting: May - June.

Flowering: Last week of April – May and Aug- Sept.

Fruiting: March - May.

Floral formula: •|• ⚥ $K_{(5)}\ C_5\ A_{9+1}\ G_{\underline{1}}$

Uses: Seeds yield commercial oil (Brandis 1906, Kocchar 1981), which is used for candle and soap making and as an illuminant and lubricant. The leaves are used as manure and fruits are eaten (Pandey 1981). Mainly used as a shade tree or for ornamental planting, it is very suitable for roadside avenues, railway stations and plateforms. Leaves drive away white ants (Randhawa 1983) while seeds and roots are used as fish poison (Sinha 1996).

Medicinally, all the parts of the tree are very important (Kurian 1995, Sinha 1996). The decoction of the root and bark is useful in expelling worms from the body, diseases of the eye, vagina and skin, tumours, piles, wounds, ulcers, itching, beri-beri, accumulation of liquid in the abdominal cavity, enlargement of the spleen, and abdomen and congestion of liver. The decoction of the tender shoots or sprouts is an appetizer, so it strengthens and tones up the stomach and checks inflammation, oedema and toxicosis. The infusion of leaves is a mild laxative and useful for digestion and expelling of worms from the body. The infustion of flowers is useful in diabetes and biliousness. The decoction of seeds is useful in earache, chest complaints, chronic fevers, hydrocoel and purification of the blood. The seed oil extracted by pressure technique is useful in rheumatic pains, leucoderma, itching, leprosy, ulcers, lumbago, urinary disorders, headaches, chronic fever, pain in the liver, wounds and various skin diseases, like herpes and scabies. Poultice of leaves is applied to ulcers infested with worms. The root extract is used for closing fistulous sores, cleansing foul ulcers and given internally in gonorrhoea. Leaf juice is given in flatulence, dyspepsia, diarrhoea, cough and leprosy.

Wood used for yokes of cart wheels, bullock cart, ploughs, rafters, oil mills, thatched cottages, furniture and small turney articles (Ambasta *et al.* 1986).

Cultivation: Through seeds and cuttings in rainy seasons (Randhawa 1983). It can also be propagated through root suckers (Brandis 1906). The seeds are collected during March-May, sown during April – May in nursery to raise seedlings, which are transplanted during July (Chundawat and Gautam, 1993).

Pongamia pinnata has a wider range of adaptability in relation to soil pH. Even in higher soil pH (10.5), when the soil is mixed with sand and FYM mixture at 2:1:1 ratio karanj gave more germination, seedling growth attributes and chemical constituents in 180 days old seedlings. Hence, their species may have greater potential for waste land afforestation programme (Masilamani *et al.* 2002).

Recently, vegetative propagation of *P. pinnata* has been achieved through hardwood stem cuttings treated with IBA 2500 ppm maintained in hydropit. This method induces 80% rooting. Besides, wedge grafting in three months old seedlings has also been achieved to the tune of 95% success for the purpose of seed production (Karoshe and Hegde 2002).

114. ALUBOKHARA

Botanical: ***Prunus domestica*** Linn.

English: Plum, Prune, Sloe, Damson

Hindi: Alubokhara, Alucha

Bengali: Alu-bukhara

Gujarati: Alubukhara, Alucha

Manipuri: Plum

Punjabi: Alucha

SYSTEMATIC POSITION

Bentham & Hooker (1883)	Engler & Prantl (1899)	Hutchinson (1959)
Phanerogams	Phanerogams	Angiospermae
Dicotyledones	Dicotyledoneae	Dicotyledones
Polypetalae	Archichlamydeae	Lignosae
Calyciflorae	Rosales	Rosales
Rosales	Rosaceae	Rosaceae
Rosaceae		

Geographical distribution: Indigenous to Europe and Western Asia where temperate climatic conditions exist and different kinds have been cultivated since time immemorial. The plum is cultivated in Kashmir, Punjab, Himachal Pradesh (Kullu), Uttarakhand (Kumaon), Nagaland, Manipur (Mao, Ukhrul) and other North-eastern hilly states.

Habit: A shrub or middle sized, unarmed or armed (spinescent) tree with pubescent young shoots.

Habitat: Hilly regions between 250-1500 meter ASL altitudes where a long spell of winter and a dry atmosphere is present for profuse blossoms and successful fruiting.

Root: Tap, branched.

Stem: Aerial, erect, branched, solid, woody, cylinderical, unarmed or armed (spinescent), glabrous, young stem pubescent.

Leaves: Simple, ovate or ovate-lanceolate, serrate, alternate, more or less pubescent beneath along the nerves, petioles shorter than greatest breadth of leaf, stipules linear, fimbriate, unicostate, reticulate venation.

Inflorescence: Solitary or fasiculate from lateral often leaf bearing buds.

Flower: Bisexual, pedicellate, pedicel slender, 3-4 times longer than calyx, actinomorphic, regular, complete, perigynous, ebractate, thalmus cupshaped, pentamerous.

Calyx: Sepals 5, gamosepalous, sepaloid (green), aestivation imbricate, deciduous in fruits, calyx tube bell shaped, 5-lobed and inferior.

Corolla: Petals 5, polypetalous, large, arranged on hypanthium, conspicuous, pink, inferior.

Androecium: Stamens many, perigynous, polyandrous, inserted at the rim of calyx tube (hypanthium), filaments equal, anthers bithecous, dorsifixed, introse, dehiscing longitudinally.

Gynoecium: Monocarpellary, ovary superior but situated within the hypanthium, unilocular, stigma capitate.

Fruit: Drupe, smooth, globose or oblong, pericarp pink-red, edible, fleshy, single seeded, enclosing an indehiscent or two valved furrowed stone.

Flowering: Jan. – Feb.

Floral formula: $\oplus$ ⚥ $K_{(5)}\ C_5\ A_{\alpha}\ G_{1}$.

Uses: Fruits are edible, delicious and vitaminized when fresh. Fruits used as dessert, cooked, canned and made into jams. Dried fruits are demulcent, laxative and refrigerant (Sinha 1996). This is one of the secondary /tertiary host plant of tropical tasar silkworm *Antheraea mylitta* D. (Nakvi and Srivastava 2003).

Kernels yield oil for lubricating fine machinery with light mineral lubricating oil. Bark contains tannins and yields a gum used in confectionary. Wood used for papier mache boxes, suitable for cabinet work, inlay and tunery (Ambasta *et al.* 1986).

Cultivation: Through seeds, seedlings and cuttings.

115. AMROOD

Botanical: ***Psiduim guajava*** Linn.

English: Guava
Hindi: Amrood, Amritphal
Marathi: Jamba
Bengali: Peyara
Gujarati: Sadaphal, Peru
Kannada: Jamphala
Malayalam: Pera, Koyya
Manipuri: Pungdol
Oriya: Pijuli
Sanskrit: Madhuphal
Tamil: Koyya
Telugu: Goyya

SYSTEMATIC POSITION

Bentham & Hooker (1883)	**Engler & Prantl (1899)**	**Hutchinson (1959)**
Phanerogams	Phanerogams	Angiospermae
Dicotyledones	Dicotyledoneae	Dicotyledones
Polypealae	Archichlamydeae	Lignosae
Disciflorae	Myrtiflorae	Myrtales
Myrtales	Myrtaceae	Myrtaceae
Myrtaceae		

Geographical distribution: Indigenous in tropical America, cultivated in most tropical countries and naturalized in India (Brandis 1906, Kocchar 1981).

Mexico to Columbia, Peru and Brazil region has been regarded as centers of origin of Guava. Portugese introduced it in India. Now, it is fourth most important fruit of India, both from the point of view of production as well as the area of cultivation (about 50,000 ha). Uttar Pradesh has the largest area (about 10,000 ha) under guava cultivation. Allahabad (U.P) and Rajahmundry (A.P) have the finest guava with a delicious taste and a luscious soft pulp (Rao 1991).

Habit: Medium sized tree with whitish bark attaining a height of 3.0-4.5 meters.

Habitat: It grows well in plains, hilly and slightly tough soil. It is susceptible to water logging and alkalinity. It flourishes well in yellow fertile soils brought by rivers in the plains.

Root: Tap, branched.

Stem: Aerial, erect, woody, branched, cylinderical, solid, glabrous, white or brown.

Leaves: Simple, oblong-elliptic, 10-15 cm long, cauline, alternate, exstipulate, petiolate, petiole 0.3-0.6 cm long, gland dotted, aromatic, entire, glaucous, light green, neatly glaborous above pubescent beneath, apex ovate, unicostate reticulate venation, secondary nerves 15-20 pairs, prominent beneath.

Inflorescence: Cymose, solitary axillary, peduncles 2.5-4.00 m long.

Flower: Pedicellate, bracteate, complete, hormaphrodite, actinomorphic, hypogynous, white, 4.0 cm across, mildly fragrant.

Calyx: Sepals 5, gamosepalous, reduced, forming a cap like structure on the corolla, superior.

Corolla: Petals 5, gamopetalous, forming a corolla cap covered by a calyx cap, the formed calyptra falls off when flower opens, superior.

Androecium: Stamens indefinite, polyandrous, attached on the rim of calyx cap, folded inwards in bud conditions, anthers dorsifixed, versatile, dithecous, small, dehisce longitudinally.

Gynoecium: Carpels 5, syncarpous, ovary inferior, pentalocular, stigma minute, terminal, placentation axile.

Fruit: Bery, edible, sweet, light green to yellow skin with pink or cream coloured flesh.

Flowering: Rainy season (June – July).

Fruiting: Sept. and sometimes up to Dec.

Floral formula: $\oplus$ ⚥ $K_{(5)}\ C_{(5)}\ A_{\alpha}\ G_{(\bar{5})}$

Varieties: "Chittidar", "Dharwar", "Harija", "Karela", "Lucknow 49", "Nasik" and "Safeda" are popular. Other promising varieties are "Allahabad" "Apple

Guava", "Banarasi", "Bangalora", "Guinea guava", "Seedless", "Red flesh" and the like with or without seeds and either cream coloured or pinkish flesh.

Uses: Fruits with delicious taste and a luscious soft pulp are eaten. Jellies and Jams are made from guava pulp. The fruit is an excellent source of vitamins C and A, besides being rich in minerals like calcium and phosporus (Rao 1991).

Medicinally, young leaves with salt are given in dysentery while decoction of bark and leaves is prescribed in skin diseases. Tender leaves are used for treatment of mouth sores, expulsion of intestinal worms and for stopping bleeding from wounds and cuts. Fruit is rich in vitamin-B complex also, besides being digestive, tonic and nutritive. The leaf –tannin and aromatic essential oil are antibacterial. The infusion of leaf is astringent and given in cholera to check vomiting. Shoot and leaf extract is febrifuge, antispasmodic and used in nephritis, epilepsy, local swellings and rheumatism. Bark extract is tonic. Flowers are cooling and given in bronchitis and infantile diarrhoea (Sinha 1996).

This is utilized as one of the host plants of wild silkworm *Attacus atlas* L. in Indonesia for commercial production of the non-mulberry elegant wild silk (Mishra 2002). The plant yields fodder and fuel also (Chundawat and Gautam 1993).

Wood used for engraving, lac turney and spear-handles.

Cultivation: Guava is cultivated either on commercial scale or in backyards of house. With little care, it grows up to 3.0-4.5 meter within three – four years from seedlings and start flowering and fruiting but the fruits are small, much seeded and less fleshy. On the contrary, the saplings raised from cuttings start flowering and fruiting within two years and fruits are more tasty. Hence, they should be propagated and cultivated through stem cuttings from commercial point of view. (Chaudhary 1997). The seeds are collected during Dec.- Jan., nursery is raised during March – April and seedlings are transplanted during June-July (Chundawat and Gautam 1993).

116. BIJA SAL

Botanical: *Pterocarpus marsupium* Roxb.

Eng: Indian Keeno tree — Hindi: Bija Sal, Sihora, Pia Sal

Bengali: Pitshal — Gujarati: Bibla — Gujer: Bio

Kannada: Honne, Valle hone — Kol: Bio

Malayalam: Venga, Keeno — Marathi: Honne, Dhorbenla

Oriya: Byasa Tamil: Vengai Telugu: Pedegu

SYSTEMATIC POSITION

Bentham & Hooker (1883)	Engler & Prantl (1899)	Hutchinson (1959)
Phanerogams	Phanerogams	Angiospermae
Dicotyledones	Dicotyledoneae	Dicotyledones
Polypetalae	Archichlamydeae	Lignosae
Calyciflorae	Rosales	Leguminales
Rosales	Leguminosae	Papilionaceae
Leguminosae	Papilionaceae	
Papilionatae		

Geographical distribution: Sub-Himalayan tract in Avadh, Gorakhpur and South-east corner of Kumaon (Bhabar) forests, Central Indian regions of Madhya Pradesh, Chattisgarh, Orissa, Maharashtra, Andhra Pradesh, Karnataka forests and Mount Abu in the Northern most point on the west side. Also found in Sri Lanka.

Habit: A large deciduous tree up to 200 ft. height with spreading branches and brown heartwood with dark streaks. Produces a straight clean bole up to 8 ft. girth under favourable climatic conditions.

Habitat: Warm tropical to sub-tropical conditions with high to moderate rainfall.

Root: Branched tap root.

Stem: Aerial, erect, woody, solid, hard, branched, branches spreading.

Leaves: Compound, alternate leaves with softly adpressed hairs when young, dark green, shining, leaflets 5-7, elliptic, obutse, coriaceous, emarginate, glabrous on maturity, sometimes shortly acuminate, exstipulate. Secondary nerves 15-20 pairs with prominent reticulate veins joining the intermediate nerves.

Inflorescence: Panicled racemes (terminal).

Flower: Hemaphrodite, zygomorphic, complete, regular, cyclic, hypogynous, yellow or white with pedicels much shorter than calyx. Calyx, pedicels and peduncles clothed with dark brown hairs.

Calyx: Sepals 5, gamosepalous, teeth unequal, turbinate, often curved, aestivation ascending imbricate.

Corolla: Petals 5, unequal, exserted with long claws, divided into standard/ vexillum, the two free lateral wings and two united petals as keel enclosing stamens and pistil, venation conspicuous and aestivation descending imbricate.

Androecium: Stamens 10, united in a single sheath (monoadelphous), sheeth deeply bifid, anthers bicelled, dorsifixed, dehiscing longitudinally.

Gynoecium: Carpel one, free, ovary superior, unilocular, style bent and short, stigma simple, placentation marginal.

Fruit: Pod, nearly orbicular, 4-5 cm across, usually two sided.

Flowering: July-Oct., sometimes earlier or later.

Fruiting: Nov - Jan.

Floral formula: ⊕ ⚥ $K_{(5)}\ C_{1+2+(2)}\ A_{10}\ G_{\underline{1}}$

Uses: It serves as one of the tertiary but recognised as one of the secondary food plants of tropical tasar silkworm, *Antheraea mylitta* D. in Central India forests especially for Andhra local, Bhandara and other local ecoraces. (Jolly *et al*. 1979)

Bark is filled with sacs containing a red astringent medicinal oily gum which is extracted by incisions in the bark and sold as East Indian Kino (Brandis 1906). Kino used in diarrhoea and dysentery. Bark used for dyeing. Paste of leaves cure skin diseases and also used for external application to pains. Bark and wood decoction is used in treatment of diabetes (Singh 1990).

The wood is very hard, close grained, heartwood yellowish - brown, durable and much used for building, agricultural implements, carts, wheelwork, boats etc (Sagreiya 1967). Also used in railway carriages, wagons, ships poles, drums, camp furniture, frames, mathematical instruments and combs (Ambasta *et al*. 1986).

Cultivation: It may be cultivated either through seeds or seedings raised in nursery. Collected dried seeds may be stored in gunny bags after mixing with BHC 10% powder. According to Regional Forest Research Centre, Rajahmundry, alternate wetting and drying of seeds for seven days gives 23% peak germination and 43% overall germination (Rao *et al*. 1998). Seedlings require 20-30% culling in secondary stage (root trainer plants) to have uniformly good planting stock in case of seedlings raised from the seeds collected from wild source.

117. DINGRITTIANG

Botanical: *Qurecus acutissima* Carr. Syn. *Q. serrata* auct. non Thunb.

Manipuri: Uyung

SYSTEMATIC POSITION

Bentham & Hooker (1883)	Engler & Prantl (1899)	Hutchinson (1959)
Phanerogams	Phanerogams	Angiospermae
Dicotyledones	Dicotyledoneae	Dicotyledones
Monochlamydeae	Archichlamydeae	Lignosae
(or incompletae)	Fagales	Fagales
Unisexuales	Fagaceae	Fagaceae
Cupuliferae		
Quercineae		

Geographical distribution: From Nepal, Sikkim and Bhutan to Arunachal Pradesh between 3000-8000 ft. in the khasi hills (Meghalaya), Manipur, Mizoram, Nagaland. Also found in Myanmar, S.W. China, Indo-China, Taiwan, Korea and Japan. It was introduced in Himachal Pradesh during 1985-86 and considerable variation of growth rate was observed between different locations. Low altitude area (790-1140 m ASL) was found most suitable for the recruitment of *Q. acutissima* in Himachal Pradesh. It has also been introduced in Garhwal and Kumaon region of Uttarakhand since 1995 for large-scale plantation purposes through NGOs.

Habit: A medium or large sized tree reaching up to 250 ft. height and acquiring a girth of 5 ft., deciduous, crown rounded but straight when grown closely (Fig. 50).

Habitat: Subtropical wet or moist habitat, hill slopes or low grounds without water logging.

Root: Branched, tap root system.

Stem: Erect, aerial woody, solid, branched, young shoots softly pubescent but glabrous on maturing, bark dark grey, 1.8-2.0 cm thick splitting horizontally as well as vertically, brown and granular on the inner side, annual rings conspicuous by numberous large pores in the spring wood.

Leaves: Simple, lanceolate, oblong-lanceolate or elliptic - lanceolate sometimes gradually obovate or spatulate, apex acute or acuminate, serrate, glabrous, shining on maturity, thinly coriaceous, young leaves clothed with soft hairs, petiole 2.5-4.0 cm long, slender, secondary nerves 10-20 pairs, prominent, parallel, each nerve terminating in a sharp serrature and forming the setae up to 0.6 cm long, base obtuse, stipules oblong.

Inflorescence: Spike, male and female spikes separate.

Male spike: Elongated, lax, drooping, rachis tomentose, bracteoles ovate-lanceolate, pubescent, much larger than the perianth.

Female spike: Short, axillary, erect, each flower is found in an involucre of numerous bracts or scales which coalesce in fruit and form a cup or a bag enclosing the nut and are often connate into concentric belts.

Flower: Unisexual, incomplete, monoecious, epigynous.

Male flower: Perianth 5-6 lobed, perianth connate at base, stamens twice the number of segments, glabrous, carpels nil.

Female flower: Perianth as in male flowers, ovary inferior, trilocular, styles 3, filiform, stigma capitate.

Fruit: Acorns, solitary or in pairs, more or less sessile found on last years shoot; cupule woody, hard, embracing 1/3 to 3/4 of the ripe glans and whole of the young glans; scales numerous, ovate - lanceolate, spreading, imbricate, hoary, serrate vertical ridges much reflexed out side the lower rows; glans globose / sub-globose / ovoid/ ellipsoid / oblong, apiculate, bitter, smooth on ripening 1.5-2.5 cm in diameter.

Leaf fall: Nov. - Jan.

Leaf sprouting: Feb. - March.

Flowering: The male catkins appear during Oct.-Nov. at the ends of leafless branches and attain a length of 2.5 cm and elongate rapidly up to 10-15 cm during Feb. The female flowers appear on new shoots in February.

Pollination: February - March.

Fruiting: The acorns hardly attain 0.3 cm in diameter and brown in colour through the ensuing winter. During second year they start increasing size in March and ripe and fall during Sept.-Oct. Thus total period between pollination to ripening is about 18 months.

Floral formulae: Male flower: Br. ⊕ ♂ $P_{(4\text{-}6)}$ $A_{4\text{-}6}$ G_0

Female flower: Br. ⊕ ♀ $P_{(4\text{-}6)}$ A_0 $G_{(\bar{3})}$

Varieties: Based on foliar morphology, lanceo-linear, lanceolata, elliptica, oblonga, obovate, oblanceolata and spatulata, varieties may be earmarked whose rearing performance is under trial (Srivastav and Singh 1997a, 1999a, 2002, Srivastav *et al.* 2000 c, d, 2002 a).

Uses: It serves as one of the primary food plants of temperate (oak) tasar silkworms viz., *Antheraea proylei*, *Antheraea pernyi* G.M. and *A. roylei* Moore in North East India and *Antheraea pernyi* G.M. in China and *A. yamamai* G.M. in Japan (FAO 1976, Mishra *et al.* 2000, Srivastav and Singh 2002). The tannin contents show a 3.5 fold increase over a period of 45 days. Hence, brusing of larvae should be made within 45-47 days of leaf sprouting. The timber is used for building purposes and as fuel (Negi and Naithani 1995).

Cultivation: *Q. actussima* may be propagated through seeds, seedlings, air layerings (Srivastav and Singh 1997b, 1999a, Srivastav *et al.* 2000 b, d, 2002 c) and even cuttings.

(a) **Seedling Propagation**: The ideal period for collection of seeds of *Q. acutissima* in North Eastern India from naturally grown trees is between Oct.-Nov. and Sept.-Oct. in North-western region (Kaunli, Dehradun) from introduced trees. (Srivastav and Singh 1997 b, Negi and Naithani 1995)

The locally collected seeds could be stored for a year in 90-cm. sq. pit with the rammed earth. Such stored seed was found in advanced stage of germination after one year in Dehradun.

If the seeds have to be stored for shorter duration between Oct.-Nov. to Dec.-March for 3-5 months they may be sprayed with the solution of Nuvan @ 7 ml/litre of water to kill the maggots of insects infesting the acorns. Subsequently, they may be dusted with Dhanustin-50 powder @ 2 gm per kg. of seeds. The seeds may be stored in dry gunny bags in dry airy places.

The weight of the seeds normally ranges from 1.5-5.0 gms. On an average 30-35 kg. of seeds are required for plantation of one hectare of land at 4'x4' spacing. Seeds weighing above 3.0 gms may be selected for germination. The seeds for germination are selected by submerging them in water and rejecting the floating seeds.

The nursery may be raised between Feb.-May in nursery beds as well as in polythene tubes. To facilitate germination the seeds may be mechanically scarified. The scarified seeds are soaked in water for two-four days while unscarified seeds may be soaked for 4 days. The imbibed seeds are spread densely in nursery beds and covered by gunny bags followed by sprinkling of water for retaining moisture. Germination starts within 10-20 days and more than 90% germination occurs. The germinating seeds may either be allowed to grow in the nursery beds or dibbled in polythene tubes containing sandy loam soil and FYM in 2:1 ratio or sand, clayey soil and FYM in 1:2:3 ratio in horizontal posture. Regular watering, weeding and insect control results in quick growth and the seedlings acquire height of 9-12" within 90 days.

The seedings are transplanted in the pits of 1'x1'x1' at 4'x4' spacing during June-Aug. The pits for transplantation are dug during March-May for proper aeration. Before transplantation, the pits may be filled with 1.5-2.0 kg FYM and bottom of polythene is removed from tube. The polythene tube is completely removed in next year during various cultural operations.

Weed control may be resorted mechanically for three years and by spraying of Glyphosate @ 7 ml/litre of water after mechanical control when sprouting starts in grasses.

(b) Propagation through air layering: Recently vegetative propagation has been made successful by the author (Srivastav *et al.* 2000b,d, 2002c) for multiplication of true to type superior varieties of *Q. acutissima* Carr. through air layering which also reduces gestation period tremendously as even 5' long branches could also be successfully rooted and transplanted which can make clones ready for rearing within two to three years instead of 7-8 years as observed in case of seedling propagation. More than 75% success has been achieved in propagation through air layering by adopting following technology.

i. Pine saw dust is moistened overnight in 0.15% carbendazim solution.

ii. 250 ppm of IBA (Indole butyric acid) solution is made by dissolving 25 mg of IBA in 5 ml. of either ethyle or methyle alcohol and then by adding 95 ml. of distilled water which makes 100 ml of said solution.

iii. The 500 ppm of NAA (Napthalene acetic acid) is prepared by dissolving 50 mg. of NAA in 5 ml. of Ammonium hydroxide solution or ethyle alcohol followed by adding 95 ml. of distilled water which makes 100 ml of said solution.

iv. 2.0-2.5 cm ring of bark is removed at the bottom of each side branch. Side branches should be selected from inner plants rather than the marginal plants in oak plantations.

v. Cotton soaked in either IBA 250 or NAA 500 ppm solution is wrapped around this ring.

vi. Saw dust moistened overnight in 0.15% carbendazim solution is wrapped subsequently around this girdle/ring and this ball of sawdust is tightened by wrapping polythese sheets and by tieing its both ends by the twining thread.

vii. When ball is full of roots, air layers may be cut and kept in bucket filled with water and subsequently transplanted in 1'x 1'x 1' pits (without FYM) with the help of powdered soil and irrigated immediately. Regular watering is done to help air layers to establish in the field.

(c) Propagation through cuttings: Some success has been achieved in propagation of *Q. acutissima* through cuttings also (Srivastav *et al.* 2002c). NAA 250 ppm has been found to induce 16% rooting in 90 days old cuttings treated with aforesaid hormonal solution for 24 hours followed by IBA 250 ppm solution, which yielded 7-8% rooting in cuttings. Girdling of side branches for carbohydrate accumulation, their blanching for etiolation and spraying of 20 ppm GA3 to foliage for induction of juvenility should be carried out 15 days before cutting preparation to facilitate rooting. Each cutting should have 1-2 leaves. Sand disinfected with 0.15% carbendazim solution may be used as the rooting media in polythene sheets for treated cuttings. The cuttings may be transplanted by removing bottom of the polythene tube after emergence of roots, which are easily visible through polythene tubes. The cuttings planted in polythene tubes containing rooting media should be kept inside pits made one feet deep in the soil under shades and white polythene sheets should cover these pits. The spraying of water every day and spraying of 0.15% carbendazim solution in pits is resorted at five-six days interval regularly to maintain the mositure and control of fungal diseases.

Fertilization: The fertilizer application schedule to *Q.acutissima* plantations at 4'x4' spacing may be resorted as hereunder on the basis of age of plants.

(a) 1-2 year old plants: 10 gm urea + 10 gm Mussooriee phos/plants/year after onset of monsoon.

(b) 3-6 year old plants: 10 gm urea + 20 gm Mussoorie-phos/plant/year after onset of monsoon or

FYM-3 metric ton/ha (1/2 kg/plant), Azotobacter - 5 kg/ha *(1.5 g/plant), Azospirillum - 5 kg/ha ** (1.5 g/plant), Phosphatica - 10 kg/ha *** (3.0 g/plant) [*, **, *** Together constitute NFM 10 PSM 10 kg/ha/dose]

(c) 7 -year onwards: Three fertilizer doses may be applied in combination with FYM @ 7 metric tons/year/ha. (Srivastav and Singh 1997b, 1999b, Srivastav *et al*. 2000a)

i. N-150, P-50, K-38 kg/ha or

ii. N-75, P-25, K-38, NFM-20, PSM-20 kg/ha. or

iii. NFM-20, PSM-20 kg/ha.

Seventh year first dose, eight and ninth year second dose and tenth year third dose should be applied for sustainable utilisation of natural resources without affecting soil texture, structure and fertility for tasar culture.

In order to reduce the application of chemical fertilizers, curtailment of soil pollution and improvement of soil structure, combined application of chemical as well as biofertilizers - N75, P25, K38, NFM-20, PSM-20 kg/ha. dose has been found to yield more or less equal leaf yield (9777 kg/ha) and improved leaf quality as compared to N150, P50, K38 kg/ha dose (10422 kg/ha) at

Regional Tasar Research Station, Imphal. However, NFM20, PSM20 kg/ha dose has exhibited the best results in terms of less inputs and highest returns (12600 kg leaf yield/ha) every third year since urea, single super phosphate/ mussoorrie-phos etc. of preceding years rendered unavailable/ insoluble in soil are solublilized and made available by the biofertilizers (NFM = Nitrogen fixing microorganisms i.e. Azotobacter and Azospirillum and PSM = Phosphate - solubilizing microorganisms i.e. phosphatica) during the third year when applied without any chemical fertilizers. All the fertilizers and manures should be applied by ring method one and half months before rearing. After pruning of plants biofertilizers should be applied during first week of Feb. along with FYM while NPK should be applied after fifteen days. Irrigation should be ensured after application of biofertilizers as well as NPK for spring crop rearing during March-April (Srivastav *et al.* 1999b).

Pruning: In order to maintain the plants at manageable height, improve leaf quality and increase leaf yield *Q.acutissima* trees are subjected to pruning every year and heavy pollarding every third-fourth year during last week of November to first week of January for undertaking spring crop rearing (Singh and Tikkoo 1989, Ghosh and Srivastava 1996, Srivastav and Singh 1997b).

For second crop rearing the phase- wise pruning of trees in finger fashion is conducted at the interval of seven days during April-May and at the interval of 4 days during July-August for summer and autumn crop rearing. The length of branches (fingers) should be more than one foot for second crop rearing and six inch for spring crop. The plants should be pruned only once in a year irrespective of the crop. The unpruned *Q. acutissima* plants normally sprout during first week of March and the foliage becomes available for rearing by the end of March in low attitude and 10 days later in the middle attitude.

Pruning during last week of Nov. to first week of January prepones sprouting in the third week of Feb. and thus synchronises silkworm rearing with low disease mortality, less uzifly infestation and high quality foliage. On such plants suitable leaves for rering becomes available from first week of March.

Pruning should be avoided during peak monsoon periods and should be conducted 45 days before undertaking rearing in phasewise manner for second crop as aforesaid (Srivastav and Singh 1999b).

118. BRE - CHUR

Botanical: ***Quercus baloot*** Griff. Syn. *Q. ilex* non Linn.

Afghanistan: Balut.

English: Holly oak, Holm oak Vernaculars: Irri, Breh, Bre-chur

SYSTEMATIC POSITION

Bentham & Hooker (1883)	**Engler & Prantl (1899)**	**Hutchinson (1959)**
Phanerogams	Phanerogams	Angiospermae
Dicotyledones	Dicotyledoneae	Dicotyledones
Monochlamydeae	Archichlamydeae	Lignosae
(or incompletae)	Fagales	Fagales
Unisexuales	Fagaceae	Fagaceae
Cupuliferae		
Quercineae		

Geographical distribution: In India, it is found in inner arid tracts of the Himalaya from Sutlej valley westward to Afghanistan at 3000-8500 ft. ASL. In Kashmir, it is common in lower slopes of Kishanganga, Karna and Jagran at 1100-2100 m ASL. It also occurs in Jhelum, Chinab, Ravi and Sutlej valleys at 1200-2600 m ASL. In Himachal Pradesh, it occurs in Kanwar, Kilba, Kailash and Chinbi ranges up to 2100 m ASL. It also occurs in Afghanistan and Mediterannean regions of France, Spain, Portugal, Italy, Tyurol, Corsica, Sicily, Greece, Syria, Morrocco, Algeria, Tunia and England.

Habit: A small evergreen tree reaching up to 40 ft or usually a large shrub, branchlets, petioles and underside of leaves densely grey pubescent with stellate hairs.

Habitat: Dry temperate habitat of North-western Himalaya where precipitation ranges from 42-100 cm in snow form chifely during winter months. The tree in shallow and rocky regions tolerates the siliceous to calcareous soils.

Root: Branched tap root.

Stem: Erect, aerial, branched, woody, hard, young shoots pubescent, trunk attaining 7-8 ft. girth in old trees.

Leaves: Simple, elliptic, oblong or elliptic-oblong, entire or with large spinescent teeth, apex obtuse, acute or mucronate, coriaceous, base rounded or sub-cordate, margin recurved, blade 5.0-7.5 cm long, lower side densely tomentose or stellately pubescent, secondary nerves 5-8 pairs, not prominent, more conspicuous in serrate leaves, petioles 1-2 mm long, stout, pubescent.

Inflorescence: Spikes.

Male spikes: Fascicled, stellately pubescent, slender, interrupted, 5-7 cm long.

Female spikes: Short, flexuous, slender, erect, few flowered, 5 cm long.

Flower: Unisexual, monoecious, epigynous, incomplete, bracteate.

Male flower: Bracteate, bracts ovate-acute to obovate, pilose externally, perianth fused, obtusely 5-6 lobed, stamens 4-12, hairy, shortly apiculate.

Female flower: Perianth fused, obtusely 5-6 limbs, styles 3-5, spreading, clavate, ovary inferior.

Fruit: Acorns, 2-3 on a peduncle 2.5-5.0 cm long, nuts upto 3.5 cm long, 1.5 cm in diameter, almost entirely enclosed in the cup in beginning but one third to one fourth sunk in a thick woody cup on maturity, scales of cup hoary, closely appressed, inbricate, pale-greyish green.

Leaf duration: One and half year or less.

Leaf sprouting: March - April.

Flowering: March - April.

Fruiting: No growth occurs in acorns until April-May of next season, afterwards they develop fastly and ripen during October. Thus, nearly one and half year elapses from the time of pollination to ripening of the acorn.

Floral formulae: Male flower: Br. ⊕ ♂ $P_{(5\text{-}6)}\ A_{4\text{-}12}\ G_0$

Female flower: Br. ⊕ ♀ $P_{(5\text{-}6)}\ A_0\ G_{(\bar{3})}$

Varieties: Two varieties consisting of serrated and entire leaf have been found to occur in nature.

Uses: It serves as one of the secondary food plants of oak tasar silkworms, *Antheraea proylei* J., *A. pernyi, A. roylei* and *A. frithii* etc. in North-Western Himalayan region. (Seth 1995b, Mishra *et al.* 2000)

The wood is locally used for agricultural implements, joinery and tool handles.

The leaves and branches without prickles are used as fodder. The prickly branches are used for fencing.

The wood yields good firewood and charcoal. The bark is used for tanning and the acorns for feeding pigs as they are edible and yield edible oil also (Streets 1962, Gamble 1922, Troup 1921).

Cultivation: It may be cultivated through seeds and seedlings and grows well on dry, stony or rocky ground and hot slopes due to its xerophytic habit and production of root suckers. It coppices well and is resistant to both drought as well as frost (Negi and Naithani 1995).

119. MORU

Botanical: ***Quercus floribunda*** Lindl. Ex A. Camus Syn. *Q. dilalata* Wall. Ex A. DC., *Q. himalayana* Bahadur

Eng: Green oak, Holly oak Hindi: Moru, Punj

Garhwali: Tilonj Haz. : Barungi,

Other Vernaculars: Chora, Kali-ring, Kilonj

SYSTEMATIC POSITION

Bentham & Hooker (1883)	Engler & Prantl (1899)	Hutchinson (1959)
Phanerogams	Phanerogams	Angiospermae
Dicotyledones	Dicotyledoneae	Dicotyledones
Monochlamydeae	Archichlamydeae	Lignosae
(or incompletae)	Fagales	Fagales
Unisexuales	Fagaceae	Fagaceae
Cupuliferae		
Quercineae		

Geographical distribution: Found in North-western Himalayan region from Nepal and Kumaon westwards chiefly between 5000-9000 ft. ASL forming a well-defined zone intermediate in altitude between the zone of *Q. semecarpifolia* above and that of *Q. leucotrichophora* (*Q. incana*) below. Common in Kumaon hills, Tehri Garhwal, Mussoorie (Uttarakhand) from Jaunsar to Kulu (Himachal Pradesh) and Kishanganga valley, the lower Jhelum valley, Keran, Kishtwar, Marwa Dachhan, Ramban and Udhampur (Kashmir). It extends through West Pakistan into Afghanistan.

Habit: Largest evergreen shade-enduring oak tree attaining up to 140 ft. height and 20 ft. girth.

Habitat: Temperate regions having fairly heavy winter snow fall under –7 to 32°C temperature and normally 110-240 cm rainfall. It avoids dry situations and moist, cool localities with deep, rich, well-drained soil are highly favourable for its best development.

Root: Deeply penetrating branched tap root.

Stem: Erect, aerial, branched, woody, solid, bark dark grey or black peeling of in longitudinal scales, youngest shoots almost glabrous.

Leaves: Simple, glabrous, shining, coriaceous, entire or with sharp spinescent teeth, blade 4-9 x 1.5-4 cm, oblong or oblong-lanceolate, accuminate, petiole 0.6-1.5 cm, lateral nerves 8-12 pairs, slender, forked and reticulate, base obtuse or cordate.

Inflorescence: Spike or catkins.

Male spike: Facicled, lax, drooping, 4-8 cm long, bracts lanceolate or ovate – acute.

Female spike: Short, few flowered.

Flower: Unisexual monoecious, epigynous, incomplete, bracteate, almost sessile.

Male flower: Perianth obtusely six cleft, hairy outside, stamens 4-8, anthers glabrous, apiculate.

Female flower: Perianth obtusely six cleft, styles 3-5, linear clavate, ovary inferior, syncarpous.

Fruit: Acorns, mostly soliary, almost sessile, 1.2-1.5 cm across, 1.7-3 cm long, ovoid or oblong, glabrous, brown on ripening, found on last years shoots, lower half enclosed in the cup, scales of cup closely appressed.

Leaf fall: May – June (some may fall earlier and some may later).

Sprouting: April – May.

Flowering: April - May.

Fruiting: Cold season (16-18 months after pollination). Ripening starts in August till middle of October.

Floral formulae: Male flower: Br. ⊕ ♂ $P_{(6)}$ $A_{4\text{-}8}$ G_0

Female flower: Br. ⊕ ♀ $P_{(6)}$ A_0 $G_{(\overline{3-5})}$

Uses: It serves as one of the primary food plants of temperate tasar silkworm *Antheraea proylei, A. pernyi* and *A. roylei.* (FAO 1976, Mishra *et al.* 2000, Srivastav and Singh 2002).

The wood is used for agricultural implements and constructional purposes. It suits well as tool handles, axe handles, sledge runners, walking sticks and carrying poles. It is found highly suitable for making strong and heavy wheel spokes, cask and barrel making. Leaves serve as good fodder and wood is also used as firewood and for charcoal (Pearson and Brown 1932, Gamble 1922). The leaves contain 9.85% and bark contain 4.99% tannin contents and are used for tanning locally (Negi and Naithani 1995)

Cultivation: It can be successfully used for afforestation at an elevation of 6000-7000 ft. ASL in Uttarakhand both by sowing of seeds as well as seedling transplantation. It prefers sheltered and hollow conditions and demands moderate light only. Seeds are collected during Aug-Sept. and sown immediately, which gives germination of 90%. For direct sowing, patches are prepared at the spacing of 1.5 x 1.5 m or 1.8 x 3.0 m and seeds are dibbled. In contour terraces, between contours spacings may be 60-90 cm and 1.8-3.0 m between plants to plants. Seed weight classes of 1.0-2.5 gm and 2.5 gm and above give significantly better germination and survival percentage, seedling growth as well as seedling dry weights than the seed weight classes upto 1.0 gm. Hence, seeds below 1.0 gm should not be used for nursery raising (Singh 1998).

Nursery is raised by sowing seeds in drills and seedlings are allowed to grow for two years there. Subsequently, they are transplanted during rainy season in pits prepared at 1.5 x 1.5 m or 1.8 x 1.8 m spacing. Gap filling is attended during next year. Transplantation during winter is not successful. Natural seedlings also yield good results. Transplantation and sowing on hot dry slopes results in failure but complete success is achieved on moist soil in cool situations (Troup 1921).

120. BANNI

Botanical: *Quercus glauca* Thumb. Syn. *Q. annulata* Smith, *Q. glauca* var. *annulata* (Sm.) A. camus, *Q. phullata* Buck-Ham.ex D.Don

English: Blue Japanese Oak | Hindi: Banni | Nepali: Phalat

Haz: Barin | Lepcha: Siri | Kumaon: Phanat

Punjabi: Bran, Banni

Other Vernaculars: Baran, Banku, Dieng-rih, Inai, Diengal-Sari-War, Imbri, Dieng-Syrtap, Musre-Phalant, Phaniant, Pharonj Siri.

SYSTEMATIC POSITION

Bentham & Hooker (1883)	Engler & Prantl (1899)	Hutchinson (1959)
Phanerogams	Phanerogams	Angiospermae
Dicotyledones	Dicotyledoneae	Dicotyledones
Monochlamydeae	Archichlamydeae	Lignosae
(or incompletae)	Fagales	Fagales
Unisexuales	Fagaceae	Fagaceae
Cupuliferae		
Quercineae		

Geographical distribution: In outer Himalayan valleys ascending to 6000 ft. from 1800 ft. ASL in between Kashmir to Arunachal Pradesh and in Meghalaya

(Khasi hills) between 3600-4500 ft. ASL. In Kumaon hills it grows below Ranikhet in Uttarakhand and between Baijnath and Dhali in Himachal Pradesh. Also found in Manipur and Nagaland in India. Also found in China.

Habit: Middle sized or large, beautiful, evergreen tree with dense, rounded crown and a cylinderical, clean bole. In Western Himalaya it usually attains height up to 15-18 m and girth up to 1.5-1.8 m. In Kumaon it may attain a height of 26 m and girth of 4 m under favourable conditions. It coppices freely.

Habitat: Typically found in moist sheltered conditions in rich soils especially in valleys, on the sides of ravines and along streams. It does not grow well in dry conditions.

Root: Branched tap root deeply penetrating.

Stem: Erect, aerial, woody, hard, branched, young shoots sparsely pilose, speedily glabrous and afterwards lenticillate, bark smooth, grey up to pole stage and subsequently fissured and rough.

Leaves: Simple, thinly coriaceous, glaucous and minutely pubescent beneath with appressed hairs when young, afterwards glaucous, dorsal side and petiole glabrous and shining, blade 7-15 cm long, oblong or ovate-lanceolate, acuminate or cuspidate, upper half sharply serrate, teeth in-curved, base entire and unequal, secondary nerves 10-15 pairs, straight, parallel excurrent in serratures, tertiary nerves slender, parallel, base acute, petiole slender, 1.2-2.5 cm long.

Inflorescence: Spikes / catkins.

Male spikes: Solitary or in small lax panicles, hairy, axillary, drooping and shorter than the leaves.

Female spikes: Solitary, axillary, 2-4 flowered, shorter than the leaves.

Flower: Unisexual, monoecious, epigynous, incomplete, bracteate.

Male flower: Perianth segments 5, lanceolate, stamens 4-5, anthers broad and glabrous, bracts lanceolate, hairy and much lower than the flowers.

Female flower: Perianth segments 5, lanceolate, Carpels-3, ovary inferior, styles separate, syncarpous, bracts long, early deciduous.

Fruit: Nut, sessile, apiculate, shining, Ovoid or sub-globose, 1.5-1.8 cm long, half buried in the hemispherical cup, cup 1.6-2.0 cm in diameter consisting of many thin, silky, crenate broad truncate concentric teeth. Spikes usually only single fruited and on last years shoot.

Leaf sprouting: March – April.

Leaf shedding: June.

Flowering: March - June.

Fruiting: 15-18 months after flowering.

Floral formulae: Male flower: Br. ⊕ ♂ $P_{(5)}$ A_{4+5} G_0

Female flower: Br. ⊕ ♀ $P_{(5)}$ A_0 $G_{(\bar{3})}$

Uses: It serves as one of the secondary/tertiary food plants of oak tasar silkworms viz, A. pernyi, *A. proylei*, *A.roylei* and *A. frithii* in jungles (Seth 1995 b, Mishra *et al.* 2000).

The wood of *Q. glauca* is used in the same fashion as that of *Q. floribunda* especially for soles for sledges, bridges and houses in Tehri Garhwal area in Uttarakhand (Gamble 1922). The tree is lopped for fodder and wood is also used as firewood and for charcoal. The bark contains as much as 12% tannins (Negi and Naithani 1995).

Cultivation: It may be cultivated through either seeds or seedlings. Seeds may be stored as long as for eight months but if exposed to sun crack and become infertile, hence seeds should be buried or become buried soon after falling or lie under shade on very moist ground.

Germination starts early in rainy season i.e., July but seeds may germinate in May in moist places. The germination percentage varies between 53-90. One kg seeds contain 570-770 nuts (Negi and Naithani 1995).

The best season for sowing directly in the field as well as in nurseries is early spring. Transplantation of seedlings may be carried out in the rainy season of next year after sowing and moist ravines and identical places should be selected for transplantation of seedlings. Sound seeds give high germination (Troup 1921 Sengupta 1937).

121. CHAKPA-UYUNG

Botanical: ***Quercus grifithii*** Hook. & Thom. Ex. Miq.

Burmese: Thitcha, Metlein — Other Vernaculars: Dieng-wah, Dingim

Khasi: Dingim — Manipuri: Chakpauyung,

SYSTEMATIC POSITION

Bentham & Hooker (1883)	Engler & Prantl (1899)	Hutchinson (1959)
Phanerogams	Phanerogams	Angiospermae
Dicotyledones	Dicotyledoneae	Dicotyledones
Monochlamydeae	Archichlamydeae	Lignosae
(Or incompletae)	Fagales	Fagales

Unisexuales	Fagaceae	Fagaceae
Quercineae		
Cupuliferae		

Geographical distribution: Occurs in Himalayan region from Sikkim and Bhutan to Arunachal Pradesh and in Manipur, Nagaland, Meghalaya and Mizoram from 2700 – 5400 ft. ASL. It has also been introduced in Darjeeling and Nainital (Bhimtal) areas of West Bengal and Uttarakhand respectively. Also occurs in Myanmar and probably in China and Japan as well.

Habit: A large, deciduous, gregarious tree, reaching up to 50 ft. height, young parts usually rusty pubescent (Fig. 51).

Habitat: Sub-temperate to sub-tropical hilly regions with high humidity, low temperature and moderate light. It regenerates freely in gaps and open spaces.

Root: Branched tap root.

Stem: Aerial, erect, branched, woody, hard, bark silvery grey, warty, deeply furrowed, branchlets pale often rusty tomentose when young.

Leaves: Simple, coriaceous, underside usually rusty pubescent when young but yellowish-grey pubescent when mature, glabrous above when mature, obovate, obovate-oblong or oblanceolate, blade 10-18 x 5-12 cm, acute or subacuminate, more or less coarsely dentate, usually entire towards the base, petiole 0.4-1.2 cm, stout, hairy, secondary nerves 12-16 pairs prominent, parallel, nearly straight, terminating in a denticulation, base sub-cordate, stipules fugacious, hairy.

Inflorescence: Short spikes.

Male spikes: 5-12 cm long, yellowish, fascicled at the base of young shoots, sub-pendulous, glabrous.

Female spikes: Clustered at the tips, few flowered.

Flower: Unsexual, monoecious, actinomorphic, incomplete, epigynous.

Male flowers: Perianth 5-6 lobed, fused, hairy, stamens 8-12, anthers hairy, basifixed.

Female flowers: Perianth 5-6 lobed, carpels 3-5, syncarpous, ovary inferior, stigma capitate, glabrous.

Fruit: Acorn, solitary or in short spikes, almost sessile in clusters of 2-5 at the end of branchlets, cup hemispherical, 1.2 cm in diameter and 0.7-1 cm deep, slightly sericeous, smooth inside, scales closely appressed, ovate, acute, nut 1.2-1.5 cm long, smooth, ovoid or elliptic, almost cylinderic.

Flowering: Oct. – March. In Arunachal Pradesh the male flowers are seen in June and young fruits appear in July.

Floral formulae: Male flowers: Br. ⊕ ♂ $P_{(5\text{-}6)}$ $A_{8\text{-}12}$ G_0

Female flowers: Br. ⊕ ♀ $P_{(5\text{-}6)}$ A_0 $G_{(\overline{3\text{-}5})}$

Varieties: Two varieties are clearly discernible within this species (Brandis 1906).

1. **Var. typica** – Leaves obovate or oblanceolate, acute or sub-acuminate, more or less coarsely dentate.
2. **Var. oblonga** – King.mss. Leaves oblong or oblong-obovate, shortly acuminate, subentire, concolorous, and puberulous or glabrate beneath. It is a very constant form of Khasia Mountains in Meghalaya.

Like foliar variations, the variations at fruit levels also exist in *Q. grifithii*. Further, as quotd by Hooker (1888), King thinks that *Q. grifithii* Hook. F. & Thomas and the Mangolian and Japanese *Q. aliena* Bl. and *Q. grosseserrata* Bl. are probably one species. Identical views had been expressed by Brandis (1906) also.

Uses: It serves as one of the primary food plants of temperate oak tasar silkworms *A. proylei, A. roylei, A. frithii* and *A. pernyi* in North–Eastern India. *A. pernyi* in China and *A.yamamai* in Japan (FAO 1976, Srivastav and Singh 2002). Its timber is used for building, agricultural implements and other purposes locally in North-Eastern India. The wood is also used as fuel and bark contains 5-10% tannins (Gamble 1922, Kanjilal *et al.* 1940). According to Shukla and Kukreti (1983), it can be used for construction work, poles and ballies, tool handles, anvil blocks, mallet heads and agricultural implements.

Cultivation: It may be cultivated through seeds, seedlings or air layerings. The seeds weigh about 620 to a kg. and exhibit germination percentage between 49-94. The germination starts after 9 days of sowing and continues up to 160 days (Sengupta 1937). The package of practices for propagation and cultivation are more or less identical to those of *Q. acutissina* Carr.

122. PHALANT

Botanical: ***Quercus kamroopi*** D. Don. Syn. *Q. lineata* Bl. var. *thomsoniana* (A.DC.) Wenzig, *Q. lineata* (non Bl.), *Q. thomsoniana* A.Dc.

Vernaculars: Phalat, Phalant, Sri-kung, Dieng-lingim

Lepcha: Siri

SYSTEMATIC POSITION

Bentham & Hooker (1883)	Engler & Prantl (1899)	Hutchinson (1959)
Phanerogams	Phanerogams	Angiospermae
Dicotyledones	Dicotyledoneae	Dicotyledones
Monochlamydeae	Archichlamydeae	Lignosae
(or incompletae)	Fagales	Fagales
Unisexuales	Fagaceae	Fagaceae
Cupuliferae		
Quercineae		

Geographical distribution: Eastern Himalaya from Nepal eastwards between 4500-8400 ft. ASL and in Meghalaya and Nagaland between 2700-5400 ft. ASL in Khasi and Naga hills respectively. Common near Darjeeling between 6400-7500 ft. ASL. Also found in Myanmar and Indo-Malayan Archipelago.

Habit: A small or middle-sized tree, crown spreading widely and young parts brown-tomentose.

Habitat: Temperate wet as well as temperate moist forests.

Root: Branched tap root.

Stem: Erect, aerial, cylinderical, woody, hard, branched, branches spreading, bark dark grey or dark brown, peeling off in rectangular flakes, young branches rusty tomentose.

Leaves: Simple, elongate – elliptic, oblong – elliptic or obovate – elliptic, acute or acuminate, brownish hairy along midrib or nerves and glaucous beneath, glabrous and shining above, secondary nerves 13-17 pairs, amastomising under the margin or ending in a spinescent serratures, base obtuse, blade 6-12 x 3.5-4.0 cm, petiole 1.0-1.5 cm long.

Inflorescence: Unisexual panicled spikes.

Male spikes: 2-6 cm long in small fasicles, sliky brown tomentose, bracts prominent.

Female spikes: Solitary, terminal, few flowered.

Flower: Incomplete, unisexual, monoecious, regular, cyclic, epigynous.

Male flower: Perianth segments-4, filaments long, stamens 4, anthers broad.

Female flower: Perianth segments-4, carpels 3-4, syncarpous, ovary inferior, stigmas sub-capitate.

Fruit: 2-6 acorns or nuts on 5-12 cm long spikes, young cupules sessile, obovoid, the apex truncate, densely fulvous-tomentose, margins of scales denticulate or crenate, cup encloses base of the nut, ripe cupules cup shaped or saucer shaped, 1.5-3.0 cm across, nut ovoid-conic, ripe glans hemispheric, depressed, apiculate, shining and smooth.

Flowering: April - May.

Fruiting: Sept. – Dec.

Floral formulae: Male flower: Br. ⊕ ♂ $P_{(4)}$ A_4 G_0

Female flower: Br. ⊕ ♀ $P_{(4)}$ A_0 $G_{(\bar{3})}$

Uses: It serves as one of the secondary food plants of the oak tasar silkworms viz., *A. roylei, A. proylei, A. frithii* and *A. pernyi* (Seth 1995b, Mishra *et al*. 2000).

The wood is used for timber and fuel and converted into charcoal. Dry leaves, twigs bark and mature bark contain 11.06%, 10.50% and 9.69% tannins respectively (Fraymouth and Pilgrim 1918).

Cultivation: It may be propagated through direct sowing of seeds or plantation of seedlings by raising nursery. The 110 fruits weigh a kg while 176 seeds weigh in one kg. The germination percentage of seeds is 40%, which may be increased by soaking in water for 2 days and scarification before soaking. The germination is completed in five months (Homfrey 1935).

In plantations, the pruning of double stems and forks should be carried out between 6-8 years.

123. BAJRANT (BUK)

Botanical: *Quercus lamellosa* Sm. Syn. *Q. imbricata* Buch.-Ham.ex D. Don., *Q. lamellata* Roxb.

Lepcha: Buk Nepali: Shalsi Budgrat

Vernaculars: Bajrant, Buk, Shalsi Bujrat, Budgrat, Pharat-Singhali

SYSTEMATIC POSITION

Bentham & Hooker (1883)	**Engler & Prantl (1899)**	**Hutchinson (1959)**
Phanerogams	Phanerogams	Angiospermae
Dicotyledones	Dicotyledoneae	Dicotyledones
Monochlamydeae	Archichlamydeae	Lignosae
(or incompletae)	Fagales	Fagales
Unisexuales	Fagaceae	Fagaceae

Cupuliferae

Quercineae

Geographical distribution: Eastern Himalaya from Nepal eastward in Sikkim and Arunachal Pradesh between 5000-9000 ft. ASL and in the hills of Manipur and Nagaland at 6000-9000 ft. ASL. In Darjeeling hills (W.B) it is common between 5500-7500 ft. ASL. This species has also been recorded from Bhutan, Myanmar, Tibet and South-west China.

Habit: A very large, evergreen, dioecious tree bearing silky buds, young shoots with tawny deciduous pubescence and very hard and heavy wood.

Habitat: Wet temperate habitat with atleast 200 cm rainfall on usually steeply sloped – moderately sloped hills. It thrives well on Northern slopes with deep, well drained fairly moist soil of clayey loam nature (Troup 1921).

Root: Branched tap root.

Stem: Aerial, erect, woody, very hard, branched, branches gregarious, young branches with fulvous tomentose apices, bark grey – brown with rough spots.

Leaves: Simple, oblong, elliptic or elongate–elliptic, acute at both ends or acuminate at apex and acute at base, sharply serrated near the base, remotely and sharply serrated towards the apex, dorsal side glabrous, ventral side glaucous except the nerves which are pulvinous when young, blade 17-23 x 6-11 cm, petiole 2.5-5 cm long, lateral nerves 10-20 pairs, straight, impressed on the dorsal but very prominent on the ventral side, base rarely rounded.

Inflorescence: Spikes, unisexual, male and female spikes on different trees.

Male spikes: Solitary, axillary, much shorter than the leaves, rachis fulvous tomentose.

Female spikes: Axillary, 3-4 flowered, very short.

Flower: Unisexual, dioecious, incomplete, epigynous, actinomorphic, sub-glomerulate, bracteate, tetramerous, tetratepalous, syntepalous, stamens 4 in male flowers only, tricarpellary, ovary syncarpous and inferior in female flowers only.

Fruit: Nut, sessile on short spikes, peduncles thick, cups very large up to 7.5 cm across and 3.5 cm deep, embracing two third or more of the glans, woody, obovoid – truncate, with 10 thin, loose, broad, lamellae which are minutely tomentose, finely serrate with erased or sub-entire margin. Glans turbinate on ripening, apiculate, nearly 3.5 cm across and 3 cm long, exposed part pubescent when young but smooth on ripening.

Flowering: April - May.

Fruiting: Acorns ripen in Nov. – Dec. during second year. A good seed year occurs once in three years.

Floral formula: Male flower: Br. ⊕ ♂ $P_{(4)}$ A_4 G_0

Female flower: Br. ⊕ ♀ $P_{(4)}$ A_0 $G_{(\bar{3})}$

Uses: It serves as one of the secondary food plants of oak tasar silkworms viz., *A. proylei, A. pernyi, A. roylei* and *A. frithii* in temperate forests (Seth 1995 b, Mishra *et al.* 2000).

The timber is used for beams, doors, window frames, posts, rafters, bridge constructions, shingles (aplit), ploughs, cart axles, naves, spokes and felloes.

It also makes an excellent fuel and used for charcoal conversion. Bark contains 12.6% tannins and is used for tanning (Negi and Naithani 1995).

Bears, pigs, deers, monkeys, squirrels and rats readily eat acorns and destroy seed crops considerably.

Cultivation: It may be propagated either by direct sowing of seeds or by nursery raising. One kg contains approx. 60 fruits and one kg consists of 100-110 seeds. The germination of seeds has been found to range between 60-80%.

124. RANJ

Botanical: ***Quercus lanata*** Sm. Syn. *Q. lanuginosa* D.Don, non Thuill, Kiani. *Q. nepaulensis* Desf.

Eng: Woolly oak

Nepali: Banga, Banja

Kumaon: Ranj, Rianj, Raibanj

Other vernaculars: Sanj, Riani

SYSTEMATIC POSITION

Bentham & Hooker (1883)	Engler & Prantl (1899)	Hutchinson (1959)
Phanerogams	Phanerogams	Angiospermae
Dicotyledones	Dicotyledoneae	Dicotyledones
Monochlamydeae	Archichlamydeae	Lignosae
(or incompletae)	Fagales	Fagales
Unisexuales	Fagaceae	Fagaceae
Cupuliferae		
Quercineae		

Geographical distribution: Extending from Kumaon (4000-8000 ft. ASL in Uttarakhand) eastward to Arunachal Pradesh at 3500-8000 ft. ASL through Nepal and Bhutan in localised patches along with *Q. leucotrichophora* / *Q. floribunda* / *Q. grifithii*. Also occurs in Myanmar and Indo-China.

Habit: A large evergreen monoecious tree bearing branchlets and ventral side of leaves densely clothed with tawny or rufous wool, hence named as woolly oak.

Habitat: Moist temperate sunny slopes or cool sheltered places having limestone and shaley ridges on poor shallow soils.

Root: Branched tap root, massive and wind-firm.

Stem: Aerial, erect, woody, branched, bark pale-grey or ashy brown, rough, exfoliating in large irregular thin woody plates, lenticellate.

Leaves: Simple, coriaceous, oblong or elliptic – lanceolate, acute or acuminate, serratures distant, upper surface rugose, lower surface dark glossy, green with greenish-white tomentum and prominent secondary and tertiary nerves, midrib usually hairy, blade 10-20 x 4.0-9.0 cm, lateral nerves straight, parallel, 9-13 pairs, petiole 1.0-2.5 cm long.

Inflorescence: Unisexual spikes.

Male spikes: Bracteolate, woolly, shorter than the leaves, crowded, interrupted oftenly, bracteoles ciliate, ovate-acute.

Female spikes: Short, tomentose, few flowered.

Flower: Unisexual, monoecious, incomplete, epigynous, actinomorphic.

Male flower: Perianth 5-6 cleft, fused, pilose externally with 7-10 glabrous anthers.

Female flower: Perianth as in male flowers, styles linear, ovary inferior, trilocular, syncarpous.

Fruit: Acorn, solitary or in short spikes in pairs, axillary, sessile, 2 cm long, matured fruit one third to one half enclosed by companulate cup, cup 1.0-1.4 cm across, with appressed, thin, woody, ovate scales, nut wooly with a short cylinderic umbo.

Leaf shedding: November - January.

Sprouting: March - April.

Flowering: March – April.

Pollination: April - May.

Fruiting: December – January (eight months after pollination).

Floral formulae: Male flower: Br. ⊕ ♂ $P_{(5\text{-}6)}\ A_{7\text{-}10}\ G_0$

Female flower: Br. ⊕ ♀ $P_{(5\text{-}6)}\ A_0\ G_{(\bar{3})}$

Uses: It serves as one of the secondary food plants of oak (temperate) tasar silkworms viz., *Antheraea proylei, A. roylei, A. pernyi* and *A. frithii* (Seth 1995b).

Acorns are eagerly devoured by bears and wood is used as fuel while twigs and leaves are lopped for fodder (Negi and Naithani 1995).

Cultivation: It can successfully be raised either by direct sowing or by planting out seedlings from nursery. The growth of the seedlings is slow and 5-12.5 cm height is attained by the end of first season and approximately 50 cm in four years. It demands more light and coppices better than Moru oak and Ban oak when stems are 7.5 – 20 cm across at breast height (Troup 1921).

125. BANJ

Botanical: ***Quercus leucotrichophora*** A. Camus Syn. *Q. incana* Roxb.

Eng: White oak, Grey oak, Ban oak		Hindi: Banj, Ban.
Garhwali: Phanat	Haz: Rhin	Kashmiri: Shiddar
Kumaun: Ban, Banj	Punjabi: Shindar, Rinj	

Other vernaculars: Tri, Sila-supari, Shinddar, Bang, Rin, Rinj, Vari, Bani, Inai, Phanat, Phen.

SYSTEMATIC POSITION

Bentham & Hooker (1883)	Engler & Prantl (1899)	Hutchinson (1959)
Phanerogams	Phanerogams	Angiospermae
Dicotyledones	Dicotyledoneae	Dicotyledones
Monochlamydeae	Archichlamydeae	Lignosae
(or incompletae)	Fagales	Fagales
Unisexuales	Fagaceae	Fagaceae
Cupuliferae		
Quercineae		

Geographical distribution: Commonest oak of North-western Himalaya exending eastwards from Kashmir to Nepal between 4000-8000 ft. ASL. Also found in Arunachal Pradesh in patches.

Habit: A moderate to large sized, evergreen, gregarious tree with rounded crown bearing massive nasled branches, grey foliage and branchlets, petioles and lower side of leaves white-tomentose. Usually 50-80 ft. high and 3.5-5.5 ft. in girth, rarely 150 ft. high and 15 ft. across in girth (Fig. 52).

Habitat: It may be cultivated at low elevations and where water is in plenty it grows in plains also under temperate climate. The minimum temperature ranges from 6-7°C and maximum seldom exceeds 35°C under normal rainfall between 100-240 cm. It frequently grows well on clayey or micaceous sandy soil but may also be found growing on variety of geological formations viz., shale, gneiss, mica, schist, quartzite and limestone (Troup 1921). Often associated with *Rhododendron arboreum, Andromeda* and Deodar.

Root: Branched tap root.

Stem: Erect, aerial, woody, hard, branched, young shoots heavily pubescent, bark dark grey, rough with cracks and fissures.

Leaves: Simple, coriaceous, elliptic – lanceolate, oblong or ovate – lanceolate, apex acuminate or mucronate, serrate, teeth sharp, tough, base acute, young ones pinkish and woolly all over, matured ones dark green and glabrous above, densely white or grey – rarely brownish tomentose beneath, blade 7.5-15.0 x 2.5 – 5.0 cm, lateral nerves 14-20 pairs, straight and parallel, petioles 1.2-1.7 cm long, stipules linear, 1.2 cm long, scarious and early caducous.

Inflorescence: Unisexual spikes / catkins, bracteate, bracts obtuse, scarious.

Male spikes: 5-10 cm long, slender, drooping, usually much interrupted.

Female spikes: Few flowered and short or may be absent.

Flower: Unisexual, monoecious, incomplete, epigynous, actinomorphic, bracteate.

Male flower: Small, perianth 4-5 lobed, fused, companulate, stamens 3-5, anthers glabrous, apiculate.

Female flower: Solitary or clustered on current years shoots, axillary, sessile, perianth 4-5 lobed, campanulate, style linear – clavate, spreading, ovary inferior, trilocular.

Fruit: Acorns, mostly solitary on current years shoots, ripe nuts ovoid – conical, 2-2.5 cm long, brown, canescent, half enclosed in the rough woody cup on maturity but enclosing them almost completely in beginning.

Leaf sprouting: March – April, young leaves have liliac and purple tinge.

Flowering: March – April, occasionally erratic.

Fruiting: Autumn, 15-17 months after flowering.

Floral formulae: Male flower: Br. $\oplus$ ♂ $P_{(4\text{-}5)}$ $A_{3\text{-}5}$ G_0

Female flower: Br. $\oplus$ ♀ $P_{(4\text{-}5)}$ A_0 $G_{(\bar{3})}$

Uses: It serves as one of the primary food plants of temperate tasar silkworms viz. *Antheraea proylei, Antheraea pernyi* and *Antheraea roylei* and secondary host plant for *Antheraea frithii* etc. in North-western regions of Indian subcontinent (FAO 1976, Mishra *et al*. 2000, Srivastav and Singh 2002).

Bears, monkeys, flying squirrels, rats and birds readily eat the acorns.

Timber sometimes used for wall construction and for agricultural implements, tools handles and building works. The foliage makes excellent fodder during winter for cattles. Bark is used for tanning due to 6-23% tannins. Wood is often used as fuel and firewood and also for charcoal making. Wood pulp may be used for high-grade hardboards with water resistance and good strength. The leaves are used as cattle fodder in the hills (Anon 1969). Medicinally, acorns are diuretic and used as an astringent in indigestion and diarrhoea in children and in gonorrhoea in adults. It also yields manna used in confectionary (Negi and Naithani 1995).

Cultivation: The growth survival and mean annual increment of *Q. leucotrichophora* on degraded lands of Doon valley has been found very encouraging (Singh *et al.* 1998). Ban oaks may be raised either by direct sowing or by planting of seedlings raised in nursery or through air layerings.

Fresh, mature and sound seeds are collected between Dec.-February and insect infested seeds are rejected. Fresh seeds may be stored without drying in sun in a cool, dry place protected from insects for at least a year without losing fertility. The stored seeds have exhibited fertility of 60%. Fresh seeds weigh 1.5 gm on drying on an average. One kg of seeds contains 400-600 seeds and lowest germination of 7-18% and maximum 96-99% germination has been recorded. The germination starts after 25 days and continues up to 5 months when seed coat is intact (Negi and Naithani 1995). The germination of seeds requires soaking in plain water for 48 hrs. Lowest germination (7-18%) occurs when seed coats are intact while maximum (96-99%) germination occurs when seed coats are opened, and in such cases germination completes within 6 to 30 days (Pandey 1999).

For nursery raising, soil is dug well during early spring (Feb.-March) and seeds are sown in drills of 22.5-30 cm apart at the depth of 1.3-2.5 cm. Irrigation is made periodically during dry seasons and weeding, cleaning and loosening of soil is attended as and when required. The seedlings become ready for transplantation during 3-4 years after attaining the height of 30-36 cm because of slow growing nature. Wire nets may be used in nursery to save seeds from rodents.

Pits are prepared at the spacing of 1.5 x 1.5 m or 1.8 x 1.8m and 2-4 years old plants are transplanted during rainy to winter seasons. According to Lal and Karnataka (1993) *Q. incana* seeds can be raised by sowing the seeds in horizontal position in the soil mixture containing FYM, sand and soil in 1:1:2 ratio in perforated polythene bags or in nursery beds for better germination.

Direct sowing of seeds *in situ* in prepared patches at the 1.2 x 1.2 to 3 x 3 m spacing has been made during early spring or winter successfully. Seeds fail to germinate on dry, hot slpoes but may succeed during wet seasons there. During hot dry situations sowing should be made under shade of bushes, boulders or grasses and should be covered with earth to a depth of 1.3 – 2.5 cm only (Negi and Naithani 1995). It can also be propagated through air layerings while untreated air layers gave only 4% success, treatment with one part rootone in 100 parts lanoline gave 28% rooting (Chandra & Mahindru 1977).

Pruning: *Q.leucotrichophora* is slow growing in nature hence it is pollarded once in three – four years at 5-6 ft. height during Dec.-Jan. When heavy pollarding is not resorted light clipping of branches and leaf plucking is resorted to get delicate quality leaves for silkworm rearing in western Himalaya. The leaf plucking is resorted during 15-20 Dec. and sprouting occurs in defoliated sets around 13th – 14th Feb and in non-defoliated sets around 23rd-25th Feb. in patches, which receive sunlight most of the day hours. On the contrary, where shade occurs most of the day hours the sprouting occurs around 25th-27th Feb. in above defoliated sets and around 22nd –24th March in non-defoliated sets (Sinha and Sinha 1992).

According to Maurya (1991) defoliation conducted during 10-11-89 to 25-11-89, 26-11-89 to 20-12-89 and 21-12-89 to 10-01-90 exhibited initiation of sprouting during Feb. 8-10, Feb.20-25 and March 1-5 and foliage became ready in such plants around 15-2-90, 3-3-90 and 10-3-90 respectively in Garhwal region. Due to quick maturity of *Q. incana* foliage during commercial crop, it is advisable to undertake rearing of *Antheraea proylei* silkworms upto III instar on *Q. incana* at 4000 ft. ASL and IV-V instar worms should be shifted on *Q. incana* at 6000 ft. – 7000 ft. ASL where maturity of foliage is slower and moderate pleasant atmosphere is available to the worms during April-May.

126. KHARSU

Botanical: *Quercus semecarpifolia* Sm. Syn. *Q. obtusiloba* D.Don., *Q. cassura* Buch. – Ham. ex. D. Don

Eng: Brown oak	Hindi: Kharsu, Karsu, Kasru	Nepali: Ghesi, Karsu
Bash: Khor	Chamba: Maru, Kreo	
Haz: Banjar	Punjabi: Banchar, Klarshu	

Other vernaculars: Ara, Ghesi, Banchar, Dieng-sheh, Dingiem, Dieng-sia-soh-dingim

SYSTEMATIC POSITION

Bentham & Hooker (1883)	**Engler & Prantl (1899)**	**Hutchinson (1959)**
Phanerogams	Phanerogams	Angiospermae
Dicotyledones	Dicotyledoneae	Dicotyledones

Monochlamydeae	Archichlamydeae	Lignosae
(or incompletae)	Fagales	Fagales
Unisexuales	Fagaceae	Fagaceae
Cupuliferae		
Quercineae		

Geographical distribution: Throughout Himalaya from Afghanistan, Pakistan to Arunchal Pradesh and east Manipur – Myanmar frontier from 8000-10000 ft ASL. Also found in South Tibet and China. Occasionally descending to 6500 and ascending to 1200 ft. ASL.

Habit: A large, gregarious, evergreen tree of 70-80 ft. height with 6-7 ft. girth and clear bole. Occasionally larger in favourable localities but stunted and gnarled on exposed ridges. Occasionally leafless for a few weeks in spring. Trees bearing fruits do not shed leaves that year whereas trees bearing flowers (male) shed leaves by middle of May.

Habitat: Moist temperate regions of heavy snowfall with moderate rainfall and short summer. Found on deep, rich, moist soil in more or less sheltered localities and also on poor, rocky ground on the crests of ridges with stunted growth. (Troup 1921, Mohan and Puri 1955). It forms pure forests along the tops and upper slopes of ridges.

Root: Branched tap root.

Stem: Aerial, erect, branched, woody, hard, bark dark-grey, cut into shallow cracks into small more or less 4- sided scales, young parts clothed with silky hairs.

Leaves: Simple, elliptic or oblong to obovate – elliptic, rigidly coriaceous, young leaves softly rusty tomentose, glabrous and dark green above, hairy on underside, entire on old individuals but dentate with long spinescent teeth on young trees, petiole very short, 0.5 mm long, lateral nerves 6-12 pairs, bifurcating and branching, base cordate.

Inflorescence: Unicellular spike / catkin

Male spikes: Fascicled, 5-10 cm long and congested on seeding year but 10-20 cm long and lax in other years.

Female spikes: Short, few flowered, at the tip of current years shoots.

Flower: Unisexual, monoecious, incomplete, actinomorphic, epigynous, bracteate.

Male flower: Perianth segments-5, ciliate and obtuse, stamens indefinite, anthers apiculate and glabrous.

Female flowers: Perianth segments-5, ciliate and obtuse, styles 3-5, linear, recurved, ovary inferior, trilocular – pentalocular, syncarpous.

Fruit: Acorns, usually solitary, on previous years wood, glabrous, globose, black on ripening, 2.5 cm across and base enclosed by the thin concave cup, 1.2 cm across.

Flowering: April – June.

Sprouting: April – June.

Pollination: May – June.

Fruiting: Aug- Sept. of next year, 15 months after flowering. Fruits are produced profusely every alternate year but occasionally some trees produce fruits every year. Occasionally 2-4 seeded acorns are found.

Floral formula: Male flower: Br. $\oplus$ ♂ $P_{(5)}$ A_{α} G_{0}

Female flower: Br. $\oplus$ ♀ $P_{(5)}$ A_{0} $G_{(\overline{3-5})}$

Uses: It serves as one of the primary food plants of temperate tasar silkworms viz., *Antheraea pernyi, Antheraea roylei, Antheraea frithii* and *Antheraea proylei* in North Western Himalayas (Gamble 1922, FAO 1976, Mishra *et al.* 2000, Srivastav & Singh 2002)

Kharsu oak yields a fine, strong timber locally used for building purposes as beams, posts, rafters, door frames, wall plates etc. and also for agricultural implements, bed steads and mule saddles. Because of having good steam bending properties it can also be used as bent wood for furniture, cart wheel-rims, barrel staves, whisky vats and roof sticks of lorry bodies (Rahman et.al 1956). Its attractive silver grains contribute magnificently in making handsome furnitures. It also serves as excellent firewood and yields charcoal of best quality (Pearson and Brown 1932), Gamble 1922). The leaves are used as fodder and bears with great interest eat acorns extensively. Bark from branches yields as much as 24% tannin whereas stem bark yields only 11.63% tannin (Troup 1921).

Cultivation: It may be cultivated through direct sowing of seeds or transplantation of seedlings raised in nursery. The matured seeds may be collected from ground between July-Sept. after falling from the trees. Since their viability is very short they should not be stored.

One kg. of seeds weigh 130-140 seeds only. The germination percentage of fresh seeds is very high to the tune of 95-100. The fresh seeds are sown in drills during July-Sept. and maintained in nursery for two years. The pits are prepared before rains at the spacing of 5'x5' or 6'x6' and two years old seedlings are transplanted in these pits during rains as winter planting is not much successful.

Direct sowing of seedlings is preferable and seeds are dibbled *in situ* in July-Aug. in open places at 5'x5' or 6'x6' spacing. It may successfully be sown along the contour lines prepared in advance before ripening of the seeds (Negi and

Naithani 1995, Negi 1985). Because of slow growth it needs weeding cleaning, protection against browsing and other cultural operations for several years.

127. SEHOP

Botanical: ***Quercus semiserrata*** Roxb.

Burmese: Ta-tu-polu, Thicha, Zagat

Assamese: Ramrotor Lushai: Sehop

Vernaculars: Sehop, Cyok, Sekho, Ramrota Phokhoh, Suchi-arong, Tu-ung

SYSTEMATIC POSITION

Bentham & Hooker (1883)	Engler & Prantl (1899)	Hutchinson (1959)
Phanerogams	Phanerogams	Angiospermae
Dicotyledones	Dicotyledoneae	Dicotyledones
Monochlamydeae	Archichlamydeae	Lignosae
(or incompletae)	Fagales	Fagales
Unisexuales	Fagaceae	Fagaceae
Cupuliferae		
Quercineae		

Geographical distribution: Assam, (Cachar), Arunachal Pradesh, Meghalaya (Garo & Khasi hills), Mizoram and Manipur deciduous forests up to 3600 ft. ASL in plains as well as hills. Cultivated in Sikkim and North Bengal. Also found in Sylhet (Bangladesh), upper and lower Myanmar and Indonesia at an altitude of 1800-2700 ft. ASL.

Habit: A middle sized evergreen tree of 60-90 ft. with straight and buttressed bole. Youngest shoots and leaves clothed with soft deciduous tomentum.

Habitat: It occurs in East-Himalayan tropical wet forests intermixed with *Q. acutissima*.

Root: Branched tap root.

Stem: Aerial, erect, branched, bark dark grey, rough, hard, 1.5 cm thick.

Leaves: Simple, coriaceous when mature, elliptic, lanceolate, oblong or oblanceolate, acuminate, coarsely serrate or semi-serrate towards the apex, lower portion entire, blade 8-20 x 2.5-6.0 cm, glabrous above, glaucous or glauscent beneath, base usually unequal, obtuse or cuneate, petiole 1.5 – 2.5 cm long, secondary nerves 10-14 pairs, straight, prominent beneath, family anastomising, close under the edge.

Inflorescence: Unisexual spikes.

Male spikes: Solitary, axillary, pendulous, 2-3 flowered, fulvous, tomentose.

Female spikes: Solitary or paired, axillary, little longer than the petiole.

Flower: Unisexual, monoecious, incomplete, epigynous, small, bracteolate.

Male flower: Perianth 5-6 lobed, anthers about 10, glabrous.

Female flower: Perianth as in male flowers, style long, divericating, bifid, stigma sub-capitate, ovary inferior, trilocular.

Fruit: Acorns, cupule annulate, hemispheric covering about two fifth of the glan, the lamellae 4-6, velvety, tomentose outside, smooth, purplish inside, glans cylinderic, ovoid 3-4 x 1.8-2 cm canescent with an annulate beak about 5 mm long.

Flowering: Hot season.

Fruiting: November and March.

Floral formulae: Male flower: Brl. $\oplus$ ♂ $P_{(5-6)}$ A_{10} G_{0}

Female flower: Brl. $\oplus$ ♀ $P_{(5-6)}$ A_{0} $G_{(\bar{3})}$

Uses: It serves as one of the primary food plants of temperate oak tasar, Antheraea *pernyi, A. roylei, A. proylei* and *A. frithii* (FAO 1976)

The timber is easy to saw and work to a smooth surface. It is used for pins with which cartwheels are joined in Myanmar (Pearson and Porown 1932).

Cultivation: It may be cultivated through either direct sowing or transplantation of seedlings. Seeds weigh about 143 to a kg (Sengupta 1937). Its cultivation is similar to that of *Q. acutissima*.

128. ORA / ORCHA

Botanical: ***Rhizophora caseolaris*** Linn. Sp. pl. Syn. *Sonnertia acida* Linn.f.

Burmese: Tabu, Tamu | Bengali: Ora, Orcha | Kannada: Kandale

Malayalam: Blatti, Thirala | Marathi: Tiwar, Chippi | Tamil: Kinnari

Oriya: Sundarignua

SYSTEMATIC POSITION

Bentham & Hooker (1883)	Engler & Prantl (1899)	Hutchinson (1959)
Phanerogams	Phanerogams	Angiospermae

Dicotyledones	Dicotyledoneae	Dicotyledones
Polypetalae	Archichlamydeae	Lignosae
Calyciflorae	Myrtiflorae	Lythrales
Myrtales	Lythraceae	Lythraceae
Lyrthrarieae		

Geographical distribution: Sunderbans (West Bengal), Deccan Peninsula, Delta of Indus, seacoast from Chittagong to Tennasserim, Sri Lanka, Malaya Peninsula and Archipelago including Myanmar (Rangoon), Thailand (Siam) and Java.

Habit & Habitat: A small evergreen, glabrescent tree up to 15 ft. height, growing in mangroove swamps flooded by sea tides.

Root: Branched tap root. Widely spreading horizontal roots send up vertical branches with soft pith like wood, which facilitates exchange of gases between roots buried in the mud of the tidal swamp and the atmosphere. These vertical branches are known as respiratory roots or breathing roots or pneumatophores and contain special minute pores situated mostly near the tips for oxygen and carbon dioxide exchange.

Stem: Erect, branched, woody, glabrous, branches usually quadrangular.

Leaves: Simple, opposite, coriaceous, entire, oblong or narrowly obovate-elliptic, petiole broad, very short or subobsolete, blade 5.0-10.0 cm long, 2.5-4.5 cm broad.

Inflorescence: Solitary flowers at the end of branchlets.

Flower: Bisexual, complete, regular, actinomorphic, hypogynous or perigynous, 5.0 cm in diameter, perfect.

Calyx: Sepals 6-8, usually 6, 2.5 cm or much longer, gamosepalous, calyx tube shallow, unribbed, lobes acute, longer than tube, valvate, persistent.

Corolla: Petals 6-8, usually 6, linear-oblong, acuminate, dark purple or dark red, free.

Androecium: Stamens numerous, arranged in 3-4 whorls, inserted at the mouth of calyx tube, inflexed in bud.

Gynoecium: Carpels usually 6, nearly free or adnate at the base to the calyx tube, hexalocular, style long, exserted, up to 7.0 cm long sometimes, stigma capitate.

Fruit: Capsule, thick walled, depressed - globular, 5-7.5 cm broad, concave at the summit, fleshy, indehisecent, supported by persistent calyx, edible, placentas thick, ovules numerous, placentation axile.

Seeds: Non-endospermic, angular.

Flowering: Rainy season.

Floral formula: $\oplus$ ⚥ $K_{(6-8)}$ C_{6-8} A_{α} $G_{\underline{6}}$ or G_{-6}

Uses: It serves as one of the tertiary but recognised as one of the secondary food plants for tropical tasar silkworm, *Antheraea mylitta* D. in jungles (Jolly *et al.* 1974). "Keora water" is obtained by distillation of *Sonneratia caseolories* (Shukla and Mishra 1970).

Fruits edible, raw ones used for flavouring curries and chutneys. Pericarp vermifuge. Vinegar prepared from fruits. Wood may be used for bridges, boat, ship building, doors, ceiling, floorings, furnitures, cabinet work and musical instruments with precautions (Ambasta *et al.* 1986).

Cultivation: It propagates through seeds in nature.

129. CASTOR

Botanical: *Ricinus communis* L.

Eng.: Castor
Hindi: Randi, Erandi, Bherand, Arind
Bengali: Bherenda
Gujarati: Diveligo
Kannada: Haralu
Malayalam: Avanakku
Manipuri: Kege
Marathi: Erandi
Telugu: Amudamuchettu
Tamil: Amanakku

SYSTEMATIC POSITION

Bentham & Hooker (1883)	Engler & Prantl (1899)	Hutchinson (1959)
Phanerogams	Phanerogams	Angiospermae
Dicotyledones	Dicotyledoneae	Dicotyledones
Monochlamydeae	Archichlamydeae	Lignosae
(Incompletae)	Geraniales	Euphorbiales
Unisexuales	Euphorbiaceae	Euphorbiaceae
Euphorbiaceae		

Geographical distribution: India and Africa have been considered as ancestral homes of Castor from where it spread to Sudan, UAR, Egypt, Burma, Sri Lanka, Malaysia, Phillippines, Indonesia, Brazil, USA and Mexico.

Andhra Pradesh, Maharashtra, Assam and Karnataka are major castor cultivating states followed by West Bengal, Bihar, Jharkhand, Orissa,

Meghalaya, Arunachal Pradesh, Manipur and Himachal Pradesh where it is used for eri culture.

Habit: A tall annual, biennial or perennial evergreen shrub, rarely a small tree, red or green in colour (Fig. 53).

Habitat: Cultivated in diverse climates and on poor sandy to rich alluvial soils. Assumes a perennial and excessively luxuriant characteristics in humid tracts. It can withstand drought and slightly acidic soil conditions but detests water logging and soil alkalinity. Black cotton soils with more than 80% clay content are unsuitable for castor.

Root: Tap root.

Stem: Erect, cylinderical, herbaceous above and woody below, hollow, or solid, glabrous and branched.

Leaves: Simple, alternate, broad, palmately lobed, venation reticulate and multicostate, lobes 7-9, petiolate, serrated, acute, glands green or reddish.

Inflorescence: Panicled terminal raceme.

Flower: Small, incomplete, unisexual, monoecious, actinomorphic, apetalous, hypogynous, pedicellate, bracteate, male and female on same inflorescence, upper female and lower male.

Male flower: Perianth 3-5, valvate, uniseriate, membranous, green, united when young, spreading in open flower. Stamens usually 5, rarely numerous, polyadelphous, arranged upon partially connate repeatedly branched filaments. Anthers, globose, introse, bicelled, basifixed and divergent, pistillode absent.

Female flowers: Perianth 3-5, uniseriate, spathaceous, gamophyllous. Staminodes absent. Gynoecium tricarpellary, syncarpous, ovary superior, three lobed, covered with numerous fleshy spines, trilocular with one ovule in each loculus, placentation axile, styles 3, bifid, long or short, spreading, feathery or papillose, stigma 6, papillose, red.

Fruit: Globose capsule or regma; generally echinate (spinous), schizocarpic, splitting into three cocci, containing one seed each.

Seeds: Albuminous, oblong, smooth, testa crustaceous and mottled with knob like carbuncles.

Germination: Epigeal.

Flowering and fruiting: Most part of the year

Floral formulae: Male flower: $\oplus$ ♂ $P_{(3-5)} A_{(5-8)} G_{(0)}$

Female flower: $\oplus$ ♀ $P_{(3-5)} A_{0} G_{(\underline{3})}$

Uses: Castor serves as the main primary food plant of eri silkworm *Philosamia ricini* and eri silk industry largely depends on cultivation of castor plants (FAO 1976, Sarkar 1988, Seth 1995c, Sarmah 2000). It can also serve as the secondary host plant for tasar silkworm *Antheraea mylitta* D. (Patil *et al.* 2000). Besides, castor seeds are also known to yield 35-55% of famous castor oil, which is thick, colourless or greenish oil. Although improved varieties of castor exist in India for oil seed production, high leaf yielding varieties from sericulture point of view are meager. Rearers prefer usually red variety, which is due to its red colour and aesthetic value only. The green variety is delicate and cannot stand adverse weather. Rearers prefer the perennial green variety to annual variety due to the quality, quantity and duration of the leaves. The perennial variety lasts for 3-4 years, branches profusely and grows almost like a tree yielding abundance of seeds and leaves.

The castor oil is an excellent lubricant for airplane engines. It is also used for manufacturing varnishes, paints, typewriter inks, aromatic perfumes, transparent soaps and textile soaps. Due to inherent water resistance, it is enormously used for protective coverings for insulation, food containers, gums, airplanes and also for coating fabrics. Besides, it is used as an illuminant and for preserving leather. The oil cake of castor seeds may be used as fertilizer (Pandey 1981, Kochar 1981).

The wood pulp is utilized for making writing and printing papers whereas stems are used for straw boards and cheap wrappings (Pandey 1981).

Medicinally, its leaves are warmed, crushed and applied to anus for treatment of bleeding piles. The castor seed oil is purgative and leaf paste is often used as poultice on sores, gout, abdominal ailments and rheumatic swellings, whereas decoction of roots is given for treatment of lumbago (Sinha 1996, Ansari 1997).

Administrative of the pulp of one or two castor seeds during the menstrual period is stated to prevent fertilization for one and two years respectively (Dash 1993). In Bundelkhand, the seed oil is burnt with a piece of leather and latex of *Euphorbia neriifolia* L. and the resulting paste is applied on eyes of cattle for 4-5 days in cataract. The leaves are also used to ward off evil spirits (Maheswari and Singh 1991).

Ladies and girls in Manipur boil the leaves of red castor with water for washing hairs.

Cultivation: Castor is cultivated through seeds all over the world. High flat or sloppy land where water does not stagnate should be selected for its plantation. The ideal seasons for castor plantation are March-April and Sept.-October. The land should be ploughed 2-3 times to a depth of 20-25 cm and leveled. Subsequently, 25x25x25 cm (LxWxD) pits are made at 1x1 meter spacing and 1 kg of FYM/pit is mixed thoroughly with soil and leveled. The seeds of castor are shown @ 2-3 seeds/pit at a depth of 3-4 cm and irrigated. Germination starts within 7-10 days. Only one plant should be kept in every pit after 15 days of germination. Hoeing and weeding should be carried out 3-4 times in a year. After one month of germination NPK @ 60:40:20 kg/ha should be applied by making ring at the base of plants. Subsequently, after three months nitrogen @ 30 kg/ha and FYM/compost @ 1 kg/pit may again be applied.

For silkworm rearings first leaf harvest may be made after three months of plantation and 2nd-5th harvests may be made after one month of earlier harvests (Anonymous 1998).

An acre of trees at 12'x12' spacing may yield about 20000 kg of leaf (Sarkar 1988).

130. KUMANTA

Botanical: ***Salix vimnalis*** Linn.

Eng: English Willow, Basket willow

Himanchali: Kumanta

Vernaculars: Kumanta, Bibsu

Punjabi : Bibsu

SYSTEMATIC POSITION

Bentham & Hooker (1883)	Engler & Prantl (1899)	Hutchinson (1959)
Phanerogams	Phanerogams	Angiospermae
Dicotyledones	Dicotyledoneae	Dicotyledones
Monochlamydeae	Archichlamydae	Lignosae
(Incompletae)	Salicales	Salicales
Ordines anomaly	Salicaceae	Salicaceae
(Anomalous families)		
Salicaceae		

Geographical distribution: Temperate Himalaya from Jammu westwards viz., Kashmir, Chamba, Lahaul between 5000-9000 ft. ASL and Baspa valley and Kunawar at 9000 ft. ASL. Also found in Europe and Northern China and Western Asia (Brandis 1906).

Habit: A dioecious, deciduous shrub or small tree. Branchlets and underside of leaves densely covered with grey silky hairs.

Root: Branched tap root.

Stem: Aerial, erect, branched, branchlets slender, flexible.

Leaves: Simple, alternate, linear-lanceolate, margins entire, revolute, apex acuminate, reticulate, silvery silky beneath, blade 10-13 cm; stipules linear-lanceolate, petiole short.

Inflorescence: Catkins.

Male: 2.5 cm long, erect, golden yellow, sessile, bracts brown or black at the apex, fringed with long silky hairs, stamens 2, free, yellow, long, excerted, disc gland narrow.

Female: Sub-sessile, 5-10 cm. long, cylinderic, style slender stigmas slender, spreading, ovary unilocular.

Flower: Unisexual, actinomorphic, perianthless, epigynous, incomplete.

Fruit: Capsule, sub-sessile, tomentose, 0.6 cm, beaked, narrowed into the style, bivalved.

Flowering: Before leaf sprouting.

Floral formulae: Male flower: Br. ⊕ ♂ P_0 A_2 G_0

Female flower: Br. ⊕ ♀ P_0 A_0 $G_{(\bar{1})}$

Variety: Variety *smithiana* with stipules and half cordate broader, leaves and shorter styles is found in Sikkim which is probably *S. eriophylla* of the Khasia mountains (Hooker 1888).

Uses: It serves as one of the secondary food plants of temperate tasar silkworm, *Antheraea proylei, A. roylei, A. pernyi* and *A. frithii* (Seth 1995b).

Bark used for tanning. Twigs used for basketry.

Cultivation: It may be cultivated through either sowing of seedlings or transplantation of seeds.

Besides, it can also easily be raised through stem cuttings. The cutting thickness of 1.1 – 1.5 cm mid diameter with uniform length of 25 cm are best suited for propagation in nursery (Mughal 1996).

EXOTIC QUERCUS SPECIES INTRODUCED IN INDIA

1. ***Quercus cerris*** Linn. - Eng: The Turkey oak: It is grown in Dharmsala, Manali (5500 ft. ASL) and Kullu in Himachal Pradesh and Nilgiris in Tamil Nadu. It is exhibiting stunted growth at Manali and suffering from snowfall in Kullu (Negi and Naithani 1995).
2. ***Quercus coccinea*** Wang. – Eng: The laurel oak, Scarlet oak: It was introduced and grown in Nilgiri hills (Negi and Naithani 1995).
3. ***Quercus robur*** Linn. - Syn. *Q. pedunculata* Ehrm., *Q. sessiliflora*. The English oak: It requires deep soil in moist hill stations otherwise becomes

stunted. It can tolerate light snowfall but suffers greatly due to heavy snow break. It is grown in Himachal Pradesh and attained 14 m girth and 18 m. height in Kullu (Negi and Naithani 1995).

4. ***Quercus rubra*** Linn. – Eng: American red oak: It is grown in Kullu and Manali in Himachal Pradesh and Nilgiri in Tamil Nadu (Negi and Naithani 1995).

5. ***Quercus suber*** Linn. – Eng: Cork oak: It is a medium sized evergreen tree with relatively a short trunk, spreading branches and a uniform crown. It is a temperate zone tree indigenous to the shore areas of the Western Mediterranean region and thrives well between 19-21°C mean annual temperature on poor soil where it produces best cork. It was introduced in Kulu (Himachal Pradesh) and Nilgiris (Tamil Nadu).

It can be propagated either by direct sowing of seeds *in situ* or transplantation of seedlings raised in nurseries.

All the above exotic *Quercus* species serve as the secondary food plants of oak tasar silkworms (Seth 1995b). Besides, *Q.suber* also yields cork of the finest texture and quality for making of bungs and stoppers for bottles and other containers. Cork is also used for making mats, soles for shoes, linings, artificial limbs, life preservers, novelties etc. Cork wastes may be used for insulating materials and manufacturing linoleum while wax extracted from the cork waste is used for making shoe pastes. (Negi and Naithani 1995).

131. SAFED-CHANDAN

Botanical: *Santalum album* Linn. Syn. *Sirium myrtifolium* Roxb.

Burmese: Natha-Phyu English: Sandalwood

French: Santal blane, Santagu German: Sandel baum Sinhalese: Sadan

Hindi: Safed-Chandan, Chandan, Chandal, Sandal

Bengali: Chandan, Srikhand Gujarati: Sukhad, Suket,

Kannada: Gandha, Srigandha, Gandada, Agaadugandha Konkani: Sriganda

Malayalam: Chandanam Manipuri: Chandan

Marathi: Chandan, Sandal Oriya: Chandono

Punjabi: Chandan Tamil: Sandanam, Chandanam, Srigandam, Sandal

Sanskrit: Chandana, Sandala, Malayaja, Bhogivallabha, Ananditam

Telugu: Chandanam, Chandanamu, Sandala

SYSTEMATIC POSITION

Bentham & Hooker (1883)	**Engler & Prantl (1899)**	**Hutchinson (1959)**
Phanerogams	Phanerogams	Angiospermae
Dicotyledones	Dicotyledoneae	Dicotyledones
Monochlamydae	Archichlamydeae	Lignosae
(Incompletae)	Santalales	Santalales
Achlamydosporeae	Santalaceae	Santalaceae
Santalaceae		

Because of the fragrant oil its heartwood contains, Sandalwood is known to the world over from ancient times. The beautiful fragrance by the wood is rupturing and Government claims the right over all the trees in India.

Geographical distribution: Indigenous to Indian peninsula chiefly in Karnataka adjoining districts of Tamil Nadu and parts of Chattisgarh and/or Madhya Pradesh.

Habit: A small evergreen glabrous tree with slender drooping branchlets, growing up to a height of 9-12 meters and a girth of 75-90 cm. The Sandal seedlings live independently for some times, the root being furnished with minute root hairs, which disappear when it has attached itself to the roots of fostering plant through its haustoria. Thus, the Sandalwood tree grows as a parasite on the roots of other trees (Brandis 1906, Sagreiya 1967).

Habitat: It thrives on well-drained loamy lateritic soils of the highlands of Southern India, particularly in the scrub forests.

Root: Haustorial roots.

Stem: Erect, branched, woody, cylinderical, branchlets slender, drooping, bark black, rough, fragrant, sapwood mostly white and scentless, heartwood yellowish – brown, strongly scented, the colour of the wood depends on age and kind of tree.

Leaves: Simple, opposite, oval or ovate – lanceolate, shining 3.5-6.5 cm long, petiole 1.2 – 1.5 cm long, glabrous on both sides, entire.

Inflorescence: Axillary or terminal panicled cymes.

Flowers: Small, hermaphrodite, pink or purple, bracteate, epigynous, actinomorphic.

Perianth: Tepals 4, gamotepalous, campanulate, limb with 4 valvate triangular segments, pink or purple.

Androecium: Stamens 4, opposite to the perianth lobes, exserted, epiphyllous, alternating with 4 rounded, obtuse scales, which may be regarded either as petals or as lobes of the disk, anthers dithecous, filament very short.

Gynoecium: Carpel-1, unilocular, ovary nearly inferior, with three ovules, placentation basal, stigma trilobed.

Fruit: Drupe, globose, 0.6-1.2 cm in diameter, dark purple or black, endocarp hard, endosperm fleshy.

Flowering: Feb.-July and sometime during Sept. – Dec.

Floral formula: ⊕ ⚥ $P_{(4)}\ A_4\ G_{(\bar{1})}$

Uses: The Sandal wood grows wild or cultivated over an area of 15000 km^2 of which 85% lies in Karnataka and the reminer in Tamil Nadu and parts of Madhya Pradesh/Chattisgarh. The sandalwood is extensively used for carving pieces of art. The chips of heartwood and the roots as also the wood unsuitable for carving are used for distilling the oil used in medicine and in perfumery and manufacture of toilet soaps locally. A large quantity is also exported as it is considered an excellent base for high-class perfumery. India controls nine-tenth (9/10) of the world production of the Sandalwood and produces nearly 85% of the world's output of Sandalwood oil (Kocchar 1981). In 1975-76 138 tonnes of oil, value of Rs. 4.25 crore rupees was exported to Japan, USSR, USA and Germany. Increase of production is clearly indicated. The total annual production of oil is more than 100,000 kg (Sagreiya 1967, V edition 1994).

The sawdust of sandalwood is used in preparation of Agarbattis and mixture of home-havan. Sandalwood oils are used for manufacturing some insecticides also.

Paste of wood/stem is applied to forehead in headache, fever, various types of inflammations including those due to pox-vaccination, skin diseases and purities. Oil from heartwood is used in dysuria, gonorrhoea, urethritis and cystitis. Both wood and oil are diaphoretic, diuretic, expectorant and eaten in heat strokes (Sinha 1996, Jain 1968). Besides, the oil is disinfectant, cooling and tonic for the heart also. The oil is used to relieve scabies, cures pimples and other skin diseases. In chronic bronchitis, 3-6 ml of wood extract in one litre of water is used in doses of 2-3 cups/day. The extracted oil, 5-15 drops can also be used thrice a day in chronic bronchitis (Kurian 1995). The Sandalwood facilitates conservation of soil also (Chandawat and Gautam 1993).

Cultivation: It may be propagated through seeds, seedlings and cuttings. Recently, tissue culture of sandalwood has also been found successful.

The seeds are collected during March-April and nursery is raised during May-Aug, or Oct.- Dec. The seeds may be directly sown during June or else seedlings as well as root cuttings may also be transplanted during June (Chundawat and Gautam 1993). It attacks some hosts extensively and other sparingly. With *Casuarina, Terminalia, Albizia, Dalbergia* and *Pongamia* the haustorial association becomes so intimate that both the roots metamorphose to a single entity, which enables Sandal to absorb nutrients through the host roots in better way. But in hosts like *Emblica, Swietenia, Delonix, Acacia, Ailanthus and Leucaena* haustoria formation takes place but it becomes non-functional due to poor anatomical association. Hence, synergism exists between Sandal and good hosts and allelopathy exists between Sandal and poor hosts (Taide *et al.* 1999). Attempts are being made to cultivate Sandal hydroponically without any host (Chauhan and Chandra 2003).

132. KORHA

Botanical: ***Sapium eugeniaefolium*** Ham. Syn. *S.hookeri* Herb. Hort. Bot. Cal.

Nepalese: Pipalpate, Phirphire

Assamese: Korha | Khasi: Dieng-soh-mrit

SYSTEMATIC POSITION

Bentham & Hooker (1883)	Engler & Prantl (1899)	Hutchinson (1959)
Phanerogams	Phanerogams	Angiospermae
Dicotyledones	Dicotyledoneae	Dicotyledones
Monochlamydeae	Archichlamydeae	Lignosae
(Incompletae)	Geraniales	Euphorbiales
Unisexuales	Euphorbiaceae	Euphorbiaceae
Euphorbiaceae		

Geographical distribution: Sub-tropical Himalaya from Kumaon eastwards 2000-4000 ft. ASL, Sikkim, Assam and Meghalaya (Khasi hills).

Habit: A glabrous glaucous tree.

Root: Branched tap root.

Stem: Erect, aerial, woody, solid, branched, branchlets glaucous (a dull blue colour), glabrous.

Leaves: Elliptic or ovate, simple, long petioled, acute or sub-acute apex, entire, alternate, blade 7-10 cm, petiole 2.5-10 cm long, very slender, 2-glandular below the blade, base acute, nerves 6-10 pairs, very slender arched.

Inflorescence: Racemes (spikes) simple, dense, terminal, solitary and densely flowered, 7-10 cm long, female flowers in lower part of spike or in separate spikes.

Flower: Monoecious, unisexual, incomplete, actinomorphic, apetalous, discless, hypogynous.

Male: Several flowers in each bract, petals and pistils absent.

Calyx: Sepals membranous, 2-3 lobed or toothed or split to the base into 2-3 valvate sepals.

Androecium: Stamens 2-3, filaments free, anther cells ovoid, distinct, contiguous, parallel.

Female: Solitary in bracts, petals and stamens absent.

Calyx: Trifid or partite.

Gynoecium: Carpels 2-3, styles joined on top of the ovary, cells single ovuled.

Fruit: Capsule, globose 0.8-1.3 cm across, obtuse, 3-valved, valves thick, woody separating from persistent broad three wings, wings thick, horned at the outer angle.

Seeds: Globose.

Floral formulae: Male flower: ⊕ ♂ $K_{(2\text{-}3)}\ C_0\ A_{2\text{-}3}\ G_0$

Female flower: ⊕ ♀ $K_{(2\text{-}3)}\ C_0\ A_0\ G_{(\underline{3})}$

Uses: It serves as one of the tertiary food plants of eri silkworm, *Philosamia ricini* (FAO 1976, Sarkar 1988, Seth 1995c).

Cultivation: It can be cultivated through seeds as in the case of *Sapium sebiferum.*

133. VILAYATI SHISHAM

Botanical: ***Sapium sebiferum*** Roxb. Syn. *Excoecaria sebifera* Muell. Arg.

Eng: Chinese tallow tree   Hindi: Vilayati Shisham

Bengali: Mom-china   Kannada: Meenadabattimara   Oriya: Ronojita

SYSTEMATIC POSITION

Bentham & Hooker (1883)	**Engler & Prantl (1899)**	**Hutchinson (1959)**
Phanerogams	Phanerogams	Angiospermae

Dicotyledones	Dicotyledoneae	Dicotyledones
Monochlamydeae	Archichlamydeae	Lignosae
(Incompletae)	Geraniales	Euphorbiales
Unisexuales	Euphorbiaceae	Euphorbiaceae
Euphorbiaceae		

Geographical distribution: A native of China, commonly cultivated in North-West India especially in the Kangra dist. of Himachal Pradesh and near Saharanpur and Dehradun in Uttar Pradesh and Uttaranchal respectively.

Habit: A medium sized, deciduous, glabrous, frost resistant tree.

Stem: Erect, woody, aerial, branched, glabrous, solid and woody.

Leaves: Simple, rhomboid, entire, on long slender petioles, resembling those of Shisham, glabrous, acuminate, turning bright scarlet in autumn and presenting a beautiful sight, alternate, penninerved, 2-glands present on petiole.

Inflorescence: Terminal, spiciform racemes.

Flower: Unisexual, regular, actinomorphic, hypogynous, monoecious (Male & Female present on same plant), incomplete, apetalous, male numerous, female solitary in axil of the bract, 2-glands present at the base of each bract, disc absent, pollination entomophillous.

Calyx: Sepals 2-3 toothed or lobed.

Androecium: Stamens 2-3, filaments free, anther cells contiguous, parallel.

Gynoecium: Absent.

Calyx: Sepals – trifid.

Androecium: Absent.

Gynoecium: Tricarpellary, ovary glabrous, narrowed into a style with 3 long spreading stigmas, ovule single in each cell.

Fruit: Capsule loculicidally trivalved.

Seeds: 3, attached to a central coloumn, which splits into three slender divisions enclosed in a thick layer of white fatty substance (Vegetable tallow), endospermic.

Floral formulae : Male flower: Br ⊕ ♂ $K_{2-3}\ C_0\ A_{2-3}\ G_0$

Female flower: Br ⊕ ♀ $K_3\ C_0\ A_0\ G_{(3)}$

Uses: It serves as one of the tertiary food plant of eri silkworm *P. ricini* (FAO 1976, Seth 1995c). Besides, it also serves as a beautiful ornamental tree. The kernels yield 17-20% edible oils, which can be exploited commercially (Jain *et al*. 1990).

Fruit yields chinesė vegetable Tallow oil and Stillingia oil. Chinese Tallow is used in making candles, cosmetics, soaps and dressing cloth. Stillingi a oil is used in paints, varnishes, plastics and for illumination. The oil is vulnerary, emetic, purgative and used for cutaneous troubles. Decoction of bark of root-tonic and given in dyspepsia. Resin from bark is purgative. Wood used for toys, furniture, pencils, cricket bats and block mounts. Leaves yield a black dye (Ambasta *et al*. 1986).

Cultivation: Seeds germinate easily. It can be cultivated easily near rivers and streams on moist land and also in parks. It resembles maples of USA due to coppery tints in autumn.

134. ASHOK

Botanical: *Saraca indica* Linn. Syn. *Jonesia asoca* Roxb; *S. asoca* (Roxb.) Dewilde

Burmese: Jhawka English: Ashoka tree

Hindi: Ashok, Sita Asoka

Bengali: Asok, Asoka, Ashok Gujarati: Ashopallav

Kannada: Asokadamara, Kenkali

Khasi: Dieng-soh-kyrkha Lushai: Husangidba

Manipuri: Ashok Malayalam: Asok, Hemapushpam

Marathi: Ashoka, Jasundi, Jasunda Oriya: Asoko

Tamil: Asogam, Asoka Telugu: Asokamu

SYSTEMATIC POSITION

Bentham & Hooker (1883)	**Engler & Prantl (1899)**	**Hutchinson (1959)**
Phanerogams	Phanerogams	Angiospermae
Dicotyledones	Dicotyledoneae	Dicotyledones
Polypetalae	Archichlamydeae	Lignosae
Calyciflorae	Rosales	Leguminales
Rosales	Leguminosae	Caesalpiniaceae
Leguminosae	Caesalpinoideae	
Caesalpinioideae		

There is a legend in Valmikis Ramayana about Asokavana comprising Asoka trees, where Hanuman met Sita at Lanka with a message from Lord Rama, while she was in captivity. Sage Vyasa gave a full description of Asoka tree in "Sabhaparva" of "Mahabaratha". The founder of Buddhism, Gautama was born under an Asoka tree, hence it is extensively planted in monastery gardens in Thailand and Burma. The great emperor Asoka after abjuring violence and upon his conversion to Buddism also planted Asoka tree as one of the "Bodhi trees" along the road sides to give shade and shelter. The tree was extensively sculptured in Buddhist temples of Kushan Mathura, Sanchi and Bharhut. Hindus also regard it as sacred, being dedicated to Kamdeva, God of Love (Rao 1991, Randhawa 1983).

Geographical distribution: Frequently planted chiefly near temples in India, Sri Lanka and Burma (Myanmar), it is supposed to be a native of India, Myanmar and Malaya. Wild in the evergreen forests of Khasi hills (Meghalaya), Northern Circars, Konkan, Kanara, Bengal and West coast of Bombay, it is extensively cultivated in gardens.

Habit: Handsome, small, evergreen, erect tree (Fig. 54).

Habitat: Sub-tropical to tropical moist habitats in various types of soils.

Root: Tap, branched.

Stem: Erect, aerial, branched, solid, woody, branches coloured, drooping and spreading in all directions forming a dense, closely leaved and shapely crown, bark smooth, dark brown or grayish brown.

Leaves: Compound, deep green, shiny, 30.0 cm long, pendent, alternate, leaflets 3-7 pairs, wavy edged, young leaves drooping, coppery red and flaccid.

Inflorescence: Corymbs, orange-red, dense.

Flower: Hermaphrodite, complete, zygomorphic, hypogynous, orange-red, gradually changing to scarlet, peduncles and pedicels glabrous, coloured, bracteoles persistent, 0.4-0.5 cm long.

Calyx: Sepals 5, gamosepalous, calyx tube 1.25-1.5 cm long, twice the length of lobes.

Corolla: Petals 5, free, orange-red.

Androecium: Stamens 10, perfect 7-8.

Gynoecium: Carpel 1, ovary superior, unilocular.

Fruit: Pod, 15-25 x 5 cm, valves hard, reticulate.

Flowering: February - April.

Floral formula: · |· ⚥ $Brl_l\ K_{(5)}\ C_5\ A_{7+3\ stn.}\ G_{\underline{1}}$

Uses: The profusion of orange red flowers amidst deep green foliage present as an startling beauty. Hence, it is one of the prettiest Indian trees, besides being important from religious point of view.

Bark is astringent, useful in menorrhagia due to uterine fibroids, leucorrhoea, internal haemorrhoids and blood dysentery and diabetes (Randhawa 1983, Sinha 1996).

Flowers used as uterine tonic, also used in billiousness diabetes and hemorrhagic dysentery. Wood used for shafts and plough and pods serve as good forage (Ambasta *et al.* 1986).

Cultivation: Through seeds and seedlings. It should be transplanted in sheltered places to save it from hot winds in the beginning. The seeds are collected during December, seedlings are raised during March-April in nursery and transplanted during June-July (Chundawat and Gautam 1993).

135. BHILAWA

Botanical: *Semecarpus anacardium* Linn. F. Syn. *Anacardium orientale*

Burmese: Thitsi bo

English: The Marking nut tree

Hindi: Bhilawa, Bhilum

Assamese: Bhala

Bengali: Bela, Bhelatuki

Gujarati: Bhilamu

Kannada: Ger, Goddgeru

Marathi: Biba, Bibha

Malayalam: Chera

Oriya: Balia

Punjabi: Bhela

Tamil: Shenkottei

Telugu: Giddu

SYSTEMATIC POSITION

Bentham & Hooker (1883)	Engler & Prantl (1899)	Hutchinson (1959)
Phanerogams	Phanerogams	Angiospermae
Dicotyledones	Dicotyledoneae	Dicotyledones
Polypetalae	Archichlamydeae	Lignosae
Disciflorae	Sapindales	Anacardiaceae
Sapindales	Anacardiaceae	
Anacardiaceae		

Geographical distribution: Sub-Himalayan tract from the Beas eastwards, ascending in the outer hills up to 3500 ft. in Punjab, Madhya Pradesh, Chattisgarh, Bihar, Jharkhand, Western Peninsula, Assam, Mizoram (Khasi hills) and Bangladesh (Chittagong). Also reported from Upper Myanmar (Brandis 1906).

Habit: A middle sized tree. Branchlets, petioles, underside of leaves and inflorescence clothed with harsh tomentum.

Habitat: Sub-tropical to tropical places in more or less all types of soil.

Root: Tap, branched, deep.

Stem: Aerial, erect, woody, cylinderical, solid, branched, bark thick, rough, grey-black, cracked, wood soft without distinct heartwood.

Leaves: Simple, alternate, entire, oblong-obovate, coriaceous, clustered near the end of branches, apex obtuse, blade 22-30 cm long, petiole thick, half round, 1.25 – 2.50 cm long.

Inflorescence: Panicles, erect, terminal, compressed, fascicled, the lower branches of panicles in the axil of leaves.

Flower: Hermaphrodite, actinomarphic, hypogynous, complete, small, numerous, with a glandular disc, 0.6-0.8 cm in diameter, nearly sessile, fascicled, bracteate, bracts lanceolate, pentamerous.

Calyx: Sepals 5, gamosepalous, 5-cleft, calyx segments deciduous.

Corolla: Petals 5, free, alternate with sepals, oblong, greenish – yellow.

Androecium: Stamens 5, alternating with petals, inserted at the base and outside of a broad annular disc, anthers dithecous.

Gynoecium: Carpels 3, syncarpous, ovary unilocular, superior, styles 3, slender, ovule single, pendulous, placentation axile.

Fruit: Drupe, 2.5 cm long, fleshy, oblique, seated on a fleshy hypocorp, purplish-black, shining, pericarp full of acrid resinous juice, hypocarp orange, edible.

Flowering: May – August.

Fruiting: Nov. – Dec.

Leaf fall: Feb. - April.

Floral formula: $\oplus$ ⚥ $K_{(5)}\ C_5\ A_5\ G_{(3)}$

Uses: A black resin obtained from the pericarp of the fruit is used as marking ink to write on cloth. The seed oil is for protection against white ants and for floor dressing. The nuts are used for tanning (Pandey 1981). The hypocarp of the fruit is edible (Brandis 1906).

The tree serves as one of the secondary/tertiary host plant of tasar silkworm, *Antheraea mylitta* Drury in jungles (Nakvi and Srivastav 2003).

Wood suitable for packing boxes, crates, match-boxes, splints and cheap and light furnitures. Kernels yield a oil used as a wood preservative and lubricant for wooden axles of carts. Gum-resin used in nervous debility and leprosy. Fruits used in asthma, ascites, epilepsy, psoriasis, rheumatism, warts and tumours. Fruit extract is effective against human epidermoid carcinoma of nasopharynx and hypoglycaemia (Ambasta *et al.* 1986).

Cultivation: Through seeds and seedlings.

136. SAL

Botanical: ***Shorea robusta*** Gaertn. f.

Arabic: Kaikhar Burmese: Enkhyen

Eng: Indian Dammer, The Sal Tree Nepali: Sakwa

Hindi: Sal, Sakhu, Salwa, Sakhua, Sarai

Persian: Lalemohari Sinhalese: Dammala

Assamese: Borsal, Hal-orang Bengali: Sal, Sakhu, Salwa, Sakhua

Garhwali: Kandar Gujarati: Rala Haryanavi: Sal

Kannada: Kabba Oriya: Sal, Sodingi, Soringhi

Punjabi: Sal, Seral Malayalam: Maramaram

Marathi: Guggilu, Rala, Sajara Lepcha: Taksal-kung

Tamil: Attam, Kungiliyam, Mulappumarutu Shalam

Telugu: Gugal, Guggilam, Saluva, Sarjamu

Sanskrit: Agnivallabh, Shal, Ashvakarna, Rala, Sala, Divyasara, Chirparna, Dirghaparna, Kaushika

Urdu: Ral

SYSTEMATIC POSITION

Bentham & Hooker (1883)	**Engler & Prantl (1899)**	**Hutchinson (1959)**
Phanerogams	Phanerogams	Angiospermae
Dicotyledones	Dicotyledoneae	Dicotyledones
Polypetalae	Archichlamydeae	Lignosae
Guttiferales	Parietales	Ochnales
Dipterocarpaceae	Dipterocarpaceae	Dipterocarpaceae

The earliest description of Sal (*Shorea robusta*) was given in Sabhaparva of Mahabharatha by sage Vyasa. Lord Buddha attained "Nirvana" (heavenly abode) in a Sal grove (Rao 1991). Hence, Buddhists regarded it as one of the "Bodhi Trees". The great emperor Asoka planted 'Sal Trees' along the roadsides to give shade and shelter to travellers after abjuring violence and upon his conversion to Buddhist during 260 BC.

Geographical distribution: In India two irregular belts separated by Gangetic plain are occupied by Sal. The northern or sub-Himalayan belt starts from Kangra valley in Himachal Pradesh (severe cold climate) to Darrang and Nowgang districts in Assam (humid climate) comprising of Palampur, Una, Sirmur (H.P), Ambala (Haryana) and several districts in U.P., Western Bengal hills, Arunachal Pradesh, Meghalaya and Tripura in the North-east while the Central and Southern Indian belts starts from Coromandel coast west to Pachmarhi hills and south to the Godavari river comprising of several places in Bihar, Madhya Pradesh, Orissa and Andhra Pradesh. Also grown widely in Sri Lanka, Myanmar, Philippines and Malay Peninsula.

Habit: A large gregarious tree upto 50 m height and 7.5 m girth with conical or elongated crown, strong branching system and clean bole, rarely completely leafless. young shoots and inflorescence stellately grey pubescent (Fig.55).

Root: Strong branching tap root system.

Stem: Erect, straight, cylinderical, young branches grey tomentose, young bark greyish-brown, smooth consisting of few deep longitudinal fissures, old bark dark brown, 2.5-5.0 cm thick, rough consisting of deep longidutinal fissures, heartwood dark brown, coarse and cross-grained, hard, heavy, strong and tough.

Leaves: Simple, alternate, stipulate, entire, tough, thinly coriaceous, glabrous, and shining when matured, 10-25 cm long, 5-15 cm broad, stipules 0.7-0.8 cm, falcate, pubescent and deciduous, petiole 1.2-2.5 cm, ovate, oblong, base rounded or cordate, apex acuminate ending in an obtuse point, lateral nerves 10-15 pairs.

Inflorescence: Axillary or terminal, racemose, unilateral panicles, 8-25 cm long with pale, velvetty pubescence, cream, pink or creamy pink coloured.

Flower: Complete, bisexual, yellowish on short pedicles, regular, pentamerous, actinomorphic, floral axis concave or saucer shaped, hypogynous, bracteolate, bracteoles 2, deciduous.

Calyx: Sepals 5, persistent, gamosepalous, calyx tube short, wing like in fruits, lobes triangular, grey tomentose outside, 2.5-3.0 mm long, imbricate.

Corolla: Petals-5, pale yellow, free, oblong or lanceolate, grey tomentose outside, four times longer than the calyx, 12-15 nerved, twisted, orange inside.

Androecium: Stamens numerous (50), shorter than petals, connective prolonged into a bearded appendage in the anthers, anthers ovate or oblong.

Gynoecium: Carpels-3, ovary trilocular, superior, syncarpous, pubescent, each locule with two ovules, style short, glabrous, subulate, stylopodium large and hairy, stigma minute, trifid at apex, placentation axile.

Fruit: One seeded nut with persistent calyx farming a wing for aiding in dispersal, wings 3-5, pubescent, unequal, brown, obtuse, oblong, ovate or spatulate, larger 5-7.5 cm long with 10-15 parallel longitudinal nerves and distinct transverse veins, fruit edible, ovoid, acute, fleshy indehiscnt.

Seed: Non-endospermic, 1.5-4.0 x 3.0-4.5 cm, cotyledons fleshy, filled with starch or fatty oil, germination hypogeal.

Sprouting: Feb - May.

Flowering: Feb-April.

Fruiting: May-July.

Floral formula: ⊕ ⚥ $K_{(5)}\ C_5\ A_{\alpha}\ G_{\underline{(3)}}$

Uses: It serves as one of the primary food plants of tropical tasar silkworm, *Antheraea mylitta* D. Some races of *A. mylitta* viz., Palma, Laria and Barharwa in Jharkhand, Railey, Laria, Sukli, Mandalla, Jahabua, Janggir in Chattisgarh and Modal, Nalia, Omerkote in Orissa are reared from hatching to spinning, on Sal exclusively and hence they are designated as "Sal Ecoraces" and Sal serves as primary food plants for these ecoraces. Other popular races of *A. mylitta* D. like Daba, Sukinda, Bhandara etc. may be reared on their primary food plants viz., *Terminalia arjuna* and *T. tomentose* in first three stages and fourth stage onwards they may be reared on Sal in case of scarcity of their primary food plants (FAO 1976, Sengupta 1986, Thangavelu *et al*. 2000).

Sal has been considered as the good fuel wood.

The Buddhists regard sal as a sacred tree.

The resin of *Shorea robusta* is astringent, detergent, given in dysentery, indigestion and gonorrhoea and also used as an aphrodisiac (Sinha 1996).

The Sal seeds are collected from the forests from tribals to feed the raw material to edible vegetable oil industries because it is highly rich with fatty oil (12.5-18.0%) (Jain *et al*. 1990).

Sal, when once thoroughly seasoned, stands almost without a rival, as a timber in strength, elasticity and durability. It is used for piles, beams, planking and railing of bridges, doors and window posts of houses, bodies of carts and above all for railway sleepers. In Assam, it is favourite wood for boat building and in the hills of Northern Bengal where it is perhaps largest in size, the trunks are hollowed out into canoes. It is also used for making furnitures (Pandey 1981).

Cultivation: It can be propagated through seeds/seedlings. The seed looses its viability within a week, hence it should be germinated just after fall. The seeds of Sal, if stored in air tight polythene bags retain viability for 60 days (Prasad and Parvej 1987). Sal seeds freshly collected start germination after 3rd day of sowing and if stored the germination was observed after 9 days of sowing. Thus properly stored seed can be made viable for longer time. Nursery may be prepared by rooting media of soil, sand and FYM in equal ratio. The rooting media is filled in perforated polythene tubes, seeds are dibbled and regular watering and weeding is attended. The seedlings become ready for transplantation within three-four months during monsoon period. It can also be propagated easily through airlayerings (Chaudhury 1963).

137. TALURA

Botanical: ***Shorea talura*** Roxb. syn. *S. laccifera* Heyne ex Wall., *Vatica laccifera* W & A., *S. robusta* Roth. nov. sp., *S. roxburghii* G. Don.

Eng: Lac tree of South India

Kannada: Jhalli, Jaluda   Tamil: Talura, Talari   Telugu: Jalari

SYSTEMATIC POSITION

Bentham & Hooker (1883)	**Engler & Prantl (1899)**	**Hutchinson (1959)**
Phanerogams	Phanerogams	Angiospermae
Dicotyledones	Dicotyledoneae	Dicotyledones
Polypetalae	Archichlamydeae	Lignosae
Guttiferales	Parietales	Ochnales
Dipterocarpaceae	Dipterocarpaceae	Dipterocarpaceae

Geographical distribution: Western Peninsula especially in Mysore, Malabar and North Canara southwards (on the hills of southern Deccan) in evergreen forests.

Habit: A large tree, perfectly glabrous except pubescent buds and petals outside.

Root: Branched tap root.

Stem: Aerial, erect, branched, woody, solid, cylinderical, branchlets with smooth pale brown bark, wood very hard, heartwood small.

Leaves: Simple, alternate, stipulate, ovate-elliptic or elliptic-oblong, apex and base, obtuse, occasionally apex acute, retuse or mucronulate and base emarginate, blade 9-12 x 5-6.5 cm, margin undulate, petiole 1.25-2.5 cm, lateral nerves 12-15 pairs.

Inflorescence: Terminal or lateral panicles arising from the axils of fallen leaves, lax or drooping, slender and filiform.

Flower: Complete, bisexual, regular, actinomorphic, hypogynous, pentamerous, pinkish white, 2.0 cm in diam, 1.5 cm long, bracteolate, bracteoles 2, pedicels 1.0 cm long.

Calyx: Sepals-5, gamosepalous, calyx lobes glabrous, deltoid, linear, margins ciliate, light coloured.

Corolla: Petals 5, free, pinkish white, pubescent outside.

Androecium: Stamens 15, in three rows, anthers oblong and with a long seta because connective terminating in a long filiform naked appendage, appendage equal to anther, scabrous at the end, anthers of 5 interior stamens longer than others.

Gynoecium: Carpels-3, ovary superior, trilocular, syncarpous, glabrous, stylopodium absent, style longer than ovary, stigma tridentate.

Fruit: Winged nut, one seeded, calyx persistent, larger wings of fruiting calyx 7.5-8.0 cm long, narrowed above the broad base enclosing the fruit completely, bases obovate, wings elliptic-linear, obtuse with approximately 10 slender parallel veins.

Floral formula: $\oplus$ ⚥ $K_{(5)}\ C_5\ A_{5+5+5}\ G_{\underline{(3)}}$

Uses: It serves as one of the secondary food plants of tropical tasar silkworm, *Antheraea mylitta* D. in the forests (FAO 1976, Seth 1995 a). Besides, as the name implies, it also yields lac in south India. It also yields timber of moderate quality used for building and construction purposes such as beams, piles and bridges and also for making furnitures. Seeds yield valuable fatty oil hence it also feeds raw materials to vegetable oil inudstries. Flowers yield an essential oil. Wood serves as excellent fuel wood.

Cultivation: It can be cultivated through seeds. The viability of seeds is very less hence fresh seeds within one week may be used for germination and raising of seedlings in polythene tubes filled with soil, sand and FYM in equal ratio. Three months old seedlings may successfully be transplanted during monsoon period.

138. KARAYA

Botanical: *Sterculia urens* Roxb.

English: Kuteera gum tree French: Gomme kuttera German: Kuteragummi

Hindi: Kulu, Galu, Karai Karaya, Gular Sinhalese: Kavali

Assamese: Hatchanda Chattisgarhi: Kulu, Karhar

Gujarati: Karai, Kariyo, Kandol

Kannada: Kempudale, Bhutali, Kempudare Kol: Tele

Konkani: Paudruk Madhya Pradesh: Kulu, Karhar

Malayalam: Thondi Marathi: Karai, Kandol, Kulu

Oriya: Kavili, Gindrola Punjabi: Gulli, Kulu

Santhali: Telhec, Telhech Tamil: Kavalam, Vellaiputtali

Telugu: Podaku, Tabsu

SYSTEMATIC POSITION

Bentham & Hooker (1883)	Engler & Prantl (1899)	Hutchinson (1959)
Phanerogams	Phanerogams	Angiospermae
Dicotyledones	Dicotyledoneae	Dicotyledones
Polypetalae	Archichlamydeae	Lignosae
Thalmiflorae	Malvales	Tiliales
Malvales	Sterculiaceae	Sterculiaceae
Sterculiaceae		

Geographical distribution: Sub-Himalayan tract from the Ganges eastwards. Commonly found in dry deciduous forests throughout the Peninsula and on the hills of Rajasthan, Madhya Pradesh and Bihar. Also found in Myanmar and Sri Lanka.

Habit: A large deciduous tree.

Habitat: Foot hills and hilly regions.

Root: Tap, branched.

Stem: Erect, aerial, branched, bark white, greenish-grey, smooth and papery, peeling off easily in large irregular flakes/plates.

Leaves: Simple, palmate (cordate, shallowly-5 lobed), tomentose beneath, crowded at the terminal ends of branches, nearly glabrous above, lobes entire, acuminate 20-30 cm long, petiole 15-20 cm long.

Inflorescence: Erect, more or less pyramidal panicles, clothed with a dense sticky tomentum of glandular stellate hairs.

Flower: Hermaphrodite, actinomorphic, hypogynous, yellow, clustered, a few bisexual, mixed with large number of male flowers, petals none, sepals 5, valvate, yellow or brown.

Male flowers: Staminal coloumn short, stamens 20 approximately.

Bisexual flowers: Gynophore short, thick, carpels 5, each with two or more ovules.

Fruit: Follicles of 4-5 sessile, radiating, ovoid, thickly coriaceous carpels, 7.0-8.0 cm long, red when ripe, covered outside with stiff stinging bristles, seeds 3-6 in each carpel, oblong, dark brown.

Flowering: Cold season.

Fruiting: Hot season.

Floral formula: Male: $\oplus$ ♂ $K_5\ C_o\ A_{20}\ G_{(\underline{4\text{-}5})}$

Female: $\oplus$ ♀ $K_5\ C_o\ A_{20}\ G_{(\underline{4\text{-}5})}$

Uses: The wood of Karaya is used for building purposes, packing and many domestic articles. The fibre extracted from bark of the tree is very strong and used for making ropes and thick clothes. The Karaya gum extracted from trunk is mixed with pulses or vegetables and eaten. Seeds are also eaten.

The Karaya/Karira gum is used medically. It has laxative properties. It is more suitable than Tragacanth gum for application on skin. It is used in diseases of throat, manufacturing of artificial teeth, preparation of sweet tablets/globules and chutneys etc. Small amount of gum mixed with honey is taken in morning for throat problems. The poultice made from pounded root bark is applied externally over wounds, fractures and inflammed testes. The small amount of pulverized bark is given to facilitate delivery in Bastar (Jain 1968, Kurian 1995).

Wood used for toys, guitors, inferior packing cases and boardings, dugouts, cement barrels, cheep match boxes, splints, pencils, picture frames and slate frames. Wood is good fuel and yields pulp. Roasted or cooked seeds eaten. Seeds yield fatty oil suitable for edible purposes and soap preparation. Bark fibre used for cordage (Ambasta *et al.* 1986).

Cultivation: Through seeds and seedlings. The maximum germination (80%) of seeds occurs when they are dibbled in horizontal and vertical orientation at 4 cm and 2 cm depth respectively. Early emergence and higher seedling viogour occurs when seeds are sown at 2 cm depth in vertical orientation (Panday and Khatoon 1999).

139. MAHOGANY

Botanical: *Swietenia mahagoni* Linn.

English: West Indian/Spanish/ Cuban Mahogany   Hindi: Mahogany

Bengali: Mahagni   Malayalam: Mahagoni   Tamil: Ciminukku

Telugu: Mahagonichettu

SYSTEMATIC POSITION

Bentham & Hooker (1883)	Engler & Prantl (1899)	Hutchinson (1959)
Phanerogams	Phanerogams	Angiospermae
Dicotyledones	Dicotyledoneae	Dicotyledones
Polypetalae	Archichlamydeae	Lignosae
Disciflorae	Geraniales	Meliales
Geraniales	Meliaceae	Meliaceae
Meliaceae		

Geographical distribution: Native of Southern Florida (Central America) and the West Indies. This is chiefly cultivated in Uttar Pradesh, West Bengal and Tamil Nadu.

Habit: A tall tree attaining a maximum height of 18.3 meter with a short swollen or buttressed trunk up to 1.4 meter in diameter.

Habitat: Hot and moist climatic conditions and nearly all types of soil.

Root: Tap, highly branched, deep.

Stem: Aerial, erect, woody, solid, cylinderical, branched, straight, symmetrical, bark bitter, heartwood pale to deep reddish brown, finely diffuse porous with distinct but not garish rays.

Leaves: Compound, paripinnate, nearly opposite, up to 17.5 cm long, leaflets elliptic-oblanceolate, untoothed, petiole with a thickened base, leaflets opposite, 3-6 pair, 5.0-10.0 cm long, secondary nerves 10-14 pair, alternating with shorter intermediate ones, tertiary nerves prominently reticulate.

Inflorescence: Cymose in lateral clusters.

Flower: Hermaphrodite, actinomorphic, regular, complete, hypogynous, small, greenish, pentamerous with triangular bracts.

Calyx: Sepals 5, polysepalous, green.

Corolla: Petals 5, obovate, narrowed in to a claw, free, contorted in bud.

Androecium: Stamens 10, united to form a cap shaped staminal tube, staminal tube urceolate, 10-dentate, anthers dithecous, filaments short.

Gynoecium: Pentacarpellary, syncarpous, ovary superior, pentalocular, seeds numerous in each cell, flat, imbricated winged at the upper end only, placentation axile.

Fruit: Capsule, up to 10.0 cm long, splitting upwards from the base into 5 parts.

Flowering: April – May.

Fruiting: July – Aug.

Floral formula: ⊕ ⚥ K_5 C_5 A_{10} $G_{(\underline{5})}$

Uses: The tree is valued for its timber. It has long been used as a decorative wood for inlay work especially in eighteenth and nineteenth century. Supplies of this have long since been exhausted its place being at first taken by another tree, the Honduras Mahogany or Bay wood (*S. macrophylla* King) of Mexico and Central America, which has also now become scarce. Most of our present day supply comes from *Khaya ivorensis* chev, *Aucoumea klaineana* Pierre, *Eucalyptus* spp. *Tieghemella heckelii* Pierre ex chav. Syn. *Mimusops heckelii* Hutch & Dalz., *Pentace burmanica* Kurz. and *Shorea, Parashorea and Pentacme* spp.

The Mahogany wood is medium fine textured with a straight to interlocked grain and is very hard and strong enough to resist indentation. The beauty of Mahogany is due to the light reflecting qualities of the rays, particularly when viewed in radial section. Manogany is reasonably easy to work and takes a good polish. It is used for furniture, aeroplane propellers, panelling, veneering, cabinet making, printer's blocks, joinery, musical instruments, ship building, caskets and turnery articles (Kocchar 1981).

Popular wild silkworms of *Attacus atlas* L. are extensively reared on Mahogany trees in Indonesia for their commercial exploitation (Mishra 2002). However, their commercial exploitation is yet to be achieved in India.

Bark contains tannins, used as tonic, astringent, antipyretic and substitute for Cinchona bark (Ambasta *et al.* 1986).

Cultivation: Through seeds and seedlings. The seeds are collected during July-Aug. and nursery is raised during December-March. The seedlings are transplanted during June-July. Seeds may also be directly sown during June-July (Chundawat and Gautam 1993).

140. SYMPLOCOS GRANDIFLORA WALL

Botanical: ***Symplocos grandiflora*** Wall. ex A. DC.

Assamese: Bumroti

SYSTEMATIC POSITION

Bentham & Hooker (1883)	Engler & Prantl (1899)	Hutchinson (1959)
Phanerogams	Phanerogams	Angiospermae
Dicotyledones	Dicotyledoneae	Dicotyledones
Gamopetalae	Sympetalae	Lignosae
Heteromerae	Ebenales	Styracales
Ebenales	Symplocaceae	Symplocaceae
Styraceae		

Geographical distribution: Meghalaya (Khasi hills) and Assam.

Habit: Middle sized deciduous tree.

Habitat: Sub-tropical humid forest patches.

Root: Branched tap root.

Stem: Woody, erect, aerial, branchlets glabrous.

Leaves: Simple, lanceolate, nearly entire, glabrous, 12-18 cm long and 4-5 cm broad, base cuneate or sub-obtuse, thick, coriaceous, midrib impressed, nerves sub parallel, secondary distinct beneath, petiole up tc 0.8 cm long.

Inflorescence: Racemes 5-9 cm, elongated, lax, almost glabrous.

Flower: Actinomorphic, bisexual, regular, pentamerous, bract 1, caducous, deciduous, bracteoles 1-3, lanceolate, subcilliate, epigynous.

Calyx: Sepals 5, calyx tube 0.2 cm long, glabrous, calyx teeth 0.15 cm long, ovate.

Corolla: Petals 5, 0.4-0.6 cm long, polypetalous.

Androecium: Stamens 80, filaments connate in several bundles.

Gynoecium: Bicarpellary, syncarpous, stigma trifid, ovary inferior, bilocular.

Fruit: 1.2 x 0.6 cm, sub-cylinderic, smooth drupe, calyx rim little narrower than fruit, embryo straight.

Seeds: Exalbuminous.

Floral formula: $Br_1\ Brl_{1-3} \oplus ⚥\ K_{(5)}\ C_5\ A_{20-60}\ G_{(\bar{2})}$

Uses: It serves as one of the tertiary food plant of muga silkworms in Assam (FAO 1976, Seth 1995d). Leaves used as a mordant in dyeing (Ambasta *et al.* 1986).

Cultivation: *Symplocos grandiflora* can be propagated through seeds and seedlings.

141. LUDH

Botanical: ***Symplocos paniculata*** Miq. Syn. *S. crataegoides* Buch.-Ham. ex D.Don

Burmese: Daukyat
English: Sapphire Berry, Sweet leaf
Hindi: Ludh
Jaunsari: Lodra
Khasi: Dieng-iong
Kumauni: Lodh
Manipuri: Lairee
Marathi: Lodh
Punjabi: Lodar, Lodh

SYSTEMATIC POSITION

Bentham & Hooker (1883)	Engler & Prantl (1899)	Hutchinson (1959)
Phanerogams	Phanerogams	Angiospermae
Dicotyledones	Dicotyledoneae	Dicotyledones
Gamopetalae	Sympetalae	Lignosae
Heteromerae	Ebenales	Styracales
Ebenales	Symplocaceae	Symplocaceae
Styraceae		

Geographical distribution: Himalaya, from Kashmir to Assam through Bhutan between 2000-8000 ft. ASL, Khasi and Naga hills and Manipur, upper Myanmar, China and Japan.

Habit: A large shrub or middle sized, deciduous tree attaining up to 40 ft. height.

Habitat: Localities in sub-tropical-temperate climate with high humidity.

Root: Branched tap root.

Stem: Woody, aerial, branched, young branches tomentose, stout, erect, solid.

Leaves: Simple, lanceolate to broadly obovate – elliptic, acuminate to sub-obtuse apex, membranous, sharply serrate, blade 5-7 x 2-4 cm, usually softly tomentose beneath, rarely glabrous, reticulating nerves prominent beneath, alternate.

Inflorescence: Branched terminal panicles, hairy or glabrous.

Flower: Bisexual, actinomorphic, regular, complete, pedicellate, bracts-1, small, linear & caducous, snow white in colour, fragrant, bracteoles 1-3, epigynous.

Calyx: Sepals 5, calyx tube adnate, small, imbricate, glabrous.

Corolla: Petals 5, polypetalous, imbricate.

Androecium: Stamens 20-60, filaments connate in 5 bundles, anthers short, oblong dehiscing longitudinally.

Gynoecium: Bicarpellary, syncarpous, bilocular, ovary inferior, ovules-2, pendulous, style filiform, stigma small, capitate, trilobed.

Fruit: Ovoid or nearly globose drupe, 0.6-0.8 cm long, calyx rim small, much narrower than the fruit, embryo curved, endocarp woody, 1-3 seeded.

Seeds: Exalbuminous.

Flowering: March – June.

Floral formula: $Br_1\ Brl_{1\text{-}3} \oplus ⚥\ K_{(5)}\ C_5\ A_{20\text{-}60}\ G_{(\bar{2})}$

Uses: It serves as the tertiary but regarded as secondary food plant of Muga silkworms in Assam (Seth 1995d). Besides, a yellow dye from leaves and bark used as mordant, is also obtained from *S. paniculata* Miq. Medicinally, its bark is usd in opthalmia and also for checking threatened abortion (Sinha 1996).

Leaves serve as fodder, seeds yield an oil and wood recommended for turning and carving (Ambasta *et al.* 1986).

Cultivation: *Symplocas paniculata* may be propagated through the seeds and seedlings.

142. LODH

Botanical: *Symplocos ramosissima* Wall. ex. G. Don

Hindi: Lodh

Nepali: Kharane

Lepcha: Tung chong

SYSTEMATIC POSITION

Bentham & Hooker (1883)	**Engler & Prantl (1899)**	**Hutchinson (1959)**
Phanerogams	Phanerogams	Angiospermae
Dicotyledones	Dicotyledoneae	Dicotyledones
Gamopetalae	Sympetalae	Lignosae

Heteromerae	Ebenales	Styracales
Ebenales	Symplocaceae	Symplocaceae
Styraceae		

Geographical distribution: Himalayan belt from Garhwal to Bhutan up to 5000-9000 ft. ASL. Occasionally in Assam and common in Meghalaya (Khasi hills) up to 4000 ft. ASL.

Habit: A small tree or shrub attaining up to 30-40 ft. high

Root: Branched tap root.

Stem: Aerial, woody, erect, branched, branchlets glabrous.

Leaves: Simple, alternate, exstipulate, oblong or lanceolate, apex acuminate, membranous, chartaceous, glabrous, serrulate, 10.0 x 3.0 cm, base cuneate, petioles 0.6 cm, nerves distant, distict ventrally and slender.

Inflorescence: Racemes lax, 1.0-2.5 cm, 4-8 flowered usually in lower leafless axils, minutely pubescent, longer than petioles, scattered.

Flower: Bisexual, actinomorphic, regular, complete, bracteate, bracts deciduous, pedicel lacking or 0.4 cm, bracteoles 1-2 mm, ovate, 1-3, pubescent, adpressed to the calyx tube, disc glabrous, epigynous, pentamerous, white or pale yellow.

Calyx: Sepals 5, 1-2 mm, ovate, gamosepalous, campanulate, superior, persistent, valvate.

Corolla: Petals 5, polypetalous, 3-4 mm, imbricate in bud, spreading.

Androecium: Stamens 20, adnate to petals, connate in more or less pentadelphous condition, anthers short, filaments slender.

Gynoecium: Carpels 3, ovary inferior, syncarpous, trilocular, ovules 2 in each cell, pendulous from inner angle.

Fruit: Drupe, ellipsoid, 6-8 mm long, calyx rim narrow.

Floral formula: $Br_1\ Brl_{1-3} \oplus$ ⚥ $K_{(5)}\ C_5\ A_{4+4+4+4+4}\ G_{(\bar{3})}$

Uses: It serves as the tertiary but regarded as secondary food plant of muga, silkorm, *Anthereae assama* (Seth 1995d).

Cultivation: It can be propagated through seeds and seedlings.

143. JAMUN

Botanical: ***Syzyzium cumini*** (Linn.) Syn. *Eugenia Jambolana* Lam.
S. jambolamun D.C. Skeels

Burmese: Thabyebyu, Thabye	Eng: Java Plum, Indian Allspice, Jambolan	
French: Jambol	Hindi: Jamun, Jaman	Nepali: Kalajam
Sinhalese: Madan		
Assamese: Jamu	Bengali: Kalajam, Jam	
Garhwali: Phalinda	Gujarati: Jambu	Kannada: Nerale
Kol: Kuda	Konkani: Jambol	Lepcha: Phoberkung
Malayalam: Naval, Naya	Manipuri: Gulamchat	
Marathi: Jaman, Jambul, Jambhul		Oriya: Jamo
Punjabi: Jammu	Sanskrit: Brahaspati, Jambu	
Santali: Birkod	Tamil: Neredam, Naval	
Telugu: Neredu, Neereedu	Urdu: Jaman	

SYSTEMATIC POSITION

Bentham & Hooker (1883)	Engler & Prantl (1899)	Hutchinson (1959)
Phanerogams	Phanerogams	Angiospermae
Dicotyledones	Dicotyledoneae	Dicotyledones
Polypetalae	Archichlamydeae	Lignosae
Calyciflorae	Myrtiflorae	Myrtales
Myrtales	Myrtaceae	Myrtaceae
Myrtaceae		

"Eugenia" is adopted after Prince Eugene of Savoy who was an enthusiastic Botanist of the seventeenth century while "Jambolana" is adopted from Portugese name of the tree. The word Syzyzium has been adopted from the Greek word 'Suzygos' which means paired. Jamun appears to have originated either in India or the East Indies (Kocchar 1981).

The tree is often planted near temples and it has acquired great reverence from "Buddhists" (Randhawa 1983). It has been regarded as one of the favourite fruit of Lord Ganesha in Hindu mythology.

Geographical distribution: Common throughout India excepting dry regions of North-western Rajasthan and Southern Punjab. sub-Himalayan tracts eastwards ascending up to 3000 ft. ASL and South Indian districts up to an elevation of 6000 ft. ASL. Also found in Sri Lanka, Myanmar and Malaya Archipelago to Australia. In India, it is a popular tree in rural areas.

Habit: Large evergreen tree with dark stem, shining, and glabrescent green foliage with dense crown. Considerably varies in shape of leaves and size of flower and fruits (Fig. 56).

Root: Branched tap root.

Stem: Erect, aerial, densely branched, branches spreading, cylinderical, smooth, woody, bark smooth, light or dark grey- brown, exfoliating.

Leaves: Simple, opposite, glabrous, entire, exstipulate, shining, leathery, ovate, elliptic, oblong or elliptic-lanceolate, acute, obtuse or more or less acuminate, blade 7.0-15.0 cm and petiole 1.2-2.5 cm long, secondary nerves many, very fine, slender and closely parallel, transluscent dots (glands) visible against light on the blade, turning red before falling.

Inflorescence: Compound trichotomous cymes (panicled) on previous years wood, rarely axillary.

Flower: Bisexual, complete, middle sized, regular, cyclic, actinomorphic, dirty white, fragrant, sessile, rarely axillary, arising below the leaves, usually tetramerous, epigynous, bracteolate, bracteoles 2.

Calyx: Sepals-4, gamosepalous, teeth 4, calyx tube turbinate, 0.3-0.6 cm long, cylinderical, base short, teeth truncate and very short.

Corolla: Petals 4, gamopetalous, united into a cap (calyptra), white, imbricated in bud, deciduous.

Androecium: Stamens numerous in many series, inserted on a conspicuous disc close to the mouth of calyx tube, anthers small, versatile, bithecous, stamens as long as calyx tube, filaments filiform.

Gynoecium: Carpels 2, syncarpous, ovary inferior, bilocular, placentation axile, ovules many in each locule, style filiform.

Fruit: Berry, green when young, pink while ripening and purplish black, succulent and shining after ripening, 1.6-4.0 cm long, juicy, edible, elliptic-oblong, calyx remains persistent at the apex, single seeded and edible.

Seeds: Non-endospermic, elliptic-oblong, embryo thick and fleshy, radicle short, cotyledons fleshy.

Flowering: March - May.

Fruiting: June-July.

Sprouting: April.

Floral formula: $\oplus$ ⚥ $K_{(4)}\ C_{(4)}\ A_{\alpha}\ G_{(\bar{2})}$

Varieties: Diploid (2x) and hexaploid (6x) varieties are known to occur cytologically (Gill 1974, See Roy *et al.* 1983). In Western Peninsula, North-western India, West Bengal, Assam, Andaman, Central India and hilly regions of Uttarakhand large fruited forms (6x) with elliptic or lanceolate leaves have been found. In Western Ghats (Travancore) and Manipur middle-sized trees with small fruits have been encountered (Brandis 1906 and Pers. experience).

Hooker (1879) has described two varieties under *E. jambolana.* var. *caryophyllifolia* (*E. caryophyllifolia* Linn.) comprises of ovate-lanceolate leaves with a long acumen, calyx shortly 4-toothed and berry globose size of large pea. The var. *obtusifolia* (*E. obtusifolia* Roxb.) comprises of obuse leaves and oblong berries.

Uses: It serves as the secondary food plants of tropical tasar silkworm, *Antheraea mylitta* D. (Jolly *et al.* 1974, Seth 1995 a).

Mainly used as a shade tree for parks and for fruits. The fruit is very delicious, rich in iodine and astringent. Leaves are used as fodder and yield a bright green oil on distillation. Fruit is used for preparing wine, preserves, jams, squashes, and jellies and vinegar. The seeds are used as fodder and bark is used for tanning. The wood also serves as good fuel. The wood is whitish, close grained and durable. The wood may be used for agricultural implements, construction, boats, plywoods, oars, tool handles, cart wheels, well curbs, troughs, sleepers, furniture, props for shafts and galleries in mines and building purposes. Besides, the bark affords brown dyes and also yields Kino, a kind of gum. Also used as green manure, leaves are good source of fodder and blossoms are source of honey. Seeds used for feeding live stocks (Randhawa 1983, Pandey 1981, Ambasta *et al.* 1986).

Medicinally, seeds are very good as a remedy against excess sugar in the blood. 0.5-1.0 gm of seed powder consumed twice or thrice a day or juice of the seeds in dose of two drops in water thrice a day normalizes the blood sugar. For dysentery, haemorrhages and leucorrhoea, the bark of tree in dose of 20 grams/litre of water is consumed daily. The bark is astringent and is used in decoction form as a mouthwash and gorgle. Fresh bark juice along with goat's milk is reported to cure the diarrhoea of children. Fruit pulp is light, stimulant, stomachic and good in colitis. Seed powder has also been reported to be aphordisiac whereas unripe fruit is diuretic and carminative. Bark-tannin is antibiotic and bark extract is used as a remedy in dyspepsia also (Sinha 1996; Ansari 1997, Kurian 1995), Jamun is a very useful fruit for burning urine. Flowers of Jamun triturated with the urine of cow are stated to cause strility in women when taken during the time of menstruration (Dash 1993).

Cultivation: It can be grown from seeds during rains. Though it is very slow initially but grows rapidly subsequently. The susceptibility of young trees for frost is gradually lost and plants attain hardiness with age (Randhawa 1983). Besides, approach grafting or inarching can also propagate it. Both the Forkert and T-budding methods can be used on *S. cumini* (FAO 1976).

144. GULAB JAMAN

Botanical: ***Syzyzium jambos*** (Linn.) Alston syn. *Eugenia jambos* Linn., *Jambosa vulgaris* DC, *Myrtus jambos* Kunth.

Burmese: Thabye | Eng.: The Raised Apple | Hindi: Gulab Jaman

Bengali: Gulab Jamb | Kannada: Pannerale | Malayalam: Jambavam

Manipuri: Gulab Jam | Marathi: Gulab Jaman | Oriyal: Golabjamli

Tamil: Perunaval, Sambunaval | Telugu: Jambuneeredu

SYSTEMATIC POSITION

Bentham & Hooker (1883)	**Engler & Prantl (1899)**	**Hutchinson (1959)**
Phanerogams	Phanerogams	Angiospermae
Dicotyledones	Dicotyledoneae	Dicotyledones
Polypetalae	Archichlamydeae	Lignosae
Calyciflorae	Myrtiflorae	Myrtales
Myrtales	Myrtaceae	Myrtaceae
Myrtaceae		

Geographical distribution: Indigenous to Sikkim Terai in India, Malay Archipelago, Upper Myanmar, Pegu and Peang but distributed from Yunan to Australia. Cultivated throughout India and Myanmar and wild in certain places. Commonly cultivated in Assam, Chhattisgarh, Bihar, Andhra Pradesh, Tamil Nadu, Maharashtra and Bengal.

Habit: A middle-sized tree, cultivated as well as wild.

Root: Branched tap root.

Stem: Erect, aerial, branched, branchlets compressed or nearly tetragonous, woody, solid, cylinderical, smooth, spreading, barks light or dark grey-brown, exfolialing.

Leaves: Simple, opposite, oblong or elliptic-lanceolate, apex acuminate or tapering to both ends, attenuated at the base into a short petiole, thinly coriaceous, blade 10-18 cm X 4-5 cm and petiole 0.6 cm long, Secondary nerves and intra marginal veins slender but distinct, midrib stout, nerves reticulate on the ventral side.

Inflorescence: Short terminal corymbs with 2-4 pairs of opposite flowers, oftenly upper most pair opening first.

Flower: White, or greenish-white, large, 5-7.5 cm in diameter, bisexual, tetramerous, epigynous, actinomorphic, sessile, fragrant, regular, complete, cyclic, pedicel as long as calyx tube.

Calyx: Sepals 4, gamosepalous, campanulate, calyx tube produced above ovary, turbinate, lobes-4, round, calyx tube 1.3 cm long, valvate, persistent.

Corolla: Petals 4, gamopetalous, united into a cap (calyptra), gland dotted, white-greenish white, deciduous.

Androecium: Stamens numerous, in many series, 2.0 cm long, anthers small, versatile, bithecous, filaments filiform.

Gynoecium: Carpels-2, syncarpous, ovary inferior, bilocular, placentation axile, ovules many in each locule, style filiform.

Fruit: Globose, subglobose or pear shaped berry crowned with 4 persistent calyx lobes, fruit 4-5 cm long, yellow or pink, edible.

Seed: Non-endospermic, elliptic-oblong, cotyledon fleshy.

Flowering: Feb-April but also at some other times depending upon regional climate.

Fruiting: May - July.

Floral formula: $\oplus$ ⚥ $K_{(4)}\ C_{(4)}\ A_{\alpha}\ G_{(\bar{2})}$

Varieties: Tetraploid (2n = 44) and hexaploid (2n = 66) forms have been reported in this species (Gill 1974, see Roy *et al.* 1983).

Uses: It serves as one of the secondary food plants of tropical tasar silkworm, *Antheraea mylitta* D.

This is also used as shade tree in parks and gardens for fruit. Also planted as avenue tree along the roads. The fruits are rich in calcium, vitamin B complex and cover phosphorus deficiency (Sinha 1996) but they are tasteless hence used for making candied fruit jellies, sauces and jams. Leaves yield an essential oil.

Wood may be used for building and construction purposes.

Cultivation: It can be cultivated through seeds or seedlings.

145. IMLI

Botanical : ***Tamarindus indica*** Linn.

Burmese: Magyi, Gondi Chita, Hitta

English: Tamarind, Indian Date Hindi: Imli, Amli, Imlee, Ambli

Assamese: Tetuli Bengali: Tentul, Amli, Taetul

Gujarati: Amli, Ambli Kannada: Chinch, Hunsemara

Malayalam: Valampulli, Amblam

Manipuri: Mangge Marathi: Chinch, Ambli, Chitz

Oriya: Tentuli, Koya Punjabi: Imbli Sanskrit: Amlika

Tamil: Puli, Amilam Telugu: Chinta

SYSTEMATIC POSITION

Bentham & Hooker (1883)	Engler & Prantl (1899)	Hutchinson (1959)
Phanerogams	Phanerogams	Angiospermae
Dicotyledones	Dicotyledoneae	Dicotyledones
Polypetalae	Archichlamydeae	Lignosae
Calyciflorae	Rosales	Leguminales
Rosales	Leguminosae	Caesalpiniaceae
Leguminosae	Caesalpinoideae	
Caesalpinoideae		

Geographical distribution: Although cultivated throughout Asia and America; common in India and Myanmar; cultivated as well as wild in the tropics of both new as well as old world, it is belived to be indigenous to tropical Africa. In India it is commonly found in Central India, South India and eastern part of Uttar Pradesh.

Habit: A large unarmed evergreen tree.

Habitat: Waste and forest lands (Fig. 57).

Root: Tap, branched.

Stem: Erect, aerial, solid, woody, branched, bark dark grey, rough, covered with longitudinal fissures and horizontal cracks.

Leaves: Compound, paripinnate with 10-20 pairs of small leaflets (Pinnae), exstipulate, pulvinate, alternate, pinnae entire.

Inflorescence: Raceme lax, terminal or at the end of short lateral branches with 10-15 flowers.

Flower: Hermaphrodite, complete, zygomorphic, hypogynous, bracteate, bracts and bracteoles boat shaped, reddish, inconspicuous, about 1.25-1.50 cm long, variegated yellow and red.

Calyx: Sepals 5, polysepalous, two posterior connate, lanceolate odd sepal anterior, imbricate.

Corolla: Petals 5, polypetalous, one posterior petal hooded, yellow striped with red, the two lateral ovate and the two anterior ones reduced to white scales, imbricate.

Androecium: Stamens 7, monoadelphous, fertile 3, alternating with 4 staminodes which are bristle like structures at the top of the sheath, anther oblong, basifixed, versatile filaments united to the middle.

Gynoecium: Carpel 1, unilocular, ovary superior, stipitate with numerous ovules, stalk adnate to calyx tube, placentation marginal, style shorter than ovary, thick, slightly bent, stigma flat.

Fruit: Tomentum, usually falcate, epicarp thin and crustaceous, mesocarp thick, pulpy acidic with 3-12 smooth, shining seeds.

Flowering: April – June and Oct. – Nov.

Fruiting: Nov. – Dec.

Floral formula: $\cdot|\cdot$ ⚥ $K_{(2)+3}\ C_{3+2}\ A_{3+4}\ G_{\underline{1}}$

Varieties: There are some varieties with sweet pulp while some are with reddish pulp which is considered to be the best.

Uses: It is an excellent tree for road side avenues and useful for its highly prized timber and fruits. Timber is very hard to work yet used for agricultural implements, tool handles, wheels, turnery, painting blocks, tent-pegs, decorative panels, furniture, beams, rafters, purlins, trusses, mallets, rice pounders, oil presses, etc. The pulp is pressed, preserved and sold by weight in the market. The pulp is used as a favourite ingradient in condiments and chutnies. Besides, it is principal souring agent for sauces, chutneys and beverages and used in general cooking also. The leaves and flowers of Imli are also edible. Seeds yield a cheap starch used for sizing cotton yarn, jute fabrics and woollens (Randhawa 1983). Seeds yield a oil like peanut for paints, varnishes and illumination (Ambasta *et al.* 1986).

Insects do not attack tamarind wood hence, it is also used for agricultural implements, domestic articles and furnitures etc. The charcoal from tamarind wood is used for preparing gunpowder. The leaves yield a yellow colour. The acidic pulp is very useful for dyeing, tanning, polishing and cleaning silver and utensils. The seeds are used in jam, marmalades and jelly industries. Powdered seeds are used for making adhesives for book binding, card board, plywood and woods also (Jain 1968).

Medicinally, Tamarind pulp is a refrigerant, carminative, anti scorbutic and laxative and given as infusion in biliousness and febrile conditions. Pulp from unripe fruit is roasted and applied to swellings due to sprain. Mixture of pulp with water is used in cough and cold. In rheumatic pain, paste made of fresh leaves is applied. Mixture of pulp with banana and milk is prescribed in dysentery while leaf juice is used as an eye tonic (Sinha 1996).

Leaves flowers and young seedlings serve as vegetables. Bark yields tannins. Leaves and flowers used in dyeing. Flowers are good source of honey and leaves serve as fodder and green manure (Ambasta *et al.* 1986).

Cultivation: Through seeds in March – April and seedlings in June – July. In some places seeds are collected during March, seedlings are raised during April and they are transplanted during July–Aug. (Chundawat and Gautam 1993).

Due to higher insect infestation, the seeds of fruit storage recorded no germination after 9 months of storage while the seeds (non-scarified) stored as such recorded more than 99% germination after 12 months of storage (Parmeshwari *et al.* 2001). Hence, storage of only non-scarified seeds should be practised.

146. YEW

Botanical: ***Taxus baccata*** Linn.

English: Common Yew Nepali: Dhingre Salla, Tcheiraygulab

Hindi: Thuno, Thuner, Birmi

Bengali: Burmie, Iichujhau Kashmiri: Birmi, Postil Khasi: Dhunu, Dingsableh

Kumauni: Thaner, Gallu Lepcha: Cheongbu, Tunsi Marathi: Barmi

Vernaculars: Barma, Barmi, Thunu, Arkhau, Yamdal, Thuner

SYSTEMATIC POSITION

Bentham & Hooker (1883)	**Engler & Prantl (1899)**	**Hutchinson (1959)**
Phanerogams	Phanerogams	Gymnospermae
Gymnospermae	Gymnospermae	Coniferales
Coniferae	Coniferae	Taxaceae

In Great Britain *T. baccata* is found in many churchyards. Specimen estimated to 3000 years old are known, the oldest of which, a mere hollow shell, stood in the churchyard of Fortingal in Perthshire, Scotland.

Geographical distribution: Temperate and warmer regions of Northern hemisphere and mountains in the tropics. Widely distributed in Europe, Asia, North Africa and North & Central America. In Asia, it is found in India, Japan, Afghanistan and Myanmar (Ruby Mine hills) above 5000 ft ASL. In India, it is found in sub-Himalayan regions between 6000-10000 ft. ASL in Himachal Pradesh (Dalhousie and Simla), Meghalaya (Khasi hills), Manipur, Nagaland and Arunachal Pradesh.

Habit: A slow growing, long lived, dioecious and evergreen shrub or tree attaining a height of 10-20 meters.

Habitat: Shady ravines.

Root: Tap root, branched.

Stem: Aerial, erect, columnar, woody, solid, cylinderical, branched, branches spreading, branchlets perfoliate, dense, bark reddish-grey, thin, smooth, peeling off in longitudinal shreds. The heartwood hard, usually reddish-brown. Branches of unlimited growth (long shoots) present, branches of limited growth (dwarf shoots) and resin canals absent.

Leaves: Dark green, flat, distichous, 2.0-3.0 cm long, linear – linear lanceolate, spiny, single nerved, broadens into a very short, flat, sheathing leaf base decurrent on the stem, apex pointed, spiral but twisted at the bases, prolonged leaf bases persist as ridges after leaf fall.

Flowers: Male and female flowers (cones) are borne in leaf axils of different trees on short axillary branchlets which are densely clothed with imbricating bracts.

Male cones: Solitary, short stalked, yellow, globose heads occur on the shoots of same season underside of the branchlets in the axils of leaves. Sterile bract scales 10, towards lower portion of central axis, spirally inserted increasing in size upwards. Fertile micro (male) sporophylls (stamens) towards upper portion of the central axis, 6-10, clustered, umbellate, peltate (consisting of stalk and a disc like head). Disc like head bears 4-7 sometimes 9 pendant sporangia (pollen sacs) surrounding the stalk. Pollen grains do not have wings.

Female cones: Ovulate strobilus (female cone) very much reduced. The female flower arises in axil of leaf of the previous year and consists of three crossed pairs of minute scales and a straight, terminal, solitary, ovule (megasporangium) surrounded by membranous disc which enlarges in to a red fleshy cup.

Seeds: Dicotyledonous; berry, poisonous, attractive; aril, a fleshy red out growth arises from the base of the ovule; germination epigeal.

Uses: All parts of plant are poisonous and used as fish poison except fleshy arils. Both leaves as well as fruits are emmengogue, sedative and antiseptic. Leaf is aphrodisiac, antspasmodic, lithontriptic and used in asthma, bronchitis, hiccup, indigestion, nervousness, hysteria and epilepsy whereas fleshy arils are eaten, stomachic, carminative and expectorant. A tincture from young shoots has long been in use for headache, giddiness, feeble and falling pulse, extreme colds, diarrhoea and severe biliousness. Hair lotions, beauty and shaving creams and dentrifices are made from extracts of various parts of the tree (Sinha 1996).

The bark of the tree grounded into powder, when boiled with water and mixed with salt and ghee known as namkeen tea mostly preferred by Bhotiya tribe in Mandakini valley in Uttarakhand. This keeps body warm during winter and also cures headache due to cold (Nautiyal *et al.* 2002).

Wood used for cabinet, work fancy articles, knives, combs, carvings, in lays, veneers, furnitures, parquet flooring, panelling, gates, fences, poles, ploughs, pencils and turnery (Ambasta *et al.* 1986).

Cultivation: Through seeds, seedlings and stem cuttings. Application of 0.75% IBA + 5% captan + 5% sucrose in August resulted in highest 72.22% rooting in the species. The girdled cuttings produced significantly higher rooting than non-girdled ones. Similarly, the lower portion of cuttings resulted in significantly better success as compared to upper ones (Kumar and Shamet 2002).

IBA, NAA, IAA (auxins) and TDZ (thiadiazuron) induced 82.0, 68.0, 76.0 and 88.09% rooting in stem cuttings of *T. baccata* at lower concentration (50 ug/ml). The combination of TDZ, IBA and NAA (50 ug/ml each) was highly effective for 96% root induction (Mishra *et al.* 2000). IBA at 10000 ppm is optimum for induction of better rooting percentage, higher number and greater length of primary roots in the juvenile stem cuttings of *T. baccata* (Khali and Sharma, 2003).

147. SAGON

Botanical: ***Tectona grandis*** Linn.

Burmese: Kyun, Kyum — Eng.: Teak — Hindi: Sagon, Sagun

Persian: Saj, Sal — Sinhalese: Tekka

Assamese: Chingjagu — Bengali: Segun, Sagun — Gond: Teka

Gujarati: Saga — Kannada: Jadi, Tega — Konkani: Sailo

Lepcha: Ripnyok	Malayalam: Thekku	Manipuri: Chigsu
Marathi: Sag, Sagwan	Oriya: Singuru	Punjabi: Sagwan
Rajputana: Saguan	Santali: Saguna	Sanskrit: Saka
Sindhi: Loheru	Tamil: Tekkumaram, Thekku	
Telugu: Adaviteekum, Teku	Urdu: Sagun	

SYSTEMATIC POSITION

Bentham & Hooker (1883)	Engler & Prantl (1899)	Hutchinson (1959)
Phanerogams	Phanerogams	Angiospermae
Dicotyledones	Dicotyledoneae	Dicotyledones
Gamopetalae	Sympetalae	Herbaceae
Bicarpellatae	Tubiflorae	Verbenales
Lamiales	Verbenaceae	Verbenaceae
Verbenaceae		

Geographical distribution: Native of India, Java, Sumatra and parts of Indonesian Archipllago. Extensive forests of teak occur in Myanmar and Thailand and teak plantations have been set out in Sri Lanka, Java and other islands of East Indies and in Central and Northern South America. In India, teak forests cover an area of about 7276000 ha. out of which 3119000 ha. occur in Madhya Pradesh and Chattisgarh, 1404000 ha. in Maharashtra, 1176000 ha. in Gujarat and 1000000 ha. in Karnataka. Rajasthan, Kerala, Tamil Nadu and Andhra Pradesh also produce teak.

Habit: A large deciduous tree, 30-32 m in height and 2.5-4.0 m in girth with approximately more than 200 years of age, crown of large leaves rounded, bole clean and cylinderical, buttressed at base in large trees (Fig. 58).

Habitat: Its best growth takes place in warm, moist forests of Kerala where it constitutes 20% of the crop. In dry deciduous forests, its growth is not good and size is not large, hence it is used as poles or small planks and battens in such places.

Root: Strong, branched tap root.

Stem: Erect, aerial, branched, branches pubescent, quadrangular and channelled with large quadrangular pith, cylinderical, woody, solid, heartwood golden yellow to golden brown when freshly sawn, but turning darker afterwards due to exposure and immune to termites and wood rotting fungi due to an essential oil in it, sapwood white and susceptible to termites and fungi. The wood scented like weather, greesy in touch, extremely durable, moderately hard, crack/warp/split resistant, easy to work, excellent in polish, 45 lb./ft.^{3} or 609-689 kg/m^3 average weight in dry state, annual growth rings distinct and contain narrow belt of large vessels, grain straight normally, texture very coarse and uneven.

Leaves: Simple, very large, opposite or whorled, broadly ovate or elliptic, entire, acute or mucronate, dorsal side rough, clothed with dense, stellate grey or tawny tomentum, blade 30-90 cm long, narrowed into petiole of 2.5-4.0 cm length, 60-90 cm long leaves in coppice shoots or seedlings.

Inflorescence: Terminal cymose panicles erect, hairy, large, branched, trichotomously 30-90 cm long.

Flower: Bisexual, white, complete, regular, cyclic, hypogynous, bracteate, bracts lanceolate at the forks, bracteoles below calyx, linear.

Calyx: Sepals 5-6, gamosepalous, campanulate, valvate, lobes 5-6, hairy, persistent, enlarged and bladder like in fruit enclosing it, ovoid, crumpled and veined.

Corolla: Petals 5-6, gamopetalous, white, lobes 5-6, unequal, imbricated, corolla tube short, limb spreading.

Androecium: Stamens 5 at the base of corolla, anthers exserted, bithecous, dehiscence longitudinal.

Gynoecium: Carpels 4, syncarpous, ovary superior, tetralocular, fleshy, style short and slender, stigma bifid.

Fruit: Drupe, subglobose, 4-celled, pericarp soft with felted stellate hairs, endocarp bony, enclosed in accrescent calyx, placentation axile, single ovul in each cell.

Seed: Non-endospermic, 1-3 rarely 4.

Flowering: July-August during rainy season.

Leaf fall: Nov - Jan.

Fruiting: Nov - Jan.

Sprouting: April - May.

Floral formula: ⊕ ⚥ $K_{(5\text{-}6)}\ C_{(5\text{-}6)}\ A_5\ G_{\underline{(4)}}$

Uses: It serves as one of the secondary food plants of tropical tasar silkworm, *A. mylitta* D. in forests (Jolly *et al.* 1974, Seth 1995a).

Teak has numerous medicinal uses. The decoction of root is given in anuria (absence of urine excretion) while infusion of flowers is given in biliousness, urinary discharges, bronchits and congestion of liver. The decoction of bark is useful as laxative, anthelmintic, stringent and expectorant and used as a remedy against piles, lencoderma, dysentery, headache, burning pain over the region of the liver and cures inflammed eye lids, indigestion, inflammatory swellings, itchiness of the skin and discomfort or pain with burning sensation in the stomach. The oil extracted from the flowers and seeds promote growth of hair and checks scabies. The flower and seeds have also been reported to be diuretic and promote the flow of urine. Leaf extract is prescribed in hysteria. The boild extract of leaves along with small fishes (*Puntius phutino*) is used for normalising blood circulation in Manipur (Sinha 1996, Kurian 1995).

It is one of the best timbers in the world and hence used as a standard for comparision of other timbers. It is seasoned naturally and highly prized for house buildings, bridges, piles, cabinet work, railway carriages and wagon, ornamental veneers, wheel spokes, felloes, shingles, turnery, carving, brush backs, toys, plywood and other decorative surfaces, flooring and green house construction. In ship building and boats making it is superior to oak and infact a class by itself (Sagreiya 1967, Kochhar 1981).

Teak is greatly durable as it contains a large quantity of fluid resinous matter, which fills up the pores and resists the action of water. Teak has been used in India widely for more than two thousand years as shown by the ruins of old temples in Southern and Western India, where timbers have still been found in good condition. One such example is that of karli caves near Poona, which is more than 2000 years old, but the teak woodwork is still perfectly good.

Even some streets are paved with blocks of teak wood and it is used for flooring and roofing tiles due to its plentifulness in Myanmar (Kochhar 1981).

Cultivation: It may be cultivated through seeds, seedlings and cuttings.

For propagation through stem cuttings maximum rooting (40%) was observed by their treatment for 24 hrs. with NAA 100 ppm during March and IBA 100 ppm in May, whereas in control (without harmonal treatment) rooting occurred during March and the percentage was very low 10% at Kerala Forest Research Institute, Peechi. Rooting (10-40%) was observed from Jan-June and Oct - Dec. in *T. grandis* with growth regulating substances viz., NAA, IBA and Coumarine in either 10 or 100-ppm doses. No rooting was observed during July-Sept., hence season has considerable effect on rooting. These findings may very well be utilised for propagation of teak through stem cuttings (Surendran and Seethalakshmi 1987).

Based on different experiments, technology for large-scale plantation through stem cuttings has been developed by Nautiyal and Rawat (1994) as hereunder:

1. Full grown healthy mother plants of 15-25 years with all desirable characters should be selected for collection of branch cuttings.

2. Branch cuttings of 20-25 cm length and 1.5-2.0 cm diameter should be prepared from lower part of the crown of mother tree.
3. IAA and IBA 100 ppm stock solutions should be prepared by dissolving 100 mg. of their powder separately in 1.5 ml. of absolute alcohol and making it to 1000 ml with distilled water.
4. The best rooting response season is April-May (Summer).
5. River sand and soil of 2:1 ratio is good medium for rooting.
6. Rooting media is filled in earthen pot, polythene bags or nursery beds.
7. Cuttings just after collection should be coated with wax at the upper end and treated for 24 hrs by dipping the 5 cm basal portion in the test solution and subsequently planted in rooting medium. Regular watering is done. The transplantation in field is done only after adequate rooting after 3 months during rainy season. As much as 60% rooting has been noticed in *T. grandis*.

148. ARJUN

Botanical: ***Terminalia arjuna*** Bedd. Syn. *T. arjuna* (Roxb.) W. & A.

Burmese: Toukhyan Eng: Arjun, White Murdah

Hindi: Arjun, Arjuna, Arjan, Kahua, Kaugach, Khawa, Koha, Kowa, Anjan, Anjani

Sinhalese: Kumbuk

Assamese: Orjun Baiga: Kahu

Bengali: Arjhan, Arjuna, Arjun, Kahu Bundeli: Kowa

Canarese: Arjuna, Billimatti, Halematti, Kakubha, Nirmatti, Toramatti

Chattisgarhi: Kahua, Kowah, Kowha Deccan: Kava

Gujarati: Arjuna sadada, Arjunasadadra

Gond: Arjun, Koha, Mangi

Kannada: Maddi, Holematti, Bilimaddi-nirmathi, Thasamatti

Kolami: Garahatana Konkani: Arjun, Nadiain, Holematti

Malayalam: Attu-maruthu, Kattukadukka, Marutu, Nirmaruthu, Vellilayer, Vellamarutu

Marathi: Anjan, Arujuna, Arujuna-sadura, Azun, Arjuna-ladada Arjun, Sadaru, Sanamadat, Savimadat, Vellamarda, Sadura

Mundari: Garahatana Oriya: Arjuno Punjabi: Arjan, Jumle

Santali: Kanha

Sanskrit: Arjuna, Chitrayodhi, Dhanvi, Dhavala, Dhananjaya, Gandivi, Indradaru, Kireeti, Indrasunu, Kakubha, Pandava, Virataru, Vairnatak, Virvriksha, Krishnasarathi, Prithaja, Vira, Savyasachi, Shivamallaka, Kaunteya, Indradruma, Partha

Tamil: Vellamatta, Nirmatti, Kakubhamu, Tellamaddi Nirmarudu, Vellaimarudu

Urdu: Arjan Telugu: Yerramaddi

SYSTEMATIC POSITION

Bentham & Hooker (1883)	Engler & Prantl (1899)	Hutchinson (1959)
Phanerogams	Phanerogams	Angiospermae
Dicotyledones	Dicotyledoneae	Dicotyledones
Polypetalae	Archichlamydeae	Lignosae
Calyciflorae	Myrtiflorae	Myrtales
Myrtales	Combretaceae	Combretaceae
Combretaceae		

The genus *Terminalia* derives its Latin name (terminus = end) from the position of the leaves which are borne in tuft at the ends of the branchlets. At least in the young stages, the principal lateral branches of the tree arise in whorls. The specific name explicits its sacredness and relationship with 'Arjuna' the great warrior, legendary hero and dextrous archer of Mahabharata period. The oldest *T. arjuna* trees having approximately 9.0 and 11.0 m girth from Jammu and 9.0 m. girth from Kollegal at Saviramathi (Karnataka) on the bank of Cauvery have been reported which shows that they are much older than sacred *Morus serrata* tree having only 6.0 m girth estimated to be more than 1250 years old. 'Kunti', the wife of king 'Pandu' gave birth to the third of Pandavas amidst the forest of "Arjuna" trees. Hence, Pandu named him "Arjuna" for the trees around.

According to another legend, the Arjuna tree lamented about the meaningless life since neither its flowers are utilised for sacred purposes nor dry woody fruits are edible and requested for salvation from Saint/Sage "Sumantak" who was meditating below an Arjuna tree in the dense forests of Askand. The Saint consoled it by blessing that "Arjuna" the famous personality of 'Aryavarta' shall take birth below Arjuna shade and Nala and Manigriba, the twins of Kubera (God of prosperity) shall take the form of Arjuna tree owing to a curse given by Devarshi 'Narad'. In due course of time, after getting a holy touch of child "Krishna" after a century the effects of the curse will go away and the tree shall be highly regarded for its medicinal properties (Patro and Mohanty 1998).

The blessings of Saint Sumantak, Devarshi Narad and Lord Krishna and its holy association with Arjuna, Nala and Manigrib in due course of time made it famous for its magnificant medicinal qualities bounteously beneficial for the mankind and for the wild silkworms.

Kalidasa specifically mentioned "Arjuna" among the beautiful flowering plants in his play 'Abhigyan Shakuntalam' (Rao 1991).

Geographical distribution: Common in Indian Peninsula from Avadh towards Southern, Western and Central India specifically in Madhya Pradesh, Chattisgarh, Orissa, Bihar, Chotanagpur, parts of Uttar Pradesh, Maharashtra, Tamil Nadu, Andhra Pradesh, Kerala, Rajasthan and Punjab, rare in Karnataka but plentiful in Tirunelvelli and on west coast. It extends northwards to the sub-Himalayan tract through Rohilkhand and Doon valley. In Himachal Pradesh, it is found upto 700 m ASL. Also, found in Sind (Pakistan), Indonesia (Jakarta) and United States of America. In Sri Lanka and tropical Africa, it occurs as *T. glabra*.

Habit: A large evergreen tree with smooth and grey bark flaking off in large thin layers, attaining the height up to 45 m (usually 60-80 ft.) and girth up to 10 m (usually 6-10 ft.), crown spreading, branchlets drooping, bole rarely long and straight, usually buttressed.

Habitat: This is chiefly found growing near streams, riverbanks, nalas and moist places. It thrives well at places where maximum and minimum temperature ranges between 30-45°C and 5-15°C respectively and rainfall ranges from 73.5 - 170 cm.

Root: Branched tap root.

Stem: Erect, aerial, branched, woody, solid, bark very smooth, pale pinkish green, pinkish grey or greenish grey exfoliating in large, thin and irregular flakes, branches drooping or horizontal, heartwood dark brown and very hard.

Leaves: Simple, opposite or usually sub-opposite, oblong, elliptic or lanceolate, rarely spatulate, lamina 10-25 x 4.5-9.0 cm, coriaceous, pale dull green and smooth or glabrescent below, dark green-dull green and glabrous above, hard, often crenulate rarely serrated or serrulate, apex obtuse, cordate, acute or sub-acute, base rounded or cordate or narrowed suddenly, petiole rarely more than 1.2 cm, two large, cylinderical glands on the ventral side of lamina present nearly at the top of the petiole on both sides of midrib, veins reticulate, lateral nerves 10-15 pairs.

Inflorescence: Panicled spikes, glabrous but may be slightly pubescent when young, 6-10 cm long, axillary or terminal.

Flower: Bisexual, incomplete, regular, cyclic, dull yellow or white, bracteate, actinomorphic, epigynous, sepaloid, small, sessile, about 0.5 cm across, bracteoles linear - lanceolate, caducous.

Perianth: Tepals 5, sepaloid, perianth tube above the ovary with a companulate mouth, limb with 5 short, valvate, triangular lobes, semideciduous.

Androecium: Stamens 10, biseriate, inserted on the perianth tube, episepalous, 5 upper longer ones between the calyx lobes alternating with 5 lower shorter ones opposite calyx lobes, anthers bithecous, filaments slender.

Gynoecium: Carpel 1, ovary unilocular, inferior, ovules 2-3, pendulous, anatropous, young ovary very shortly covered with curled brown or profuse hairs, style long, filiform with pointed or capitate stigma, receptacle tube bearing a disc clothed with yellowish hairs.

Fruit: Winged nut, hard, ovoid, obovoid, oblong or ovoid-oblong, drupaceous samara, 3-6 cm long, indehiscent, dark brown or reddish, almost glabrous, woody when mature with 5-10, usually 5, thick, woody projecting wings, wings less than 1.5 cm wide, closely veined which are not spreading horizontally but tend to curve upwards.

Seeds: Non-endospermic, fleshy, oily, germination epigeal.

Flowering: March-June rarely July.

Fruiting: Feb-March of next year.

Floral formula: Br. $\oplus$ ⚥ $\overbrace{P_{(5)}\ A_{5+5}}\ G_{\bar{1}}$

Varieties: *Terminalia arjuna* exhibits extensive continuous rather than discontinuous variations in foliar, stem/bark, and fruit characters and hence species/varietal delimitation has become extremely difficult. Foliar variations exist at the level of shape, size, apex, base, margin and texture, while fruit variations are found in weight, length, breadth and number of wings. Hence, *T. arjuna* has been regarded as a species complex rather than simply a species (Hooker 1878, Parker 1925, Parkinson 1936, Srivastav *et al.* 1997 a, b). Exhaustive exploration of genetic resources conducted by the author in the forests of Chattisgarh the erstwhile Madhya Pradesh, Bihar and Orissa revealed existence of following morphotypes/biotypes (Srivastav and Thangavelu 1995, 2000, Thangavelu *et al.* 2000).

1. ***T. berryi* W. & A. syn. *T. arjuna* var. *angustifolia***: Leaves narrow, elongate-oblong, suddenly narrowed into petiole with drooping branches. L/B ratio >4, leaves nearly opposite from lanceolate to linear-oblong, glabrous with two glands on the margins of the apex of the petiole, bark smooth and spikes terminal, somewhat panicled. It is distributed along Balaghat mountains and banks of Kalindi and Gutaprabha rivers. It is also found in Southern Peninsula and Konkan.

2. ***T. glabra* W. & A. syn. *T. arjuna* var. *arjuna***: Leaves often cordate, obtuse or very shortly acute with horizontal branches. L/B <4, oblong, 12-

20x3.0-5.0 cm, abrupt at both ends, slightly hairy on the veins, petioles short, fruit obovate. It is common in Deccan tableland, sub-Himalayan tracts of the North-West provinces and Sri Lanka (Fig. 63).

3. **Small leaf form**: Leaves elliptic to lanceolate, L/B = 3.0-3.5.
4. **Putative hybrids closer to *T.arjuna***: Bark grey, slightly rough, leaves broader than those of *T.glabra* and not perfectly smooth or glabrescent, veins not buldging.

All varieties of Arjuna are suitable for tasar silkworm rearing.

Uses: Arjuna serves as one of the primary food plants of tropical tasar silkworm, *Antheraea mylitta* D and most of tasar produce comes from Arjuna plantations raised in Chattisgarh, Orissa, Jharkhand, West Bengal, Andhra Pradesh, Maharashtra and Uttar Pradesh, where more than 7845 ha. Plantation has been raised under Inter-State Tasar Project (ISTP) funded by Indian and Swedish Governments (Shetty and Samson 1998). Besides, Union of India and various State Governments had raised Arjuna plantations before ISTP also. Most of the eco-races of tropical tasar silkworm, *Antheraea mylitta* D. are reared on Arjuna. The prominent among such races are Daba and/or Piprai, Sukinda, Bhandara, Bankura, Bhopalpatnam and/or Kurud, which are reared in Jharkhand, Orissa, Maharashtra, West Bengal and Chattisgarah respectively (Sinha 1998).

Ayurvedic medicines like "Arjunarishta" prepared from Arjuna tree are available in market. Medicinally, bark and fresh leaves are used as a remedy against many diseases (Kurian 1995, Srivastava 2000). The decoction of the bark is useful as a tonic styptic, antidysenteric, expectorant and a good febrifuge and used for fractures, ulcers, piles, blood diseases, dropsy, congestion of the liver, diseases of the heart, painful or difficult urination, leucoderma, anemia, tumours and asthma. It also relieves fatigue and used as a wash in ulcers. The juice of the fresh leaves is a good remedy for earche. Fruit is tonic and deobstruent. The powder of bark with milk is given in hypertension, dropsy and cirrhosis of liver.

The fundamental methods for using Arjuana as remedy in different diseases are detailed below (Patro and Mohanty 1998).

1. **Sprains**: Paste prepared from crushing and grinding of Arjuna bark with unboiled milk or Arjuna bark with black pepper is applied externally.
2. **Bile disorders**: Extract of Arjuna bark after boiling in water may be taken twice daily with powdered sugar candy.
3. **Diabetes** (Madhumeh) and Prameh: Bark extract prepared by boiling in water or bark powder is mixed with honey and eaten.

4. **Fractures**: Bark powder is taken with ghee and honey for quick joining of bones after proper bandage and dressing.
5. **Heart diseases**: Extract of bark in cow milk and water is cooled, filtered and taken.
6. **Congestion**: Bark of root is boiled in goat milk and taken.
7. **Wound**: Cow-ghee applied Arjuna leaves are roasted and tightly tied over wound.
8. **Lice**: Leaf or flower is grinded with milk and paste is applied to skin for destroying lice.
9. **Poision**: Boiled bark extract with sugar is taken to subside the action of poison.
10. **Dropsy**: 30 gm bark of Arjuna is boiled in 250 ml milk and this decoction is taken every morning empty stomach.

[Ayurvedic physicians may be consulted for appropriate dosage in various diseases.]

Fresh leaves are good fodder for goats. (Prakash *et al.* 1991, Pers.experience). It is an excellent shade tree and is often planted on roadsides. The bark and fruits are used for tanning and dyeing. The bark shows the presence of calcium and arjunalic acid, saponin and (+) leucodelphindin have been isolated from it (Chopra *et al.* 1974).

The timber from *T. arjuna* is used for agricultural implements, building of boats and carts and tool handles. Its wood is also used for construction works (Singh *et al.* 1983, Randhawa 1983), water troughs, plywood, water traps, masts, poles and jetty pics (Ambasta *et al.* 1986).

Terminalia arjuna extensively helps in soil conservation, reclammation and pollution control. It is a fast growing tree species and grows successfully in water logged areas and under various soil conditions ranging from pH 5.0 to 7.0 without any adverse effect on growth and foliage production and helps in soil conservation by adding organic matter to the soil in the form of leaf and silkworm litter every year. Subsequent to establishment of *T. arjuna*, grasses and other vegetation also get established in highly alkaline soils which indicates reclammatory response of such soils. *T. arjuna* plantations have also been raised successfully for the first time on coal ash bed in Korba (Chattisgarh) where it was observed that apart from sericultural activities, it has also controlled pollution, minimised the menace of fly ash from National Thermal Power Corporation, Korba, improved conversion of CO_2 into O_2 and also increased humous microbial activities by adding the litter to soil (Srivastav *et al.* 1998a, Prasad *et al.* 1992, Khare *et al.* 1991).

Cultivation: According to Central Tasar Research & Training Institute, Ranchi (Annual Report 1976-79), Arjuna seeds are collected during April-May and soaked in cold water for 3 days in May, which gave 88-90% germination. On the contrary, according to Regional Forest Research Centre, Rajamundry alternate wetting and drying of seeds for 15 days gives 50% peak germination and 72.5% overall germination (Rao et *al.* 1998). Subsequently, seeds are heaped under trees and covered with moist gunny bags followed by sprinkling of water to maintain the moisture. The seedlings may be raised in either polythene tubes of 11" x 4" or nursery beds of convenient size keeping one-foot gap between each bed. Rooting media is prepared by mixing FYM, soil and sand in 3:2:1 ratio and filled in the polythene tubes or nursery beds. The germinating seeds are either dibbled singly in polythene tubes or planted in nursery beds to a depth of one cm at 15 cm spacing between them and irrigated regularly. The nursery is kept under shade of tree or temporary thatched roofing is erected to avoid the impact of scorching sun. The three months old seedlings are transplanted during August in the pits of 1x1x1 ft. size manured with 0.5 ft^3 FYM. The pits may be prepared at 4x4, 6x6 or 8x8 foot spacing in which 6724, 3025 and 2025 seedlings are transplanted per ha. respectively. In these spacings after four years approximately 18.22, 10.6 and 7.3 mt. leaf may be produced to sustain nearly 500, 400 and 250 disease free layings of tasar silkworms respectively if properly maintained and fertilised. The plants grow up to 100 cm in three years. The seeds exhibit 50-90% germinability within 7-21 days and more than 60% survival has been noticed after transplantation. Seeds may be stored up to one year.

The nursery technique has been detailed by Tewari *et al.* (1978), while plantation technique and agronomical practices have been detailed by Singh (1982) and Prasad and Khare (1989).

Vegetative propagation: Arjuna can also be propagated vegetatively through air layering, stem cuttings or stumps. IBA at 200-300 ppm concentration has given the highest rooting percent (46-65%) and survival percent (44-60%) in air layerings followed by IAA at 300 ppm conc. which gave 61% rooting and 55% survival (Srivastava *et al.* 1993). Misra and Jaiswal (1994) also found 60% rooting in *T. arjuna* air layers with 20000 ppm IBA.

Cuttings have been found to show 20% success in rooting for which basal treatment with IBA 1000-ppm dose is required for 24 hrs. For best rooting percentage one-year-old stem cuttings should be prepared and treated during the month of July. Nearly 89% post transplantation survival of such cuttings has been recorded (Isa *et al.*1993)

It may also be propagated through stumps of 1.27-2.54 cm diameter, which should be transplanted in the end of July.

Recently, *T. arjuna* has also been propagated through leaf node cuttings, juvenile cuttings and semi-hardwood cuttings at Central Tasar Research and Training Institute. The rooting in leaf node cuttings was improved up to 97.5% against 61.6% and 19.2% rooting in juvenile and semi hardwood cuttings respectively during April – June 2000 (Anonymous 2000-01). Sinha *et al.* (2001) identified mechanical barrier in the form of sclerenchymatous and collenchymatous tissues as a factor hindering root emergence and achieved 86.6% root emergence in *T. arjuna* by countering this hinderance either by using softer cuttings from young bushes or by removing hard tissues from hard cuttings. Approach grafting has also been found extremely successful for establishment of clonal orchards of superior trees of *T. arjuna* at CTR & TI, Ranchi (Anonymous 1999-2000, Kumar *et al.* 2003).

Besides, micropropagation (tissue culture) of *T. arjuna* has also been achieved through leaf explants (Anonymous 2000-01), axillary buds (Ramesh *et al.* 2002) and cotyledonary nodes (Pandey and Jaiswal 2002).

Pruning: In order to maintain height, balanced crown, thinning out dense growth, and improving quality of foliage for silkworm rearing, the pruning of Arjuna plants is resorted between the period of leaf fall and sprouting i.e. Feb-March at 5 ft. height in the plants having branches of 2.5-3.5 cm diametre girth in finger like fashion at 4.0-6.0 inch length from the main trunk having 30 cm girth. When plants are very old and having unmanageable height, top pollarding may be done at 6 ft. height from the ground level. After pruning the cut ends should be covered with a mixture of soil, cowdung and Bordeaux mixture. Pruning during February, April and July-August may be resorted for I, II and III crops respectively but the plants should be pruned once in a year or only alternate year.

Spraying of 0.05% Rogar during sprouting and development of leaves may be done at an interval of 10 days to prevent gall insects and attack by other insects. For chawki rearing plants are pruned at 3 ft.height.

Fertilization: For maintenance of economic plantations in 4'x4' spacing following recommendations were made regarding application of manure and fertilizers.

2nd and 3rd year: Three trucks of FYM, 50 kg aldrin (5%) dust and 162 kg urea, 155 kg single super phosphate and 42 kg Murate of potash are required (NPK ratio 75:25:25) for maintenance of arjuna plantations during second and third year in each hactare.

4th year onwards: 600 cft (3 truck) FYM, 75 kg aldrin (5%) dust and 217 kg urea, 312 kg single super phosphate and 83 kg Murate of potash are required/ha. @ 100:50:50 kg NPK/ha/year.

Above 10 years: 600 cft FYM, 75 kg aldrin (5%) dust and 325 kg urea, 312 kg single super phosphate and 83 kg Murate of potash @ 150:50:50 kg NPK/ha/year are required for maintenance and sustenance of arjuna plantations for tasar rearing.

Application of FYM should be done during December-January in alternate years along with aldrin dust and the NPK fertilizer should be applied in two split doses during July-Sept. by deep placement method. Fertilisation is not possible in forests, so 1.5% urea (15 gm/litre of water) solution should be sprayed 30 days before rearing on arjuna leaves.

Digging, weeding, bunding of land and basin formation around the plants should be attended during June-July and November-December as inter-cultural operations.

149. BAHERA

Botanical: ***Terminalia bellerica*** Roxb.

Burmese: Thitsein | Eng.: Belleric Myrobalan

Hindi: Bahera, Bhaira, Vibhitaka | Unani: Bahere, Atriphal

Bengali: Bahera, Bhairah | Kannada: Tare | Marathi: Balra, Beheda

Malayalam: Thani | Oriya: Bhara

Sanskrit: Bibhitaki, Vibhitaki | Tamil: Chona-Atthi Tani, Thani

Telugu: Tandi, Thandra, Tanikyalu

It is legendary belief that Nala and Ashwaparna exchanged aswabridaya and akshabridaya under Thanni (Rao 1991).

SYSTEMATIC POSITION

Bentham & Hooker (1883)	Engler & Prantl (1899)	Hutchinson (1959)
Phanerogams	Phanerogams	Angiospermae
Dicotyledones	Dicotyledoneae	Dicotyledones
Polypetalae	Archichlamydeae	Lignosae
Calyciflorae	Myrtiflorae	Myrtales
Myrtales	Combretaceae	Combretaceae
Combretaceae		

Geographical distribution: Found in sub-Himalayan tract from near the Indus eastwards. Usually found throughout India, Nepal, Pakistan, Indonesia, Sri Lanka, Japan, Malaysia and Myanmar. Arid regions of Sind, Western Rajputana (Rajasthan) and the Southern Punjab usually lack it. Prevalent in West Bengal, Madhya Pradesh, Chattisgarh, Uttarakhand, Uttar Pradesh, Maharashtra, Tamil Nadu, Andhra Pradesh, Karnataka, Kerala and occasionally in Rajasthan, Punjab, Assam and Manipur.

Habit: A large deciduous tree attaining 25-30 meters height (Fig. 60).

Root: Branched tap root.

Stem: Erect, aerial, woody, wood hard and solid, yellowish-grey, without heartwood, branched, bark bluish to grey.

Leaves: Simple, alternate, clustered at the ends of branchlets, broadly obovate-elliptic or elliptic from an acute base, apex obtuse rounded or rarely sub-acute, minutely puberulous when young but glabrous when matured, blade 10-18 cm long, petiole very long, 3.8-7.5 cm leaves usually punctate on upper surface, punctations much more permanent than in other species, deciduous in cold season.

Inflorescence: Simple spikes on new shoots in axils of leaves or fallen leaves, upper flowers male, lower hermaphrodites.

Flower: Greenish-yellow, strongly offensive smell, male and hermaphrodite mixed, sessile, small, epigynous, bracteate, bracts linear, caducous.

Calyx: Sepals 5, gamosepalous, calyx tube campanulate, constricted above the ovary, woolly inside with long brown hairs, deciduous, limb 5-toothed, valvate.

Corolla: Absent.

Androecium: Stamens 10, arranged in two whorls, inserted inside the calyx tube, anthers bithecous, dehiscing longitudinally.

Gynoecium: Carpel-1, ovary tomentose when young, unilocular, inferior, ovules 2, rarely more on large funicles, pendulous.

Fruit: Ovoid or globular, grey, velvety, 1.5-3.0 cm long, 1.2-1.8 cm broad, suddenly narrowed into a short smooth stalk, covered by fulvous tomentum, obscurely 5-angled after drying.

Leaf fall: Feb.- March.

Sprouting: April, fresh foliage is copper or tan coloured.

Flowering: March - May.

Floral formula: Male flower: Br. ⊕ ♂ $\overbrace{K_{(5)}\ C_0\ A_{5+5}}\ G_0$

Bisexual flower: Br. ⊕ ⚥ $\overbrace{K_{(5)}\ C_0\ A_{5+5}}\ G_{\bar{1}}$

Varieties : According to Hooker (1878) *T.bellerica* is very allied to *T. mollucana* Miq., *T. nitens* Pres. Epimel and *T. bialata* kurz.

T. mollucana differs from *T. bellerica* by its short petioles.

T. nitens is very near to *T. bellerica* but differ by its long subpersistent bracteoles.

T. bialata is much mixed with *T. bellerica* in absence of fruit (which are 3-4 x 5-9 cm) but may be distinguished due to absence of puntations on leaves.

T. bellerica consists of three varieties:

Var. 1. *typica*: No glands are found at the apex of the petiole.

Var. 2. *bellerica* Roxb: Two glands are found at the apex of the petiole on ventral side.

Var. 3. *laurinoides Miq*: Distributed in Indonesia and Malaysia. Its leaves are obovate or obovate-elliptic, shortly acuminate, much thinner than in typical *bellerica*.

Diploid and tetraploid cytotypes have been reported in *T. bellerica*. (Gill *et al*. 1982, 1990, Janaki-Ammal 1962, Khosla and Sareen 1978, Roy *et al*. 1983).

Uses: It serves as one of the secondary food plants of tropical tasar silkworm, *Antheraea mylitta* D. in jungles. Only new foliage is suitable for feeding of larvae in wild conditions. (Jolly *et al.*1974, FAO 1976, Seth 1995a)

It is also grown as an avenue tree. The fruits are used for dyeing, tanning and for making ink and also used for increasing the potency of spirits (Jain 1968, Singh *et al*. 1983). The Kernel yields 25% oil (Jain *et al*. 1990). The oil from Kernel is used for preparing the soaps. Wood pulp is suitable for manufacturing of wrapping papers (Singh *et al*. 1983). Fruits contain 5-17% tannins (Chopra *et al*. 1965). They also contain β-sitosterol, gallic acid, ellagic acid, ethyl gallate, galloyl glucose, chebulagic acid, mannitol, glucose, galactose, fructose, rhmose etc. Fatty acids viz., palmitic, stearic, linoleic, oleic and belleric oil and oxalic acid have also be found in fruits (Anonymous 1976). As much as 12-19% oxalic acid has been found in water-extracted bark.

Medicinally, Kernels are eaten with belelnut and leaf against dyspepsia. Ripe fruit is astringent, tonic and laxative and is used with chebulic myrobalans for treating dropsy, piles, diarrhoea and fever. Oil of half ripe fruit is purgative. Fruit pulp is mixed with honey and applied in opthalmia. Fruit extract is antibiotic against *E. coli* and *Micrococcus pyrogenes* var. *aureus*. Oil is good tonic for hairs and substituted for ghee in M.P. (Chattisgarh) and applied as remedy for rheumatism (Anonymous 1976). While gum is considered to be demulcent and purgative bark has been found to be diuretic (Sinha 1996). A decoction/infusion of 15-40 gms of the pulp of fruit is administered in dropsy, piles, diarrhoea, leprosy, biliousness, dyspepsia and fever. A mixture of pulp of the fruit alongwith salt, long pepper and honey is used for cough, hoarseness of voice, sore throat and dyspepsia. While half a ripe fruit may be used as a purgative, the fried fruit roasted after covering with a wheat flour paste is household remedy for sore throat, cough and catarrh (Dastur 1962). Kernel is considered as narcotic.

The timber keeps well under water and used for boats and miscellaneous agricultural tools etc. (Jain 1968). It is also used for temporary construction work, general-purpose class-II plywood, black boards, heavy packing cases, rough shafts, carts, pit-props, rafters, dug-outs, catamarans, ploughs, turnings, slate frames and boxes. It is non-durable and treatable but complete penetration is not always obtained (Najma *et al.* 1981).

Cultivation: It may be propagated through either seedlings or stump transplantation. The seeds are collected during Nov.-Feb., seedlings are raised during March-May and they are transplanted during July normally. 24 hrs. soaking of seeds in plain water facilitates germination, which is nearly 38-69%, and germination period varies from 14-30 days. One-year-old seedlings are more suitable for transplantation than direct sowing of seeds or stump plant.

Pulp of the fruit is removed and seeds are dried in the sun before storing. Seeds remain viable up to one year after which germinability may be lost from 40-50%. Proper weeding and tending may give survivality of 80-95% after seedling transplantation (Anonymous 1976).

According to Bharadwaj and Chakravorty (1994), the overall performance of *T. bellerica* seedlings is better when seeds are collected during first fortnight of January and sowing is done in the last week of March. Pre-treatment of seeds with con. Sulphuric acid for 12 minutes gave better germination and growth to seedlings.

The cuttings planted during July, October and March failed irrespective of month, hormone treatment and portion of the cuttings. Hence *T. bellerica* may be categorized as under obstinate to root under field condition (Bharadwaj *et al.* 1993).

150. DESHI BADAM

Botanical: ***Terminalia catappa*** Linn. syn. *T. procera* Roxb.

Burmese: Bardam pin Eng: Indian Almond tree

French: Amandier des indes German: Catappanbaum

Hindi: Deshi Badam, Jangali Badam

Singhalese: Kottamba Unani: Badame-hind

Bengali: Bangla-badam Kannada: Badami mara Konkani: Hatbadam

Malayalam: Admaram Sanskrit: Grahadruma Tamil: Natvadom

Telugu: Vedam

SYSTEMATIC POSITION

Bentham & Hooker (1883)	**Engler & Prantl (1899)**	**Hutchinson (1959)**
Phanerogams	Phanerogams	Angiospermae
Dicotyledones	Dicotyledoneae	Dicotyledones
Polypetalae	Archichlamydeae	Lignosae
Calyciflorae	Myrtiflorae	Myrtales
Myrtales	Combretaceae	Combretaceae
Combretaceae		

Geographical distribution: Naturally found in the beach forests of the Andamans and Malay Peninsula. It is a widely distributed littoral species in tropical countries viz. Java (Indonesia), tropical Africa, Japan, Malaysia, Myanmar and India. It is cultivated throughout tropical India in West Bengal, Uttar Pradesh, Assam, Tamil Nadu, Rajasthan, Karnataka, Kerala, Gujarat and Punjab.

Habit: A tall tree attaining 25-30 meters height, branches in horizontal whorl and stem often buttressed, deciduous.

Root: Branched tap root.

Stem: Erect, aerial, woody, solid, branched, branches horizontal, whorled, trunk buttressed, bark grey-brown, rough, thick.

Leaves: Simple, alternate, clustered at the end of branchlets, glabrous, shortly petioled, petiole 0.6-1.5 cm, stout and channelled, blade obovate with very narrow but cordate base, softly hairy when young, hairy or usually glabrous when adult, 12-25 cm long, shining and the two glandular depressions near the base of the midribs on the ventral side are usually obscure or absent.

Inflorescence: Solitary, axillary, simple spikes, slender, shorter than leaves, grey or rusty tomentose or pilose, male flowers at the top, bisexual flowers below with minute bracts, male flowers most numerous.

Flower: Sessile, small, white or greenish white, male or bisexual, incomplete, epigynous, calyx tube campanulate, constricted above the ovary, glabrous or nearly so inside as well as outside, deciduous, limb 5-toothed, petals absent, stamens 10 in two rows, carpel 1, ovary inferior, unilocular, ovules 2 or more on large funicles, pendulous, young ovary glabrous or hairy.

Fruit: Ellipsoid drupe, glabrous, somewhat compressed, so as to show two ridges on sides, keeled all round, 5.0 cm long, pericarp fibrous and fleshy, endocarp hard.

Seed: Non-endospermic, edible, oily.

Flowering: Feb.-May (July-Aug in Bhopal).

Floral formulae: Male flower : $\oplus$ ♂ $\overbrace{K_{(5)}\ C_0\ A}_{5+5}\ G_0$

Bisexual flower: $\oplus$ ⚥ $\overbrace{K_{(5)}\ C_0\ A}_{5+5}\ G_{\bar{1}}$

It mostly resembles *T. procera* Roxb. which differs from it in having obscurely 5-sideed compressed fruit, higher height, smaller petiole and longer leaf.

Uses: It serves as one of the secondary food plants of tropical tasar silkworm, *Antheraea mylitta* D. in jungles (FAO 1976, Seth 1995a). Kernels are eaten and bark and fruits are used for tanning. Fruits are also source of black dyes. Wood is used for house building and construction work, especially suitable for posts, beams wheel weights, general carpentary work, scantlings and rafters (Singh *et al*. 1983).

Medicinally, leaves and kernels are very useful. The decoction of leaves is used as a remedy for dysentery, diarrhoea, fevers, worms and inflammation of mucous membranes of the urethra and vagina. The juice of the young leaves is used as an ointment for rashes, itching, scabies and other skin diseases and also applied on the forehead to relieve the headache. The sap of the young leaves boiled in the oil of the kernel is used for curing leprosy (Kurian 1995). The bark is mild diuretic and potent cardiotonic. Tannin is used as an astringent in dysentery. Methanolic extract of trunk bark exhibits physiological activity on blood pressure and intestine.

The kernels contain myristic, palmitic, stearic, oleic and linoleic acids. The leaves and fruits contain corilagin, gallic acid, ellagic acid and brevifolin corboxylic acid whereas the bark and the wood contain ellagic acid, gallic acid, catechin, epicatechin and leuco-cyanidin (Anonymous 1976).

Cultivation: Seeds are collected during June-July and seedlings are raised during July. The seedlings are either raised in nursery beds or in polythene tubes containing rooting media of FYM, soil and sand in 3:2:1 ratio. The seedlings are transplanted during rains of following year. The seeds may be directly sown in spots also. The germination percentage of seeds ranges between 25-63%. In one kg approximately 176 fruits are found. Survival of seedlings is nearly 25% after transplantation (Anonymous 1976).

151. HARRA

Botanical: ***Terminalia chebula*** Retz.

Burmese: Panga
Eng: Chebulic Myrobalan
Hindi: Harra, Har, Harir, Haritaki
Unani: Haleela
Assamese: Silikha
Bengali: Haritaki
Gujarati: Hardo
Malayalam: Kadukka
Manipuri: Haritok
Marathi: Hirda, Habra
Oriya: Haridra
Punjabi: Hara, Harar
Tamil: Kadukai
Telugu: Nella karaka, Karakyalu
Sanskrit: Haritaki, Abhaya

SYSTEMATIC POSITION

Bentham & Hooker (1883)	Engler & Prantl (1899)	Hutchinson (1959)
Phanerogams	Phanerogams	Angiospermae
Dicotyledones	Dicotyledoneae	Dicotyledones
Polypetalae	Archichlamydeae	Lignosae
Calyciflorae	Myrtiflorae	Myrtales
Myrtales	Combretaceae	Combretaceae
Combretaceae		

Geographical distribution: Sub-Himalayan tract from the Sutlej eastwards, ascending to 5000 ft., common in deciduous forests of both the Peninsulas especially in Punjab, Uttar Pradesh, Madhya Pradesh, Bihar, West Bengal, Maharashtra, Assam, Tamil Nadu, Karnataka and Kerala. Also, found in Sind (Pakistan), Nepal, Myanmar and Japan.

Habit: A large deciduous tree attaining the height up to 28-35 meter (Fig. 61).

Root: Branched tap root.

Stem: Erect, aerial, branched, woody, hard, young branchlets covered with soft, shining, rusty hairs, bark rough, grey coloured.

Leaves: Simple, elliptic or ovate, distant, sub-opposite, not clustered, usually acute, never acuminate, rounded at the base, secondary nerves 6-8 pairs, arching, prominent, blade 8-20 cm long, petiole about 2.5 cm often with two glands near its summit, young leaves pubescent.

Inflorescence: Terminal, panicled spikes, sometimes-simple spikes and in the axils of uppermost leaves.

Flower: Bisexual, sessile, dull white or yellow with an offensive smell, incomplete, epigynous, 0.4-0.5 cm across, sepaloid, bracteolate, bracteoles pubescent, conspicuous in young spikes, deciduous.

Calyx: Sepals 5, gamosepalous, campanulate, calyx teeth-5, acute, triangular, hairy inside.

Corolla: Absent.

Androecium: Stamens 10, biseriate.

Gynoecium: Carpel 1, ovary inferior, unilocular, style slender, stigma small, capitate, ovules 1-2, pendulous.

Fruit: Obovoid from a cuneate base, sometimes ovoid or nearly globose nut, more or less distinctly 5-angled, 2.5-4.0 cm long, glabrous.

Seeds: Non-endospermic.

Leaf fall: Feb- March.

Sprouting: April.

Fruiting: Oct. - Jan.

Flowering: April-Aug., two distinct flowering periods April-May and July-Aug. found in Central provinces.

Floral formula: Brl. $\oplus$ ⚥ $\overbrace{K_{(5)}\ C_0\ A_{5+5}}\ G_{\bar{1}}$

Varieties: Hooker (1878) recognised following six varieties under *T. chebula*.

Var. 1. *typica*: Adult leaves nearly glabrous beneath or the interstices of ultimate nerves bear minute white tomentum, young ovary pubescent externally, calyx teeth glabrous externally. Found in Deccan, Sri Lanka, Myanmar and Malay Peninsula.

Var. 2: Differs from Var. 1 only in having the young ovary entirely glabrous, ovate fruit and round based leaves. Common from Kumaon to Bengal and Chota Nagpur.

Var. 3: Adult leaves densely tomentose ventrally, fruit much smaller usually 1.8 cm long, found on the summit of Parasnath Bihar at 4000 ft. altitude.

Var. 4. *tomentella* Kurz: Young leaves densely coppery pubescent on ventral side, adult leaves glabrescent or pubescent ventrally, young ovary glabrous, fruit ovoid, hardly 2.5 cm. Found in Pegu up to 2000 ft. altitude.

Var. 5. *gangetica* Roxb: Adult leaves pubescent with brown - red silky hairs on both surfaces, very unlike the typical *chebula* by the brown silky hairs which cover the twigs, leaves and inflorescence. It may be a good species viz. *T. gangetica* Roxb. but the fruit is similar to that of *chebula* found on the bank of Ganges.

Var. 6. *parviflora* Thwaites: Calyx tubes pubescent outside.

On the basis of fruit types following grades have been isolated (Anonymous 1976):

(a)	Bhimlies	-	Madras (T.N)	(f)	Salem-I (T.N)
(b)	Jubbulpores	-	Jabalpur (M.P)	(g)	Salem-II (T.N)
(c)	Rajpores	-	Kolhapur (M.S)	(h)	Survari harde
(d)	Vingorlas	-	Bombay (M.S)	(i)	Bala harde
(e)	Madras coast	-	(T.N)	(j)	Java harde

Cytotypes: Diploid (2x), tetraploid (4x) and hexaploid (6x) varieties have been isolated so far (Gill *et al.* 1982, 1990 Janaki - Ammal and Sobti 1962).

Uses: It serves as one of the secondary food plants of tropical tasar silkworm, *Antheraea mylitta* D. in the jungles. Only delicate semi-matured leaves are suitable for feeding of silkworms (FAO 1976, Seth 1995a).

The fruit is the chebulic myrobalan of commerce and rich source of tannin (30-32%). The fruit extract is largely used for internal treatment of locomative feed waters, and as additive in oil drilling compositions (Singh *et al.* 1983)

The tree yields gum, timber is used for construction work as beams, scantlings and planks and for carts and tool handles. It is extensively used for mathematical engineering and drawing instruments. The wood is perishable, moderately durable under cover while heartwood is only partially treatable (Singh *et al.* 1983, Najma *et al.* 1981).

Medicinally, the main purgative component of Ayurvedic preparation "Triphala" (Harra, Bahera, Amla) is cheulic myrobalan along with belleric myrobalan and Embellic myrobalan.

The chebulic myrobalan is a mild safe and efficacious laxative, astringent, stomachic, tonic and alterative (Sinha 1996). The pulp of the fruit is useful in piles, chronic diarrhoea, dysentery, costiveness, flatulence, asthma, urinary disorders, vomiting, hiccup, intestinal worms, ascites and enlarged liver and spleens etc. In constipation, as a mild laxative a decoction of six fruits and either one drachm of cloves or cinnamon or the pulp of 2-3 fruits with a little salt is given at bedtime daily for 3-4 days. In dysentery or diarrhoea one drachm of the unripe fruit is given twice a day along with honey and aromatics. In case of carious teeth and bleeding ulceration of the gums a fine powder of the fruit is used as a dentrifice or as a dust over chronic ulcers and wounds. While a paste of the fruit made with carron oil is very good dressing for burns, scalds and piles, the bark is a useful cardiotonic. The decoction of the fruit is used as a gorgle in stomatitis, as a douch in case of vaginal discharge and also as a wash for bleeding piles (Sinha 1996, Anonymous 1976, Singh *et al.* 1983, Pandey 1980, Jain 1968, Kochar 1981, Dastur 1962).

The kernel yields as much as 36.4% oil (Jain *et al.* 1990). Bark is diuretic and cardiotonic.

Cultivation: It may be cultivated through seeds, seedlings, air layerings, stem cuttings, grafting and stump transplantation.

The seeds are collected during Nov.-March and nursery is raised during March-May while transplantation of seedlings is made during July. For germination, the seeds are fermented or the hard front wall at the broader end of the seeds is clipped and subsequently the seeds are soaked in cold water for 36 hrs and sowed in the nursery beds. Within 15-30 days 80% of the seeds germinate (Anonyumous 1976). According to Regional Forest Research Centre, Rajahmundry, alternate wetting and drying of seeds for seven days yields 20% peak germination and 57.5% overall germination (Rao *et al.* 1998). When seedlings are three months old, they may be transplanted. The seeds may also be directly sown at the spot of transplantation after depulping and clipping. The collected seeds after sun driage can be stored up to one year without much affecting their germination (Anonymous 1976).

Misra and Jaiswal (1994) studied the effect of IBA at 2500-20000 ppm on the rooting of air layers of *T. chebula* and found that IBA at 10000-ppm dose induced highest rootings (73%), hence it can also be propagated through air layers.

According to Bharadwaj and Chakravorty (1994), the seeds collected during first fortnight of Jan., pretreated with con. Sulphuric acid for 12 minutes and sown during third week of March yielded better results with respect to germination and growth of seedlings. The cuttings of *T. chebula* planted in July and October failed completely while in March the middle portion of the cuttings with combination of 4000 ppm IBA produced better results (Bharadwaj *et al.* 1993). Subsequently, Jose and Thomas (1998) found that 100% of the juvenile cuttings placed in pots containing river sand in mist chamber exhibited rooting without any special treatments within 30 days of planting.

Ramesh Singh *et al. (2003)* achieved 67-78% rooting in stem cuttings of 1-2 year old juvenile seedlings. Besides, peg grafting (Ramesh Singh *et al.* 2003), approach grafting and cleft left grafting (Srivastava 2000) and patch bud grafting (Sharma *et al.* 1995) have also been found successful in *T. chebula* for clonal propagation.

152. KINJAL

Botanical: ***Terminalia paniculata*** Roth.

Eng: Murdah Hindi: Kinjal, Kindal

Kannada: Honal, Huluve, Bilimatti Malayalam: Pilamurda

Marathi: Kinjol, Kindal Tamil: Puluvai, Pekadukkai

Telugu: Puli maddi, Nimiri

SYSTEMATIC POSITION

Bentham & Hooker (1883)	**Engler & Prantl (1899)**	**Hutchinson (1959)**
Phanerogams	Phanerogams	Angiospermae
Dicotyledones	Dicotyledoneae	Dicotyledones
Polypetalae	Archichlamydeae	Lignosae
Calyciflorae	Myrtiflorae	Myrtales
Myrtales	Combretaceae	Combretaceae
Combretaceae		

Geographical distribution: West coast and Ghats, Nilgiris, Namalais, Javadis and other hill ranges in the Southern parts of Peninsula viz., Maharashtra, Tamil Nadu, Karnataka and Kerala. Also found in Sind (Pakistan).

Habit: A large tree, hardly quite leafless.

Root: Branched tap root.

Stem: Erect, aerial, branched, branches spreading, woody, hard, bark grey, rough.

Leaves: Simple, glabrous, oblong or elliptic, apex acute, base rounded often unequal (oblique), pale brown beneath, the upper leaves alternate, lower often opposite – sub-oppsite, two glands present at the apex of midrib ventrally, blade 10-17 cm, petiole 0.6-1.2 cm long, innovations rusty, tomentose, secondary nerves 10-15 pair.

Inflorescence: Rusty pubescent panicles, ample, compound, dense.

Flower: Bisexual, zygomorphic, small, bracteate, sepaloid, reddish-brown, epigynous, bracts pubescent, acuminate.

Calyx: Sepals 5, gamosepalous, campanulate, limbs 5, glabrous, reddish-brown, inside clothed with long brown hairs.

Corolla: Petals absent.

Androecium: Stamens 5+5 (10), biseriate, anthers bithecous, filaments slender.

Gynoecium: Carpel 1, ovary with five rounded ridges, inferior, villous, style slender, stigma capitate.

Fruit: Brown-red, rusty villous, sessile with one very broad (1.2 cm) and two narrow wings 0.3 cm broad, fruit 0.6 cm long, one seeded.

Seeds: Non-endospermic.

Flowering: Aug-Dec.

Fruiting: Hot season.

Floral formula: Br ⊕ ⚥ $\overbrace{K_{(5)}\ C_0\ A_{5+5}}\ G_{\bar{1}}$

Varieties: Diploid (2x) and tetraploid (4x) cytotypes have been observed in nature (Srivastav *et al.* 1994).

Uses: It serves as one of the secondary food plants of tropical tasar silkworm *Antheraea mylitta* D. Tribals of three states in south India undertake their rearing on this species (Jolly *et al.* 1974, FAO 1976, Seth 1995a).

The wood is non-durable to moderately durable, frequently substituted for teak. The heartwood is partially treatable. It is used as beams, rafters, planks, tea chests, commercial grade plywoods, black boards, agricultural implements, boat building, railway sleepers and lorry bodies (Singh *et al.* 1983, Nazma *et al.* 1981, Agarwal 1986). Timber is resistant to fire and may be used in fire proofed buildings.

Bark and fruits yield tannins (14%). A new glucoside, dimethyl ellagic acid has been isolated from the heartwood. Bark and fruits are used in tanning (Chopra *et al.*1994, Agarwal 1986).

The pulp strength from woods is very good and can be used in mixtures with bamboos and hardwoods in proportion upto 20% for making various types of papers (Saraf *et al* 1980).

The bark juice is cardiotonic and diuretic. In parotitis bark juice is applied with ghee and rock salt (Anonymous 1976).

Cultivation: It can be propagated through seedlings and stump plantings. The seeds may be collected between March-April. As the seeds are highly infertile they should be sown very thickly in nursery beds under shade and transplantation of seedlings as well as stumps is done at the end of the year (Anonymous 1976).

153. ASAN

Botanical: *Terminalia tomentosa* W&A.

Burmese: Taukkyan Eng: Laurel

Hindi: Asan, Saja, Sain, Asain

Assamese: Amari, Kusshi-bom-phang Bengali: Asan, Usan

Gond: Maru Gujarati: Hadri, Sadri, Sadar

Kannada: Banappu, Karimaddi, Karimatti, Matti, Madati, Sadada

Malayalam: Karimaruthu, Karumaridu, Thembava, Thenpavu

Marathi: Ain, Madat, Yen, Sadada, Saj, Sajad

Oriya: Kala-sahju, Sahaja, Sajo Punjabi: Aisan, Sain

Tamil: Karramavda, Anemui, Karimaddi, Karumanithu, Karimarudu

Telugu: Maddi, Malla, Maddinumaddi, Tani

SYSTEMATIC POSITION

Bentham & Hooker (1883)	**Engler & Prantl (1899)**	**Hutchinson (1959)**
Phanerogams	Phanerogams	Angiospermae
Dicotyledones	Dicotyledoneae	Dicotyledones
Polypetalae	Archichlamydeae	Lignosae
Calyciflorae	Myrtiflorae	Myrtales
Myrtales	Combretaceae	Combretaceae
Combretaceae		

Geographical distribution: Sub-Himalayan tract from Ravi eastwards up to 4000 ft. ASL. Barring arid regions of Punjab and Rajasthan, it is common throughout India and Myanmar. It occurs from Kangra in Himachal Pradesh eastwards to Goalpara in Assam and southwards throughout the greater part of the Peninsula. Abundant in Uttar Pradesh, Madhya Pradesh, Chattisgarh, Jharkhand, Orissa, Gujarat, Karnataka, Kerala, Tamil Nadu and less frequently in Andhra Pradesh. Also found in Sind of Pakistan.

Habit: A large deciduous tree up to 150 ft. in height and 18 ft. in girth with a long clean bole of 45 ft. with spreading branches and a heavy crown.

Habitat: Thrives best in heavy clay soil but lateritic, red loamy and black clayey cotton soils also suit under subtropical to subtemperate climatic conditions. Tropical dry/moist deciduous forest types provide ideal climate for their growth.

Root: Branched tap root.

Stem: Aerial, erect, branched, branches spreading, woody, hard, glabrescent or pubescent especially when young, pubescence yellowish-brown, bark dark grey, brown or black, often tinged blue with fungus, rough, 2.5-4.0 cm thick, with deep transverse cracks and longitudinal fissures, resembling like a crocodile back, exfoliating in to thick rectangular flakes, heartwood hard, dark purplish-brown mottled with darker streaks, often almost black.

Leaves: Simple, elliptic, oblong, or ovate, sometimes obovate-oblong, nearly opposite, the uppermost usually alternate, 13-25 x 6-10 cm, occasionally longer in young plants, coriaceous, entire or occasionally nerves excurrent as teeth, apex obtuse, emarginate or slightly pointed, base usually rounded or cordate, glabrescent above, more or less villous (hairy) beneath or glabrescent both sides, petiole 0.6-1.3 cm long, 1-2 large, turbinate, stalked glands present near base of mid-rib, nerves many, prominent parallel, stipulate, secondary nerves 10-20 pairs.

Inflorescence: Erect, axillary or terminal, densely pubescent or glabrescent spikes, 8-20 cm long, terminal ones panicled.

Flower: Small, bisexual, sessile, incomplete, dull yellow, 0.6 cm across, epigynous, bracteolate, bracteoles linear.

Calyx: Sepals 5, petaloid, gamosepalous, campanulate, calyx tube 0.2-0.3 cm across, produced over ovary with a constricted mouth, teeth triangular, ovate, acute, 0.1 cm long, glabrous or hairy, hairs yellowish to brown.

Corolla: Petals nil.

Androecium: Stamens 10, biseriate (5+5), inserted on calyx tubes.

Gynoecium: Carpel 1, ovary inferior, unilocular, usually angular, angles equal to sepals and alternating with them, style long, filiform, stigma pointed or capitate, ovules 2-5, occasionally more, anatrapous, pendulous, funicles joining ovules from the top of the ovary long and united. The receptacle tube bears a disc, disc occasionally hairy.

Fruit: Drupaceous samara, leathery, 3.5-7.5 cm x 5-6 cm, with usually 5 equal, broad, irregularly crenate coriaceous wings, wings occasionally upto 10, usually 1.3-2.5 cm broad, marked with close, parallel, horizontal striations running from axis to the edges, green initially turning to brown on maturity, one seeded, rarely two seeded.

Seed: Non-endospermic, fleshy, oily.

Germination: Epigeal.

Leaf fall: Feb-April.

Leaf sprouting: March – April.

Flowering: May - July (Summer-Rainy season).

Fruiting: Feb-April of ensuing year.

Floral formula: Brl. $\oplus$ ⚥ $\overbrace{K_{(5)}\, C_0\, A_{5+5}}\, G_{\bar{1}}$

Varieties/Species: A lot of controversies exist regarding taxonomic status of *T. tomentosa* like *T. arjuna* alongwith which it constitutes section "*Pentaptera*" of genus *Terminalia*. Both the species have been regarded as "Species complexes" rather than simple species since both the species exhibit various distinct types distributed in different regions, which may be different species, subspecies/varities and natural hybrids/ecotypes. While Hooker (1878) has divided *T. tomentosa* into 3 varieties, viz., *typica* (*alata* Roth.), *crenulata* Roth. and *coriacea* Wight and Arnott, Parkinson (1936), Bahadur and Gaur (1980) and Srivastav *et al.* (1997 a,b) have treated them as distinct species of *Terminalia*. Likewise, *T. arjuna* has also been divided into two varieties viz., *arjuna* and *angustifolia*, subsequently upgraded to the level of species as *T. glabra* and *T. berryi* respectively (Hooker 1878, Parkinson 1936, Srivastav *et al.* 1997 a, b).

While Parkinson (1936) used foliar, bark and fruit characters, Bahadur and Gaur (1980) used geographical distribution pattern also in addition to above characters for species delimitation. According to them while *T. crenulata* is widely distributed in South-west India and is common along Western Ghats, *T. coriacea* is more commonly found in South-east India, especially in the Eastern Ghats. The typical *T. alata* has a rather restricted distribution in Central India, but its var. *nepalensis* is far more widely distributed in submontaneous Himalayan tracts in the foothills of Northern India. Distribution of these taxa, however, overlaps in several areas particularly Central India, where obviously their occurrence converges from all directions.

Central India may be regarded as the "Centre of Origin" of section "Pentaptera" of genus *Terminalia* despite of the fact that centre of origin of genus *Terminalia* may be Malaya (Malaysia) since as many as 40 species of *Terminalia* have been reported from there followed by 27 species reported from India and 11 species in Tropical Africa. Hence, Indo-Malayan region may be regarded as the "ancestral home" of genus *Terminalia* as a whole from where it spread to all other biogeographical regions of the world since each, at one or the other time, has some land connection with another, except for the Australian region, across which the plants and animals could pass. Conversely, all the 18 Australian species of *Terminalia* appear to be endemic with the exception of *T. microcarpa* which occurs in Timor also. Hence, genus *Terminalia* probably differentiated independently in two or more regions and appears to have bitopic or polytopic rather than monotopic centres of origin (Srivastav *et al.* 1997a). The distinguishing features of taxa under "tomentosa complex" together with their distribution are given below (Srivastav and Thangavelu 1995, Thangavelu *et al.* 2000).

1. *T. alata* Heyne ex Roth.

Bark dark grey to black with deep vertical fissures and transverse cracks exfoliating in irregular flakes. Leaves hairy beneath. Leaves cordate or suddenly narrowed into petioles. Twigs, inflorescence, hypanthium and young ovary more or less hairy. Fruits glabrous (Fig. 59).

(a) ***T. alata* var. *alata* Heyne ex Roth**. (Saj of Central India)

Leaves elliptic - oblong or ovate-oblong, approximately 15 x 8 cm, thinly woolly beneath especially on nerves. Distribution in Chotanagpur and Central India (from Chatarpur, erstwhile Ganjam to the Vindhya mountains and Vishakapatnam to Pune and also in lower Myanmar).

This variety has also been described as β-*typica* Clark., often with serrate leaves especially when young, secondary nerves usually fewer than 20, inflorescence with a looser longer indumentum but flowers 0.4-0.5 cm across with the sepals inside, shortly villous outside, deltoid, hypanthium broader than long, densely pubescent. The fruit is usually glabrous and very large.

L/B ratio of leaves = 2.0-2.5

(b) ***T. alata* var. *nepalensis* (Haines) Fernandez**. [Hindi: Ujar-ki-Asan]

The leaves oblong-elliptic to oblong-lanceolate, 18-30 x 4-10 cm, strongly and closely tomentose beneath due to which under surface appear, white or grey. Distributed in Northern India (Sub- Himalayan tracts in the foot hills, Tarai of Northern area entering Nepal and Bhabar from Punjab eastwards to Assam) and Upper Burma.

Secondary nerves more than 20 with oblique scalriform tertiaries, inflorescence tomentose or closely shortly pubescent, bracts linear-subulate. Flower 0.35-0.45 cm across, hypanthium tomentose, slightly longer than broad, calyx outside thinly pubescent or almost glabrous at tips, sepals triangular, mucronulate, inside with long villi and large lobed disc densely villous.

Wood is lighter colour and superior to that of ordinary Saj. L/B ratio of leaves > 3.

2. ***T. crenulata* Henye ex Roth.**

Bark dark gray, exfoliating in rectangular flakes, leaves, inflorescence, twigs, hypanthium and fruits glabrous. Leaves obovate-elliptic or narrowed into petiole, glabrous beneath (Fig. 62).

Distributed in Southern, Western, Central and Eastern India and Myanmar. Common in Western Ghats.

It may further be differentiated into big leaf forms, small leaf forms and kahvi, which bear obovate, oblong and elliptic leaves respectively.

3. ***T. coriacea* (Roxb.) W. & A.**

Bark dark coloured and cracked like crocodile skin, exfoliating in squarish flakes, leaves cordate or suddenly narrowed into petiole, more or less like that of *T. alata* but beneath with a close, hard fulvous tomentum rather than villous, fruits minutely tomentose especially between the wings. Distributed in Southern and Central India, mostly in drier and hotter parts and in lower Myanmar. Common in Eastern Ghats.

It may further be differentiated into elliptic, oblong or obovate leafy forms.

4. **Putative hybrids**

Two types of putative hybrids have been found between *T. arjuna* and *T.tomentosa* complexes:

(a) **Hybrids closer to *T. arjuna*:** Bark grey, slightly rough, leaves broader than those of *T. glabra* and not perfectly smooth or glabrescent, veins not buldging.

(b) **Hybrids closer to *T. tomentosa* :** Bark greyish black, rough and slightly cracked, leaves bigger than those of *T. glabra*, glabrous but not perfectly smooth, veins slightly bulding.

All varieties/species of *T. tomentosa* complex are suitable for rearing of tropical tasar silkworm but *T. crenulata* and putative hybrids are superior for rearing than other forms because larval duration is reduced and cocoons of superior quality with higher effective rate of rearing are obtained if rearing is conducted on them.

Uses: *Terminalia tomentosa* serves as one of the primary food plants of tropical tasar silkworm, *Antheraea mylitta* D. in tropical dry/ moist deciduous forests. Ecoraces of *Antheraea mylitta* D. *viz.,* Daba and Sarihan in Bihar and/or Jharkhand (Chotanagpur), Bhandara in Maharashtra, Andhra Local in Andhra Pradesh, Sukinda in Orissa, Bankura in West Bengal and Kurud, Bhopal Patnam, Multai and Piprai in Chattisgarh are reared on *Terminalia tomentosa* varieties/species by tribals (Sengupta 1986, Sinha 1998, Thangavelu *et al.* 2000).

T. alata leaves contain gallic, ellagic, chebulinic acids, corilagin and β-sitosterol. Its gum consists of D-galactose, L-arabinose, D-xylose, L-rhamnose and L-glucuronic acid (Anonymous 1976).

The bark of *T. alata* is bitter, styptic, diuretic and cardiotonic. Decoction of bark is used in dyspepsia, atonic diarrhoea and used locally as an application to weak indolent ulcers. Gel like gum is purgative (Sinha 1996). According to Dastur (1962), the bark of *T. tomentosa* is useful in ulcers, haemorrhages, fractures, bronchitis, leucorrhoea, gonorrhoea, diarrhoea, dysentery, heart disease and calculi. A decoction of the bark is given in retention of urine, gonorrhoea, leucorrhoea, ecchymosis, spitting of blood from chest, diarrhoea and dysentery. In heart diseases, a powder of the bark is given with ghee and gruel made of wheat flour, medicated ghee prepared with the powder is equally efficacious. In fracture bark is given with milk. Locally a decoction of bark is used for washing ulcer contusions, suppurating ears and ecchymosis.

The bark of *T. coriacea* is a potent cardiotonic and heart stimulant (Sinha 1996).

The bark of *T. crenulata* exhibits physiological activity on blood pressure. The decoction of bark is also used in Chattisgarh for painless delivary by tribals. (Anonymous 1976, Srivastav 2000).

The wood from *T. tomentosa* is normally known as lawrel. It is 25-30% heavier than teak and in hardness it stands 50-60% higher than teak. In shock resistance it is usually well above teak. It is good, strong, tough wood suitable for most constructional work. It is easy to saw and work and finishes to quite a good surface (Pandey 1980).

It has been used for plywood chests in Assam. It is also used for ceiling boards, other domestic purposes, cheap utility plywood, match boxes and splints. It is also used for rice pounders, ship and boat building, furniture, cabinet making and paneling purposes. The wood is also used for use as electric transmission poles, paper pulp, fire proof building, fuel wood and charcoal (Pandey 1980).

Fruit and bark yields tannin upto 20% for the industry (Agarwal 1986).

Twigs are used as excellent cattle feeds (Khanna *et al.* 1998). Bark colour is used for dyeing fabrics and gum as adhesive (Agarwal 1986, Negi *et al.* 1985).

The best export quality laurel comes from *T. crenulata* (Bahadur and Gaur 1980). It also serves as a good fuel wood (Khanna *et al.* 1998). The pulps from *T. tomentosa* can be used in mixtures with bamboos and hardwoods in proportion upto 20% (Saraf *et al.* 1980).

Frequency of *T. tomentosa* as a dominant tree species in forests reaches upto 93% in litter production which is very high and helps in soil conservation by adding every year large quantities of organic matter as leaf and silkworm litter which serves as the fuel for nutrient recycling specially in soils which are poor in nutrients (Khare *et al.* 1998, Prasad and Mishra 1985).

T. tomentosa has capacity to thrive on infertile waste land which occupies 20-25% landscape in India. Subsequently, it helps in the reclamation of such soils (Khare *et al.* 1991).

The tannins obtained from the bark of *T. alata* serves as promising material for soil stabilization because mud wall made with bricks containing 5% of the spray dried powder prepared from the water extract of this bark can stand a rainfall of 298 cm while the control wall is washed away (Bhatia *et al.* 1987).

Cultivation: It can be propagated through seeds, stumps, entire seedling transplants or cuttings. Following technologies were developed at Central Tasar Research and Training Institute, Ranchi.

Nursery technique: Seeds of *Terminalia tomentosa* are collected during April-May from trees and heavier (1.0-5.0 gm) seeds are selected, packed in gunny bags and sunk in water for two days during May. Soaked seeds are subsequently heaped on sand bed under tree shade and covered with moist gunny bags or paddy straw and regularly sprinkled with water. Germination starts after 8-10 days and continues for 20 more days. Mixture of FYM, soil and sand in 3:2:1 ratio is filled in polythene tubes and germinating seeds are dibbled in such rooting media. Nursery is maintained by weeding, watering and insect control by spraying Rogor 0.09% twice fortnightly for three months. This technique ensures 60% germination as compared to 25-30% in control (Tewari *et al.* 1985, Isa and Thangavelu 1992). On the contrary, according to Regional Forest Research Centre, Rajahmundry, the alternate wetting and drying of seeds for 9 days gives 24% peak germination and 57.5% overall germination (Rao *et al.* 1998).

Clonal propagation: Recently, rooting response in juvenile cuttings taken from very young plants indicated 62.3% rooting with first node cutting and 50.8% rooting with second node cuttings for the first time in *T. tomentosa*. Besides, wounding could also induce 29.6% rooting in semi hardwood cuttings during Oct.-Dec. only (Anonymous 1999-2000). Leaf node cuttings exhibited as high as 70% rooting in 25.4 days while juvenile cuttings exhibited as high as 65% rooting in rooting period as less as 21.4 days during April-Dec. (Anonymous 2000-01). In *T. tomentosa,* approach grafting has also been found successful for raising clonal orchards at Central Tasar Research and Training Institute, Ranchi (Kumar *et al.* 2003).

Transplantation: 8-10" high seedlings are transplanted in 1'x1'x1' pits filled with 2 kg of FYM and 10 gm of 5% aldrin dust after cutting polythene tubes laterally.

Spacing: 4x4, 8x8 and 12x12 feet spacings were studied during 1985-87. Subsequently, 4x4, 4x6 and 4x8 feet spacings were also studied at C.T.R. & T.I., Ranchi and 4'x6' spacing was concluded as better for *T. tomentosa*. Recently, 6'x6' spacing has been recommended for plantation of *T. tomentosa* seedlings.

Seeds may also be directly sown in the spots/pits in FYM: Soil: Sand rooting media in 3:2:1 ratio. The survival of seedling progenies is 30-50% after transplantation. For areas prone to water logging, stump plantings from one-year-old seedlings after first shower in April is suggested. Seeds may be stored up to one year in gunny bags in airy space.

Fertilization: No systematic studies were conducted on fertilization of *T. tomentosa* plantations in economic as well as natural plants. Hence, the fertilizer doses of NPK worked out for *T. arjuna* 150:50:50 kg/ha. had been consistently practiced for *T. tomentosa* also on per plant basis by ring method or by making 3-4 holes around the plants after 4 years of their transplantation as seedlings. In jungles, only 1.5% urea (15 gm/litre water) solution may be sprayed on *T. tomentosa* trees (foliage) 30 days before rearing.

Pruning: Training and pruning of *T. tomentosa* is subjected to maintain low bushes as well as medium trees. The plants pruned during Feb.-March produce maximum number of new shoots and subsequently exhibit increase in leaf yield. Significant gain in leaf yield and branches per plant has been achieved when *T. tomentosa* plants are subjected to biennial pruning at 7" height from base. Heavy pruning during March is better for the first crop rearing and yield maximum leaf yield whereas pruning during April has been found to be suitable for second crop rearing.

T. tomentosa plants also respond best when they are pollarded every four years at 6 ft. height followed by light pruning in the 2nd and third year. The best recipe for wound dressing was found to be a mixture of resin and linseed oil with some fungicide (See Srivastav *et al.* 1998a).

154. TEJBAL

Botanical: ***Zanthoxylum alatum*** Roxb. Syn. *Z. armatum* D.C., *Z. planispinum* Sieb. & Zucc.

Nepali: Timur, Timuri, Sunguru-kung

Hindi: Tejbal, Tejphal, Nepali Dhaniya

Bengali: Gaira, Tambul

Garhwali: Timuri, Tejabal, Tejmal

Oriya: Tundo-poda

Punjabi: Timbar, Timal

SYSTEMATIC POSITION

Bentham & Hooker (1883)	Engler & Prantl (1899)	Hutchinson (1959)
Phanerogams	Phanerogams	Angiospermae
Dicotyledones	Dicotyledoneae	Dicotyledones
Polypetalae	Archichlamydeae	Lignosae
Disciflorae	Geraniales	Rutales
Geraniales	Rutaceae	Rutaceae
Rutaceae		

Geographical distribution: Outer Himalaya from Indus eastward ascending to 7000 ft. ASL, Khasi (Meghalaya) and Naga (Nagaland) hills and Manipur foothills in the North-east and hills of Vizagapattam in Andhra Pradesh and Ganjam in Orissa.

Habit: A shrub, sometimes a small tree, dioecious.

Root: Branched, tap root.

Stem: Almost entirely glabrous, wood white, close-grained, prickles shining, strong, broad and flat on branchlets, those on older branches often on a conical corky base.

Leaves: Compound, leaflets opposite, 2-4 pair, sessile, entire, lanceolate, secondary nerves indistinct, petioles and midrib prickly.

Inflorescence: Sparingly branched, axillary panicles.

Flower: Unisexual, incomplete, yellow, pedicellate, actinomorphic, hypogynous, bracteate.

Fruit: Ripe carpels rugose, red, solitary, rarely 2-3.

Flowering: October - November.

Uses: In addition to serving as tertiary food plants of eri as well as muga silkworms (Seth 1995 c, d), walking sticks are made from its wood and branches are used as toothbrush. Bark, fruits and seeds are carminative, stomachic and anthelmintic. Fruits and seeds are aromatic, tonic and given in fever and dyspepsia. In dental troubles and scabbies fruits are used. Bark is used for intoxicating fishes and for repelling insects. Seed oil has been considered as deodorant, antiseptic and antibacterial (Sinha 1996) and may be used for commercial production of cosmetic soap, dental preparations and edible oil as seeds contain 20% oil (Jain *et al.* 1990).

Cultivation: *Zanthoxylum alatum* may be propagated through seeds.

155. BADRANG / BAJRAMANI

Botanical: ***Zanthoxylum rhesta*** DC. Syn. *Z. limonella* (Dennst.) Alston, *Z. budrunga* Wall. ex DC., *Fagara budrunga*

Burmese: Maya-nin   Eng: Assembly tree   Hindi : Badrang

Assamese: Bajramoni   Bengali: Bajnali   Gujarati: Tejbala

Kannada: Jummina   Khasi: Dieng-soh-mirik

Malayalam: Mulillam, Katta Murakku

Manipuri: Nagang   Marathi: Tirphal, Chirphal

Mizo: Chingit   Sanskrit: Atitejani, SutejasiTamil: Iratchai

Telugu: Rhesta Maum

SYSTEMATIC POSITION

Bentham & Hooker (1883)	**Engler & Prantl (1899)**	**Hutchinson (1959)**
Phanerogams	Phanerogams	Angiospermae
Dicotyledones	Dicotyledoneae	Dicotyledones
Polypetalae	Archichlamydeae	Lignosae
Disciflorae	Geraniales	Rutales
Geraniales	Rutaceae	Rutaceae
Rutaceae		

Geographical distribution: Western ghats and coast from Konkan southwards chiefly in moist evergreen forests and on seashores. Also found in Assam, Khasi hills (Meghalaya), Dhaka and Chittagong (Bangladesh) and lower Myanmar.

Habit: A large deciduous tree, height up to 75 ft. and girth up to 4-5 ft. crown large, prickles numerous on trunk branches and petioles.

Root: Branched tap root.

Stem: The prickly stem resembles like that of Bombax, prickles on branches curved while those on the old trunk have a solid conical corky base, bark corky and pale yellow outside and light brown inside which turns to dark chocklate colour on exposure.

Leaves: Compound, exstipulate, paripinnate or imparipinnate, crowded at terminals of branchlets, 1-2 ft. long rachis terete and glabrous, leaflets opposite, 8-20 pairs per leaf, the terminal usually absent, entire, ovate or ovate-lanceolate, caudate- acuminate, very oblique (both sides unequal), larger half with rounded base, smaller half cuneate and above the base of other half, length 9.5-12 cm, breadth 2.5-3 cm, glabrous, membranous or thinly coriaceous, shining above, pale and glaucous beneath, main lateral nerves up to 23 in larger half, 2-3 lesser in smaller, petiole 0.5-1.2 cm long, turns to yellow before falling.

Inflorescence: Largely crowded terminal puberulous panicles or cymes with opposite branches up to 22 cm long.

Flower: Small, 0.4 cm diameter, unisexual, incomplete, tetramerous, pedicels short, prickly sometimes, bracts minute, caducous, actinomorphic, hypogynous, light green – pale yellow in colour.

Calyx: Sepals 4, small, persistent, acute, triangular, polysepalous (free), imbricate.

Corolla: Petals 4, polypetalous (free), elliptic or elliptic-obovate, aestivation valvate.

Androecium: Stamens 4, polyandrous (free), reduced to scales in females, anther bilobed, introse, versatile.

Gynoecium: Bicarpellary, syncarpous, bilocular with 2 ovules in each locule, placentation axile, ovary superior, styles sub-lateral, stigma capitate, ripe carpels globose, double valved. Ovary rudimentary in males.

Fruit: Hesperidium, fleshy, globose

Seeds: Solitary, sub-globose, blue black, 0.5 cm in diameter, seeds identical to black pepper.

Floral formulae: Male flower: $Br_1 \oplus ♂ K_4 C_4 A_4 G_0$

Female flower: $Br_1 \oplus ♀ K_4 C_4 A_0 G_{\underline{(2)}}$

Uses: Besides serving as secondary food plants of muga silkworm, *Antheraea assam* Ww. and eri silkworm *Philosamia ricini* Bois (Sarkar 1988, Thangavelu *et al.* 1988, Seth 1995c,d, Das *et al.* 2000), it has timber value also because wood is used in making shuttles. Aromatic oil is also extracted from its seeds and its fruits are used as spice (Pandey 1981).

Medicinally unripe fruits are given in dyspepsia, asthma, bronchitis, heart troubles and rheumatism. Pericarp is astringent, stimulant and digestive. Seed tincture is given in cholera and essential oil of fruits called Mullilam oil has been found to have antiseptic, disinfectant and anti-inflammatory properties and hence used in cholera and applied locally on skin swellings (Sinha 1996). In Mizoram, young shoots and leaves are boiled and used as vegetables during March – June (Laha 1998).

Bark is pickled and fruits are appetizing and digestive. Tender leaves used as condiments. Bark is aromatic, tonic, diuretic, cholinergic, hypogycaemic and spasmolytic. Wood used for furniture, decorative panelling, carving, turnery, construction and building purposes, boards, plywood etc. (Ambasta *et al.* 1986).

Cultivation: *Z. rhesta* can be propagated through seeds.

156. BERI

Botanical: *Zizyphus jujuba* Mill syn. *Z. sativa* Gaertn., *Z. vulgaris* Lam.

Eng: Chinese date
Hindi: Beri, Banber, Pitni Ber
Bengali: Boroi
Kashmiri: Phitni
Manipuri: Boroi
Marathi: Ranbor, Unab
Punjabi: Sinjli, Amlai

SYSTEMATIC POSITION

Benthan & Hooker (1883)	Engler & Pranth (1899)	Hutchinson (1959)
Phanerogams	Phanerogams	Angiospermae
Dicotyledones	Dicotyledoneae	Dicotyledones
Polypetalae	Archichlamydeae	Lignosae
Disciflorae	Rhamnales	Rhamnales
Rhamnales	Rhamnaceae	Rhamnaceae
Rhamnaceae		

Geographical distribution: Cultivated in the Punjab, wild and cultivated in the North-west Himalaya ascending up to 6500 ft. ASL. Indigenous and cultivated in Japan, China, Western Asia, North Africa and South Europe.

Habit: A small monoecious, deciduous, glabrous, thorny tree or hedge.

Habitat: Subtropical to tropical agroclimatic localities.

Root: Tap root.

Stem: Hard, woody, solid, heartwood brown, branches flexuose, branchlets frequently fasiculate, thorns unequal, one straight up to 3.3 cm long, the other straight when young, afterwards hooked and bent down, glabrous.

Leaves: Simple, alternate, petiolate, blade obliquely ovate - lanceolate, mostly obtuse bluntly serrate.

Inflorescence: Short axillary nearly sessile cymes.

Flower: Greenish yellow, bract 1, bracteoles 2, pedicel short, flower complete, regular, actinomorphic, bisexual, pentamerous, cyclic, perigynous.

Fruit: Drupe, dark red or black, ovoid or oblong, obtuse at both ends 1.3-1.9 cm long, stones tuberculate.

Floral formula: ⊕ ⚥ $Br_1\,Brl._2\,K_5\,C_5\,A_5\,G_{(3)}$

Uses: Leaves serve as secondary food for tasar as well as muga silkworms, whereas fruits are edible. Japla, Nagaon, Jiribam and Shivalika races of *A. mylitta* are exclusively reared on it (Sengupta 1986, Sinha 1998, Thangavelu *et al.* 2000).

Medicinally, leaves are laxative and consumed in scabies and throat trouble. Fruits are considered as emolient, cooling, anodyne and tonic. They are given as an antidote to aconite poisioning and consumed in nausea and vomiting. Seed powder is sedative (Sinha 1996). It is stated that if a woman takes badari (*Z. jujbba*) and lac boiled in gingelly oil after menses she would not conceive (Dash 1993). It serves as one of the important host plants of lac insect *Kerria lacca* Kerr. in the forests (Sagreiya 1967).

Stones of fruits are strung up as beads.

Cultivation: *Z. jujuba* is cultivated through seeds.

157. BER

Botanical: ***Zizyphus mauritiana*** Lam. syn.*Z. jujuba* Lam. non Mill.

Burma: Ziben	Chinese: Date	Eng: Indian jujube
Hindi: Ber, Baer		
Assamese: Bogori	Bengali: Boroi	Gujarati: Bor
Kannada: Jelachi	Malayalam: Elentha	Manipuri: Boroi
Marathi: Bor	Oriya: Barkoli	Tamil: Yelandai
Telugu: Regu	Sanskrit: Ajapriya, Badara, Kuvala, Madhurphal	

SYSTEMATIC POSITION

Bentham & Hooker (1883)	Engler & Prantl (1899)	Hutchinson (1959)
Phanerogams	Phanerogams	Angiospermae
Dicotyledones	Dicotyledoneae	Dicotyledones
Polypetalae	Archichlamydeae	Lignosae
Disciflorae	Rhamnales	Rhamnales
Rhamnales	Rhamnaceae	Rhamnaceae
Rhamnaceae		

Geographical distribution: It is indigenous and naturalised throughout India, China and Myanmar ascending in the outer Himalaya up to 4500 ft. ASL. Cultivated in Punjab, Uttar Pradesh, Bihar and Rajasthan.

Habit: A large shrub or middle sized tree with drooping branches up to 30 ft. height and 4-5 ft. girth, monoecious, almost evergreen (Fig. 64).

Habitat: All sort of tropical and sub-tropical climatic places.

Root: Branched tap root with spreading tendency.

Stem: Branches armed with stipular spines, equal or one straight and the other bent, sometimes spineless. Branches clothed with bright tawny or almost white tomentum, bark dark green or nearly black in old stem with long vertical cracks, reddish and fibrous inside.

Leaves: Simple, alternate, petiole 0.2-2.0 cm long, blade shape variable from ovate-oblong to nearly orbicular, obtuse or acute apex, margin entire or serrulate, 2.5-6.5 cm long and 2.5-4.0 cm broad, dark green, glabrous and shining above, bright tawny or white tomentose, beneath, strongly 3-nerved, base obliquely rounded.

Inflorescence: Short axillary nearly sessile cymes.

Flower: Greenish yellow, bract 1, bracteoles 2, pedicel short, complete, regular, actinomorphic, bisexual, pentamerous, cyclic, perigynous.

Calyx: Sepals 5, polysepalous, glabrous inside, oblong, concave or hooded, caducous, valvate.

Corolla: Petals 5, polypetalous, narrow, concave, reflexed, clawed, valvate.

Androecium: Stamens 5, polyandrous, antipetalous, hidden within concave petals, anthers bilobed and basifixed.

Gynoecium: Tricarpellary, syncarpous, ovary superior but mimics like inferior, trilocular, each locule with one erect basal ovule, stigma trilobed.

Fruit: Drupes varying in size and shape, mostly 1.2-2.5 cm long, globose or ellipsoidal or obovoid - acuminate with a fleshy, mealy aromatic acidic or sub-acidic pulp, orange or red when ripe. Stones tuberculate, bony, irregularly furrowed, mostly 2 celled.

Seeds: Albuminous with large and straight embryo.

Flowering: Usually April- Sept.

Fruiting: Usually Dec.- March.

Sprouting: The young foliage sprouts when old one are falling. Sometimes second flush appears in July and August.

Floral formula: $\oplus$ ⚥ $Br_1\ Brl._2\ K_5\ C_5\ A_5\ G_{\underline{(3)}}$

Uses: Its foliage serves as primary food for tropical tasar silkworm *Antheraea mylitta* D. as well as secondary host plant or eri silkworm, *Philosamia ricini* and muga silkworms, *Antheraea assama*. Japla, Nagaon, Jiribam and Shivalika races of tropical tasar silkworm, *Antheraea mylitta* D. are exclusively reared on Ber in Jharkhand (Japla, Palamu), Assam (Nagaon), Manipur (Jiribam) and Jammu & Kashmir (Shivalika) in forests or they themselves thrive well on Ber food plants in these areas (Sengupta 1986, Sinha 1998, Seth 1995 a,c,d, Thangavelu *et al.* 2000).

It also serves as host plant of lac insect (Sharma and Jaiswal 1997). Besides, its fruits rich in vitamin-C, B and iron are edible. The fruit is used fresh, dried or preserved. The bark is used for tanning, whereas leaves are utilised for cattle feeding. Fruits are cooling and anodyne. In cough mixture dry fruits are used as ingradient. Kernel is sudorific, sedative whereas seeds are used in hard boils as poultice. Leaves are eaten with catechu as an astringent, leaf juice is used in typhoid. Seeds also serve as an antidote to aconite poisoning and used against the abdominal pain in pregnancy and diarrhoea. If a woman takes badari (*Z. mauritiana*) and lac boiled in gingilly oil after menses she would not conceive for birth control (Dash 1993).

Cultivation: *Z. mauriana* is cultivated through seeds. Besides, it can also be propagated through air layerings, cuttings (Jauhari 1960), cleft graftings and T-budding methods (Singh 1963).

158. CHURNA

Botanical: *Zizyphus rugosa* Lam.

Hindi: Suran, Churna, Bhand

Lower Myanmar: Myauksi
Upper Myanmar: Tabu, Mintha Tabu

Malayalam: Todali
Marathi: Turan, Churna

Oriya: Simu koli
Sanali: Sekra

Tamil: Katti landai
Telugu: Pinduparighamu

SYSTEMATIC POSITION

Bentham & Hooker (1883)	Engler & Prantl (1899)	Hutchinson (1959)
Phanerogams	Phanerogams	Angiospermae
Dicotyledones	Dicotyledoneae	Dicotyledones
Polypetalae	Archichlamydeae	Lignosae
Disciflorae	Rhamnales	Rhamnales
Rhamnales	Rhamnaceae	Rhamnaceae
Rhamnaceae		

Geographical distribution: Sub-Himalayan tract from the Ganges eastwards in Bihar, Central India, Western Peninsula asecnding up to 2000 ft. ASL. Also found in Shan hills up to 4000 ft. in Myanmar and Sri Lanka.

Habit: A spreading deciduous monoecious shrub, usually climbing rarely a small tree, branchlets, young spines and under side of leaves densely clothed with ferruginous or grey tomentum.

Root: Tap root.

Stem: Solid, woody, erect, aerial, trunk often 2-3 ft. girth, armed with short recurred usually solitary spines, bark dark brown to black exfoliating as rectangular scales.

Leaves: Simple, alternate, ovate or elliptic, petiolate, petiole 0.6 cm, blade 5-12 cm long, basal nerves-3, rarely 4 or 5, prominent, secondary nerves found along midrib as well as on the outside of the lateral nerves.

Inflorescence: Cymes large, pedunculate, axillary as well as terminal forming a long compound generally drooping panicle.

Flower: Hermaphrodite, complete, actinomorphic, petals absent, regular, hypogynous.

Flowering: Nov. - March.

Fruit: Drupe 0.8 cm long, stone crustaceous, 1-celled and 1-seeded.

Fruiting: June - September.

Uses: It serves as one of the tertiary but recognized as one of the seconday food plants of tropical tasar silkworm *Antheraea mylitta* D (Seth 1995 a).

Fruits eaten. Bark used in diarrhoea and flowers used in menorrhagia. Wood used as fuel (Ambasta *et al.* 1986).

Cultivation: *Z. rugosa* can be propagated through seeds.

159. KATHBER

Botanical: *Zizyphus xylopyrus* Willd.

Hindi: Katber, Kathber, Bhander, Goti — Kannada: Challe

Marathi: Goti, Ghot — Oriya: Got, Gotoboro

Tamil: Kottai — Telugu: Goti

SYSTEMATIC POSITION

Bentham & Hooker (1883)	Engler & Prantl (1899)	Hutchinson (1959)
Phanerogams	Phanerogams	Angiospermae
Dicotyledones	Dicotyledoneae	Dicotyledones
Polypetalae	Archichlamydeae	Lignosae
Disciflorae	Rhamnales	Rhamnales
Rhamnales	Rhamnaceae	Rhamnaceae
Rhamnaceae		

Geographical distribution: Found in foot of North-west Himalaya, Central India and the Western Peninsula.

Habit: A large straggling shrub or small tree. Branchlets inflorescence and underside of leaves pubescent or tomentose and spineless on good soil.

Leaves: Simple, alternate, broad-ovate or orbicular, finely serrulate with cordate base usually. Basal nerves 3, rarely 4-5, midrib with 1-2 pair of secondary nerves while lateral basal nerves 3-5 secondary nerve on the outside, petiolate, petiole 0.6-0.8 cm, blade 3.0 - 7.0 cm long.

Inflorescence: Axillary cymes.

Flower: Regular, actinomorphic, bisexual, complete, hypogynous.

Fruit: Globose drupe, grey-tomentose, almost dry, 0.7-2.5 cm in diameter, stone furrowed, 3-celled, 3 seeded.

Flowering: April - May.

Uses: It serves as tertiary but recognized as secondary food plants of tropical tasar silkworm *Antheraea mylitta* D (Seth 1995a). It also serves as the host plants of lac insects (Sharma and Jaiswal 1997, Sagreiya 1967).

Leaves serve as fodder. Wood used for cart wheels, handles, beading, panelling, frames, inlay works and ornamental and fuel purposes (Ambasta *et al.* 1986).

Cultivation: *Z. xylopyrus* is mainly propagating through seeds.

REFERENCES

Agarwal, K.(2003). Pau Barah Kisano Ki. Sarita 1165: 59-61.

Agarwal, V.S.(1986). Economic plants of India. M/s Kailash Prakashan, Calcutta.

Ambasta, S.P., Ramchandran, K., Kashyapa, K. and Chand, R. (1986). The useful plants of India. P.I.D., C.S.I.R., New Delhi.

Ambad and Jambbale (1997). Response of single nodded cuttings of bamboo (*Bambusa vulgaris* L.) to vegetative propagation. Jour. Non-Timber For. Prod. 4(¾): 165-166.

Anonymous 1949-1979 (Revised 1985 todate). The wealth of India. Raw materials. I-XI Vols. Pub. & Inform. Director, CSIR, Hillside Road, New Delhi.

Anonymous (1962). The wealth of India, VI: L-M, CSIR, New Delhi, pp 429-439.

Anonymous (1969). The wealth of India (Raw materials) III, CSIR, New Delhi, pp 349-357.

Anonymous (1976). Wealth of India, CSIR, New Delhi.

Anonymous (1976). Medicinal plants of India I. Eds. Satyawati, G.V. *et al.*, ICMR, New Delhi.

Anonymous (1981). Improved crop production for drought prone areas of Maharastra. Mahatma Phule Agriculture University Tech. Bull., 1981: Dry Farming Research Station, Solapur.

Anonymous (1985). Wealth of India I (Revised). PID, CSIR, New Delhi.

Anonymous (1998). Package of practices in sericulture. Regional Development Office, CSB, Guwahati.

Anonymous (1999-2000). Annual Report. Central Tasar Research & Training Institute, Ranchi.

Anonymous (2000-01). Annual Report. Central Tasar Research & Training Institute, Ranchi.

Ansari, A.A. (1997). Medicinal plants of Madhulia forest of Gorakhpur. Jour. Non-timber For. Prod. 4 (34) : 138-150.

Arora, G.S. and Gupta, J.J. (1979a). Taxonomic studies on some of the Indian Non-mulberry silkmoths (Lepidoptera: Saturiniidae). Mem. Zool. Sur. of Ind. 16 (1) : 25-28.

Arya, S. and Arya, I. (1998). Micropropagation of Neem (*Azadirachta indica*) by adventitious organogenesis. Ann. For. 6(2): 123-129.

Bahadur, K.N. and Gaur, R.C. (1980). A note on the *Terminalia tomentosa* complex. Ind. Jour. For. 3: 367-369.

Barah, A, Isa, Md., Devnath, M. and Samson, M.V. (1992). A new host plant of the muga silkworm, *Antheraea assama* Ww. Sericologia 32(3): 529-530.

Benchamin, K.V. (2000). Muga and eri culture: Subsistence to substantial farming. Ind. silk 38 (9&10) : 51-54.

Benson, L. (1957)). Plant classification. (original not consulted).

Bentham, G. and Hooker, J.D. (1862-1883). Genera plantarum, 3 volumes, London (Original not consulted).

Bharadwaj, S.D., Chakraborty, A.K. and Joshi, N.K. (1993). Vegetative propagation of *Terminalia bellerica* Roxb. and *T. chebula* Retz. By stem cuttings. Ind. For. 119(5): 360-366.

Bharadwaj, S.D., Chakraborty, A.K. (1994). Studies on the time of seed collection, sowing and pre-sowing seed treatments of *Terminalia bellerica* Roxb. and *T. chebula* Retz. Ind. For. 120(5): 430-439.

Bhatia, K., Lal, J., Gulati, A.S. and Ayyar, K.S. (1987). Tannins from *Terminalia* bark - A strengthening material for soil. Ind. For. 109 (8) : 521-523.

Brandis, D. (1906). Indian trees. Periodical experts book agency, Delhi and International book distributors, Dehradun.

Chandra, J.P. and Mahendru, J.L. (1977). Propagation of *Quercus incana* by the air layering. Ind. For. 103: 539-541.

Chatturvedi, N. (1999). A new larval food plant for tasar silkmoth, *A. paphia* (Linn.). Ind. For. 125(2): 233.

Chaudhary, M. (1997). Aahaar hi aushadhi hai. Navneet Publications (India) Ltd., Ahmedabad / Mumbai.

Chaudhury, N.R. (1963). Air layering in Sal (*Shorea robusta* L.). Ind. For. 89: 69-71.

Chauhan, P.S., Joshi, N.K., Bist, H.S. and Dhiman, R.C. (1994). Effect of growth regulators in rooting performance of stem cuttings of some shrub species of Western Himalayas. Ind. For. 120(2): 105-109.

Chauhan, S.V.S. and Chandra, S. (2003). Cultivation of Sandal (*Santalum album* L.) hydroponically without host. Ann. For. II(1): 85-87.

Chinnaswamy, K.P., Doreswamy, C., Devaiah, M.C. and Venkatesh, M. (1993). Pharmaceutical value of mulberry. Ind. Silk 31 (11) : 47-48.

Chopra, R.N., Badhwar, R.L. and Ghose, S. (1965). Poisonous plants of India. ICAR, New Delhi.

Chopra, R.N., Chopra, I.C. and Verma, B.S. (1974). Supplement to Glossary of Indian Medicinal Plants. PID, CSIR, New Delhi.

Chopra, R.N., Nayar, S.L. and Chopra, I.C. (1956). Glossary of Indian Medicinal Plants. CSIR, New Delhi.

Choudhury, S.N. (1965). Studies on the propagation of Som (*Machilus bombycina*) by stem cuttings with the aid of growth regulators. Indian Agriculturist 19(1): 82-83.

Choudhury, S.N. (1984). Mulberry Silk Industry. Dibrugarh, Assam.

Chundawat, B.S. and Gautam, S.K. (1993). A textbook of Agroforestery. Oxford & IBH Pub. Co., New Delhi.

Das, P.K., Sahu, A.K. and Babulal (2000). Management of muga silkworm and host plant germplasm. Base paper in: National Workshop on Management of sericultural germplasm for posterity, CSGRC, Hosur.

Dash, B. (1993). Ayurvedic cures for common diseases. Hind pocket books Pvt.Ltd., Delhi.

Dastur, J.F. (1962). Medicinal plants of India and Pakistan. D.B.Taraporewala Sons & Co., Bombay.

Devanath, M., Isa, Md., Samson, M.V., Sengupta, A.K. and Rajaram (1992). Potential of propagation through leaf bud cuttings of Som, *Machilus bombycina* King (Lauraceae). Sericologia 32 (3): 457-464.

Drew, R.A. (1993). Clonal propagation of Neem by tissue culture. World Neem Conf., Bangalore. Abs., pp 61.

Dwivedi, N.K. (1988). Importance of white mulberry as a fodder tree. Ind. Silk 27 (6) : 18-20.

Engler, A. and Prantl, K. (1887-1915). Die naturlicten Pflanzenfamilian, 23 volumes, Leipzig (Original not consulted).

Evans, W.C. (1989). Trease and Evans Pharmacognosy (13th edition). English Language Book Society/Balliere Tindall, London.

FAO (1976). Manuals on sericulture. FAO Agriculture Service Bulletins. FAO, Rome I-IV (Reprinted 1987 by Central Silk Board, Bangalore).

Gala, D., Gala, D.R. and Gala, S. (1997). Rasahaar ke dwara tandursti aur rogemukti. Navneet Pub. (I) Ltd., Ahmedabad.

Gamble (1922). A manual of Indian timbers, London, pp 671 (original not consulted).

Gauthier, D., Diem, H.G., Doomergues, Y.R. and Ganny, F. (1985). Soil. Biol. Biochem. 17: 357-379.

Ghosh, M.K. and Srivastava, R.C. (1996). Effect of pruning time on the nutritive value of *Q. serrata,* leaves, the primary food plants of *Antheraea proylei* J. and its impact on oak tasar silkworm rearing. Sericologia 36(1): 139-141.

Gill, B.S. (1974). Advancing frontiers in cytogenetics in : Mehra Comm. Vol. Ed. Kachroo, P., pp 348 – 356.

Gill, B.S., Bir, S.S. and Singhal, V.K. (1982). Cytogenetics of some timber species of *Terminalia* Linn. Proc. Ind. National Science Academy. 48 B : 779 – 790.

Gill, B.S., Singhal, V.K., Bede, V.S. and Bir, S.S. (1990). Cytological evolution in the woody taxa of Pachmarhi hill. J. Cytol. & Genet. 25: 308 – 320.

Goor, A.V. and Barney, C.W. (1976). Forest tree planting in arid zones. The Roland Press, New York.

Homfrey, C.K. (1935). Nursery and plantation notes for Bengal. IV Edition, pp 162 (original not consulted).

Hooker, J.D. 1872–1897 (Reprinted 1973). Flora of British India. B.S.M.P.S., Dehradun & Periodical Experts, Delhi.

Hutchinson, J. (1959). Families of Flowering Plants (Original not consulted).

Isa, M., Devnath, M., Ram, R. and Samson, M.V. (1989). Nursery technique for raising Som and Soalu seedlings. Ind. Silk 28 (2): 22 – 23.

Isa, M. and Thangavelu, K. (1992). Nursery technique for raising Asan seedlings. Ind. Silk 31(8) : 30-31

Isa, M, Srivastava, D.P., Goel, A.K. and Thangavelu, A.K. (1993). Vegetative propagation of *Terminalia arjuna* Bedd. through auxins treated stem cuttings. Ind. J. Seric. 32(1) : 115 – 116.

Jain, A.K. and Kumar, R. (1989). Mulberry species and their distribution in North-eastern India. In: Genetic resources of mulberry and utilization. Eds. Sengupta, K. and Dandin, S.B., CSR&TI, Mysore, pp 37-42.

Jain, P.P., Suri, R.K., Mathur, K.C. and Goel, C.K. (1990). Scope of utilisation of oil seed from the Garhwal region. J. Econ. Bot. & Phytochem. 1 (2-4) : 34-37.

Jain, S.K. (1968). Aushdhiya Paudhe. National Book Trust, New Delhi.

Jain, S.K. (1968). Medicinal plants. National Book Trust, New Delhi.

Janaki-Ammal and Sobti, S.N. (1962). Polypoloidy in the genus *Terminalia.* Sci. & Cult. 28(8): 378-388

Jauhari, O.S. (1960). Some observations on vegetative propagation of *Z. mauritiana* Lam. (Ber) from gootees and cuttings with the aid of growth regulators. Current Science 29 : 30 – 31.

Jolly, M.S., Sen, S.K. and Ahsan, M.M. (1974). Tasar culture. Ambika Publishers, Bombay.

Jose, P.A. and Thomas, J. (1998). An account of the vegetative propagation in *Terminalia chebula* Retz. Ind. For. 124 (5) · 357 – 359.

Joshi, M.S. and Thengane, S.R. (1993). Potential application of *in vitro* propagation of neem (*Azadirachta indica* A. Juss.). World Neem Conf., Bangalore, Abs., pp 56.

Kaja, Maideen, S., Jacueline, Selvaraj, A. and Vinaya Rai, R.S. (1997). Effect of geographic source and age of mother tree on seed viability and vigour in *Casuarina equisetifolia* F.R.&G. Forst. Ind. J. For. 20(3): 220-222.

Kanjilal, U., Kanjilal, P.C., De, R.N. and Das, A. (1940). Flora of Assam 4: 304-325, Govt. of Assam.

Karoshi, V.R. and Hegde, G.V. (2002). Vegetative propagation of *Pongamia pinnata* (L.) Pierre : hitherto a neglected species. Ind. For. 128(3): 348-350.

Ketkar, C.M. (1999-2000). Commercial products from Neem. Jour. Eco. Bot. & Phytochem 10-11: 49-53.

Khali, R.P. and Sharma, A.K. (2003). Effect of Phytohormones on propagation of Himalayan Yew (*Taxus baccata* L.) through stem cuttings. Ind. For. 129(2): 289-294.

Khanna, P., Pacholi, R.K. and Singh, V.P. (1998). Fuelwood and leaf fodder yield tables of *Anogeissus latifolia* and *Terminalia tomentosa.* Ind. For. 124 (3) : 198 – 205.

Khare, R., Isa, M., and Thangavelu, K. (1991). Soil conservation through tasar culture. Proc. Workshop on Soil Moisture conservation in mulberry, 4-5 Oct., 1991, CSR&TI, Mysore.

Khosla, P.K. and Sareen, T.S. (1978). Chromosomal Conspectus of Indian hardwood timber. Ind. J. For. 1 (3) : 169 – 178.

Kocchar, S.L. (1981). Economic botany in the tropics. Macmillan India Ltd., Delhi.

Kumar, R., Joshi, M.C., Gargi, Beck, S. and Gangopodhyay, A. (2003). Tasar host plant germplasm: Establishment, management and cataloguing. Ind. Silk 42(3): 19,20,24.

Kumar, V. and Shamet, G.S. (2002). Vegetative propagation of Himalayan Yew (*Taxus baccata* L.) through branch cuttings. Jour. Non-Timber For. Prod. 9(½): 32-36.

Kurian, J. (1995). Plants that heal. Oriental Watchman Publishing House, Pune, India.

Laha, R.C. (1998). Edible forest vegetables of Kolasib forest division in Mizoram. Jour. Non-timber For. Prod. 5 (3/4): 173 – 175.

Lakshmisubramanian, S., Muthuraman, C., Ravichamy, P., Vanitha, K.L., Sujatha, S., Nirmala, K., Nalini, K., Stephen John David, A. and Murali, P.M. (1999-2000). Large scale plantation of tissue culture neem plantlets in Madhya Pradesh. J. Econ. Bot. & Phytochem. 10-11: 1-4.

Maheshwari, J.K. and Singh, H. (1991). Ethno-botanical notes from Banda district, Uttar Pradesh. J. Econ. Bot. & Phytochem. 2 (1-4) : 16-20.

Mascarenhas, A.K., Nadigar, A.L., Thengane, S.R., Phadke, C H., Khuspe, S.S., Shrirgurkar, M.V., Parasharami, V.A. and Nafagauda, R.S. (1990). Potential application of tissue culture for propagation of *Dendrocalamus strictus*. In: Rao, I.V.R., Gnanaharan, R. and Sastry, C.B.(eds). Bamboos Current Research. Proc. of the Intl. Bamb. Works., Cochin, India : 159-166.

Masilamani, P., Annadurai, K. and Sarvanapandian, P. (2002). Effect of soil pH and organic matter on germination and growth attributes of Pungam (*Pongamia pinnata* Roxb.). Ind. J. For. 25(4): 415-419.

Masilamani, P. and Vadivelu, K.K. (1997). Effect of growth regulator and nutrient on viability and vigour of preconditioned seeds of Anjan (*Hardwickia binata* Roxb.) Ind. Jour. For. 20 (3) : 223-226.

Maurya, K.R. (1991). Optimum period for defoliation of *Quercus incana* for preponed crop in the Garhwal region (Uttar Pradesh). Ind. Silk 30(5): 33-34.

Mishra, P.N., Sharma, K.K., Kimothi, R.C. and Thangavelu, K. (2000). Maintenance of genetic resources of *Quercus*. Base paper in: National Workshop on Management of sericultural germplasm for posterity, CSGRC, Hosur., pp 1-15.

Mishra, S.N. (2002). Sericulture industry in Indonesia, Indian Silk 41 (7): 24-25.

Mishra, S.N. (2003). Aamkoni athva aamresham. Ind. Silk 42 (7): 45.

Misra, K.K. and Jaiswal, H.R. (1994). Effect of Indole butyric acid on rooting of air layer of Arjun and Chebulic myrobalan. Ind. For. 120 (2) : 183 – 186.

Misra, K.K. and Jaiswal, H.R. (1997). Effect of planting season and storage period on survival and growth of bare root seedlings of Bael (*Aegle marmelos* Correa) and Karaunda (*Carissa carandas* L.). Jour. Non-Timber For. Prod. 4(¾) : 170 – 172.

Misra, K.K. and Ranvir Singh (1990). Effect of growth regulators on rooting and survival of air layers of Karunda (*Carissa carandas* L.). Ann. Agric. Res. 11(2): 208-210.

Mughal, A.H. (1996). Effect of stem cutting thickness on growth and survival of *Salix* species in nursery. Ind. For. 122(9): 834-836.

Mukuntha kumar, S., Mathur, J., Nair, P.K.K. and Mathur, S.N. (1999). *In vitro* morphogenetic responses of nodal explants of golden yellow Bamboo – *Bambusa vulgaris* Schrad. Jour. Non-Timber For. Prod. 6(½): 39-43.

Mutha, N., Burman, U., Kumar, P., Aggarwal, R.K. and Harsh, L.N. (1997). Effect of flyash substitution in nursery mixture on growth of *Hardwickia binata*. Ind. Jour. For. 20(3): 213-219.

Nadagoudar, B.S. and Nataraja, K. (1999). Studies on approach grafting, air layering and root suckers as a means of vegetative propagation in Neem. Jour. Non-Timber For. Prod. 6(3/4): 183-186.

Nagpal, R. and Sehgal, R.N. (1985). Propagation of some agroforestry species by air layering. Ind. J. For. 8: 160-165.

Nair, K.K.N. and Jayakumar, R. (1998). Ethnobotany of Muthuva tribe in the context of biodiversity rehabilitation at Chinnar Wildlife sancutary, Western Ghats of India. Jour. Non-timber. For. Prod. 5 (3/4) : 159-172.

Najma, Ganapathy, P.M., Sasidharan, N., Bhat, K.M. and Ganaharan, R. (1981). A handbook of Kerala timber, K.F.R.I. Res. Rep. 9: 201-207.

Nakvi, A.H. and Srivastava, A.K. (2003). Ushnakatibandhiya reshamkeet ke anuvanshik sansadhano ke sarankshan ki avashyakta. Resham Vani 22: 6-7, CTR&TI, Ranchi.

Nalini, K.,Vanitha, K.L., Nirmala, K., Sujatha, S., Hemalatha, S., Muthuraman, C., Stephen, John David, A., Sakthivel, R. and Murali, P.M. (1999-2000). Effect of growth hormone on micro propagation of Neem. J.Eco. Bot. & Phytochem. 10-11: 25-30.

Narayan, P. and Jaiswal, V.S. (1985). Plantlet regeneration from leaflet callus of *Azadirachta indica* A. Juss. J. Tree Sci. 4: 65-68

Nautiyal, S., Rao, K.S., Maikhuri, R.K., Negi, K.S. and Kala, C.P. (2002). Status of medicinal plants on way to Vashukital in Mandakini Valley, Garhwal Himalaya, Uttaranchal. Jour. Non-Timber For. Prod. 9(3/4): 124-131.

Nautiyal, S. and Rawat, M.S. (1994). Macropropagation of teak (*Tectona grandis* L.f.). Ind. For. 120(2): 146-151.

Negi, K.S., Tewari, J.K. and Gaur, R.D. (1985). Economic importance of some common trees in Garhwal Himalaya: An ethnobotanical study. Ind. J. For. 8(4): 276-289.

Negi, S.S. (1985). Handbook of Silvicultural Systems. Dehradun.

Negi, S.S. and Naithani, H.B. (1995). Oaks of India, Nepal and Bhutan. International Book Distributors, Dehradun.

Palaniswamy, K., Ansari, S.A. and Mandal, A.K. (1995). Standardization of vegetative propagation technology for teak, sisso, neem, karanj and bamboos. Proc. Inl. Work. For. Res. Meth., Jabalpur, pp 18-19.

Paliwal, D.P., Jaiswal, J. and Rao, B.V.S. (1991). Propagation of muga food plants. Ind. Silk 30(5): 13-14.

Pandey, A.K. and Khatoon, S. (1999). Effect of orientation of seed placement and depth of sowing on seedling emergence in *Sterculia urens*. Roxb. Ind. For. 125(7): 720-724.

Pandey, B.P. (1980). Economic Botany. S. Chand & Co. Ltd., New Delhi.

Pandey, B.P. (1981). A text book of Botany: Angiosperms. S. Chand & Co. Ltd., Delhi.

Pandey, R.K. (1999). Seed coat dormancy in Ban oak. Ind. Silk 38(2): 22-23.

Pandey, S. and Jaiswal, V.S. (2002). Micropropagation of *Terminalia arjuna* Roxb. from cotyledonary nodes. Ind. J. Expt. Biol. 40(8): 951-953.

Parker, R.N. (1925). Hybrid *Terminalia* (Arjun + Asan) and general remarks on tree hybrid. Ind. For. 51 (12) : 599-603.

Parkinson, C.E. (1936). Indian *Terminalias* of the section *Pentaptera*. Ind. For. Rec. 1 (1): 1-27.

Parmeswari, K., Srimathi, P., Malarkodi, K. and Sasthri, G. (2001) Fruit storage of Tamarind for seed purpose. Jour. Non-Timber For. Prod. 8(1/2): 91-94

Parthiban, K.T., Narayanan, R., Rai, R.S.V., Surendran, C. and Ravichandran, V.K. (1997). Callogenesis and organogenesis in *Casuarina equisetifolia* F.R.& G. Forst. Ind. J. For. 20(3): 227-230.

Patel, B.H., Agarwal, B.J. and Patel, P.J. (2003). Dyeing of silk with Babool. Indian Silk 42(4): 17-21.

Patil, G.M., Kulkarni, K.A. and Patil, R.K. (2000). Scope of shifting tasar silkworm rearing from forest trees to castor for species conservation. In: Abst. Bharat Ratna Shri M. Vishveshvaraiya Mem. Natl. Sem. on Env. Pollu. & Manage., National Pollution Control & Enviromental Conservation Committee, Mysore, Nov. 24-25, 2000, pp. 6.

Patro, K.B. and Mohanty, A. (1998). A legend: Arjuna – More than a tasar food plant. Ind. Silk 36(12): 23-24.

Pattanaik, S. and Vijayakumar, M. (1997)). *In vitro* propagation of *Eucalyptus globules*. Ind. J. For. 20(3): 236-238.

Pearson, R.S. and Brown, H.P. (1932). Commercial timbers of India 2 : 975, Calcutta.

Phillip,T. (1989). Mulberry fruit - an ideal food. Ind. Silk. 28 (8) : 13-14.

Pradhan, N.R. (1998). Important minor forest produce of Mizoram (A status paper). Jour. Non-Timber. For. Prod. 5(3/4): 102-106.

Prain, D. (1898). Jour. Asiatic Soc. Bengal 66: 508 (original not consulted).

Prasad, D.A., Reddy, K.J. and Subba Reddy, C.J. (1974). Chemical composition and nutritive value of Anjan. (*Hardwickia binata* Roxb.) leaves. Ind. J. Animal Sci. 44 (3): 150-152.

Prasad, D.N., Sinha, A.K. and Bania, H.R. (1992). *Terminalia arjuna* can be grown in various soil pH. Ind. Silk 30 (12) : 37-38.

Prasad, R. and Mishra, G.P. (1985). Litter productivity of dry deciduous teak forest stands of Sagar (M.P.) Ind. For. 111(6): 359-367.

Prasad, R. and Parvel, Z. (1987). Observation on the seed viability of Sal (*Shorea robusta*). Ind. For. 113 : 89-94.

Punam, Sharma, S. and Atul (2002). Effect of scarification, temperature and storage conditions on the germination of two ecologically important *Albizia* species of Himalayas. Ann. For. 10(2): 262-267.

Puri, S. and Nagpal, R. (1988). Effect of auxins on air layers of some agroforestry species. Ind. J. For. 11(1): 28-32.

Radwankski, S.A. and Wickens, G.E. (1981). Vegetative fallows and potential value of the neem tree (*Azadirachla indica*) in the tropics. Econ. Bot. 35: 398-414

Rahman, M.A., Jaikishan and Kukreti, D.P. (1956). The stem bending properties of Indian timbers. Ind. For. 82 (9) : 469-478.

Raja Ram, Goswami, B.C. and Kumar, S. (1993 b) Vegetative propagation of *Litsaea salicifolia*. Ind. Silk 32(2): 38-50.

Raja Ram, Kumar, S. and Goswami, B.C. (1995). *Litsaea salicifolia* : A muga silkworm food plant : Propagation and rearing technique. In: Proc. Training Course in Seri. IV. Ed. Agarwal, H.O. and Sethi, M.K., H.P. Univ., Shimla, pp. 915-919.

Raja Ram and Samson, M.V. (1991a). *Cinnamommum glanduliferum* Meissn. as a secondary host plant of eri silkworm, *Samia cynthia ricini* Hutt. Ind. J. Seri. 30(1): 64-65.

Raja Ram and Samson, M.V. (1991b). *Cinnamommum glanduliferum* Meissn as an additional primary host plant for *Antheraea assama* Ww. Ind. J. Seri. 30(1): 68-69.

Raja Ram, Sengupta, A.K. and Devnath, M. (1992). Studies on the germination behaviour and longevity of Som (*Machilus bombycina*) seeds. Sericologia 32(3): 523-525.

Ramesh, M., Pavan, U., Shyam Prasad, S., Rao, A.V. and Sadanandam, A. (2002). *In vitro* regeneration of plants from mature nodal segments of *Terminalia arjuna* Bedd. Sericologia 42(1): 75-80.

Ramesh, M., Pavan, U., Shyam Prasad, S., Rao, A.V. and Sadanandam, A. (2002). *Terminalia arjuna* : a breakthrough in micropropagation. Ind. Silk 40(4): 17.

Randhawa, M.S. (1983). Flowering Trees. National Book Trust, New Delhi.

Rangaswamy, N.S. and Promila (1972). Morphogenesis of the adult embryo of *Azadirachta indica* A. Juss. Z. Planzenphysiol 67: 377-379.

Rao, A.M., Reddy, S.K., Rao, V.R.M. and Bhalla, J.K. (2001). Clonal propagation of date palm through somatic embryogenesis. Ind. For. 127(4): 463-465.

Rao, G.M., Rao, A.R., Acharyulu, M.V.S.N. and Shyam Prasad, N. (1998). Improvement of planting stock through cutting in root training nursery. Ind. For. 124(9): 739-742.

Rao, H.S. (1958). Vegetative propagation and forest tree improvement. Ind. For. 42: 199-204.

Rao, Manibhushan, K. (1991). Text book of Horticulture. Macmillan India Ltd., Delhi.

Ravichamy, P., Lakshmisubramanian, S., Nalini, K., Vanitha, K.L., Muthuraman, C., Stephen John David, A. and Murali, P.M. (1999-2000). Effective nursery management of tissue culture Neem plantlets. J. Eco. Bot. & Phytochem. 10-11: 19-24.

Rawat, J.S., Banerjee, S.P. and Singh, T.P. (1999-2000). Silviculture and management of Neem (*Azadirachta indica* A. Juss.). Jour. Eco. Bot. & phytochem. 10-11: 31-48.

Ray, Indrajit (1990). Defy diabetes with mulberry. Ind. Silk 28 (10) : 34-35.

Rembold, H. (1989). The Azadirachtins: Their potential for insect control. In: Economic and Medicinal Plant Research, Vol 3. Eds. Wagner, H. *et al.*, Academic Press, London, pp 57-71.

Roy, R.P., Saran, S. and Dutta, B. (1983). Cytogenetics of other flowering plants. In: Genetical Research in India. Pp: 61-91.

Sagreiya, K.P. (1967). Forests and Forestry. National Book Trust, New Delhi.

Saikia, S. and Goswami, B.C. (1997). Mejankari: The unexplored Assam Silk. Ind. Silk 35 (12): 33-55.

Sanjappa, M. (1989). Geographical distribution and exploration of the genus *Morus* L. (Moraceae). In: Genetic resources of mulberry and utilisation. Eds. Sengupta, K. and Dandin, S.B., CSR&TI, Mysore, pp: 4-7.

Saraf, G.P., Pai, N.M., Meshramkar, P.M., Deb, U.K. ans Jaspal, N.S. (1980). Utilization of a mixture of *Xylia xylocarpa, T. tomentosa* and *T. paniculata* for pulping. Ippta 17(2): 1-10.

Sarkar, D.C. (1988). Ericulture in India. Central Silk Board, Bangalore.

Sarker, D.D. (1998). The silkworm : Biology, Genetics and Breeding. Vikas Pub. House Pvt. Ltd., New Delhi.

Sarmah, M.C. (2000). Management of eri silkworm and host plant germplasm. In: National Workshop on Management of Sericultural Germplasm for Posterity. CSGRC, Hosur.

Sengupta, A.K., Yadav, G.S., Raja Ram, Das, R., Devnath, M. and Basumatary, B.K. (1991). Genetic resources of muga host plant and its propagation. Base paper in : Workshop on muga culture, 19th Sept., 1991, CSB, Guwahati, pp: 1-20.

Sengupta, J.N. (1937). Seed weight, plant percents in India. Ind. For. Rec. (n.s.) Silv. 2(5): 175-221.

Sengupta, K. (1986). Tasar silk industry in India. Central Tasar Research & Training Institute, Ranchi, pp: 1-28.

Seth, M.K. (1995a): Food plants of Tasar silkworms. In Proc.: Training course in sericulture IV. Ed. Agarwal, H.O. and Seth, M.K., H.P.Univ., Shimla, pp: 787-206.

Seth, M.K. (1995b). Food plants of oak tasar silkworms. In Proc.: Traing. Course in Seri. IV. Eds. Agarwal, H.O. and Seth, M.K., H.P.Univ., Shimla, pp 852-859.

Seth, M.K. (1995c). Primary and secondary food plants of eri silkworms. In Proc.: Traing. Course in Seri. IV. Eds. Agarwal, H.O. and Seth, M.K., H.P.Univ., Shimla, pp 886-893.

Seth, M.K. (1995d). Food plants of muga silkworms. In Proc.: Traing. Course in Seri. IV.Eds. Agarwal, H.O. and Seth, M.K., H.P. Univ., Shimla, pp 894-901.

Sharda Devi, A., Katyayini, V.K.L.T. and Sumathy, B.S. (2002). Amaltas – a natural dye for silk. Ind. Silk 41(3): 19-20.

Sharma, K., Thakur, S., Badiyala, S.D. and Sharma, N.K. (1995). First report on the propagation of *Terminalia chebula* Retz. through patch budding. Ind. For. 121(8): 760-761.

Sharma K.K. and Jaiswal, A.K. (1997). A record of lac host plants from Haryana. Jour. Non-timber For. Prod. 4 (3/4): 167-169.

Sharma, L.K. and Pandey, O.N. (1999). Effect of plant growth regulators on rooting behaviour of *Dalbergia latifolia* Roxb. and *D. sissoo* Roxb. Ind. For. 125(4): 421-456.

Sharma, S.K., Verma, S.K. and Uniyal, D.P. (2002). A preliminary study on the air layering in *Acacia nilotica* (L.) Del var. *indica*. Ind. For. 128(12): 1376-1378.

Sharma, V.N. and Saxena, K.P. (1959). Spermicidal activity of sodium nimbinate. Ind. J. Med. Res. 47: 322.

Shetty, K.K. and Samson, M.V. (1998). Non-mulberry sericulture in India. Ind. Silk 37(6&7): 21-25.

Shivaganam, K., Vinaya Rai, R.S., Swaminathan, C. and Surendran, C. (1989). Studies on rooting response to growth regulators in *Azadirachta indica* A. Juss. Proc. Semi. Veg. Prop., Coimbatore, July 27-28, 1989.

Shukla, N.K. and Kukreti, M.C. (1983). Physical and mechanical properties of *Quercus grifithii* from Khasi Hills Division (Meghalaya). Ind. For. 109(4): 237-245.

Shukla. P. and Mishra, S.P. (1979). An introduction to taxonomy of angiosperms. Vikas Pub. House Pvt. Ltd., New Delhi.

Shyam Lal 1997. Nursery technique for raising *Ficus* tall plants. Ind. For. 123 (11): 1078-1079.

Siddiqui, A.A., Basumatary, B.K., Lal, B. and Das, P.K. (1996). Juvenile shoot cutting and large scale plantation of Som. Ind. Silk 35(6 & 7): 15-17.

Singh, C., Vishwanathan, M.K., Kumar, N. and Agarwal, M.C. (1998). Growth survival and mean annual increment of *Quercus leucotrichophora* (Ban Oak) on degraded lands of Doon Valley. Ind. For. 124 (9) : 732-738.

Singh, K. (1963). In: Fruit culture in India. Ed. Kurup, C.G.R., ICAR, New Delhi pp. 225-229 (original not consulted).

Singh, O. (1998). Effect of seed weight on germination survival and initial growth of *Quercus dilalata*. Ind. For. 124(11): 959-961.

Singh, P. (1990). Ethnobotanical studies on some forest trees of Jabalpur Forest Division (Madhya Pradesh). Jour. Econo. Bot. & Phytochem. 1 (2-4): 43-47.

Singh, R.V. (1982). Fodder trees of India. Oxford & IBH Pub. Co. New Delhi.

Singh, Ramesh, D.R., Dhir, K.K., Vij, S.P., Nayyar, H. and Singh, K. (2003). Study of genetic improvement techniques of *Terminalia chebula* Retz. – An important multipurpose medicinal tree species of India. Ind. For. 129(2): 154-168.

Singh, U., Wadhwani, A.M. and Johri, B.M. (1983). Dictionary of economic plants in India. ICAR, New Delhi.

Singh, U.N. and Tikoo, B.L. (1989). Effect of pruning on sprouting and growth behaviour of *Quercus serrata*. Ind. Silk 28(8): 36-37.

Singhvi, N.R. and Sinha, A.K. (1991). Mulberry : The Kalpavriksha of the twenty first century. Ind. Silk 30 (2): 24-26.

Sinha, B.R.R.P. (1998). Ecoraces of Indian Wild Silk Moths. Indian Silk 37(647): 38.

Sinha, P.S., Joshi, M.C., Beck, S., Khare, R., Gangopadhyay, A., Sinha, A.K., Sinha, B.R.R.P. and Thangavelu, K. (2001). Identification and tackling the factor hindering the regeneration of *Terminalia arjuna* (Bedd.) through cutting – A breakthrough. Sericologia 41(2): 291-295.

Sinha P.S. and Sinha, B.R.R.P. (1992). Technique to minimize the process of defoliation in oak tasar. Indian Silk 31(8): 39-40.

Sinha, S.C. (1996). Medicinal plants of Manipur. MASS & Sinha, Imphal.

Sreekumar, S., Sudhakaran Nair, Appasamy, P., Vijayaraghavan, K. and Thiagarajan,V. (1994). Now mulberry on your dining table. Ind. Silk. 32 (10) : 45.

Srivastav, A., Geetika, Prakash, D., Singh, R., Siddiqui, A.A., Das, P.K. and Raghuvanshi, S.S. (1998). Nutritional status of muga food plant – *Litsaea polyantha*. Sericologia 38(4): 693-695.

Srivastava, D.P., Isa, M. and Thangavelu, K. (1993). Comparative effects of IAA, IBA and NAA on rooting and survival of air layers of *Terminalia arjuna* (Bedd.). Ann. For. 1(1): 27-32.

Srivastav, P.K. (2000). Propagation and conservation of genetic resources in medicinal *Terminalias* of India. Adv. Pt. Sci. Res. XII: 1-24.

Srivastav, P.K., Srivastava, D.P. and Thangavelu, K. (1994). Spontaneous polyploidy in *Terminalia paniculata* Proc. IV All Ind. Conf. Cytol. & Genet., Bangalore, 4: 104-105.

Srivatsav, P.K. and Thangavelu, K. (1995). Genetic resources and utilisation in *Terminalia arjuna* and *T. tomentosa*. In: Proc. Training course in Seri. UGC Workshop, H.P. Univ. Shimla, Eds: Agarwal, H.O. and Seth, M.K., pp: 318-328.

Srivatsav, P.K., Ranjan, P. and Sinha, S.S. (1996). Conservation of *Terminalia* genetic resources : The prinicipal source of non-wood forest products in India. For. Genet. Resources No.24 : 54-58. FAO, Rome.

Srivatsav, P.K. Beck., S., Siddiqui, A.A., Brahmachari, B.N. and Thangavelu, K. (1997a). Genetic divergence in leaf characters of *Terminalia arjuna* (Roxb.) W & A. and *T. tomentosa* W & A. Jour. Non-timber For. Prod. 4 (3/4) : 103-111.

Srivastav, P.K., Srivastav, D.P. and Thangavelu, K. (1997b). *Terminalia* (section: *Pentaptera*) descriptor. Sercologia 27 (2): 289-299.

Srivatsav, P.K. and Thangavelu, K. (1997) Genetic diversity and conservation in *Terminalia arjuna* and *T. tomentosa*. Jour. Non-timber For. Prod. 46 (1/2) : 9-16.

Srivatsav, P.K. and Singh, N.I. (1997a). Genetic resources and conservation of biodiversity in *Quercus serrata*. Intl. Symp. Sustainable utilization of biodiversity, SAI Instt. Environ. Dehradun, pp 77.

Srivatsav, P.K. and Singh, N.I. (1997b). Package of practices of oak tasar food plants in North-east India. Proc. Ind. Acad. Seri. 3(2): 1-6.

Srivatsav, P.K., Thangavelu, K., Khare, R. and Singh, R.N. (1998a). Treatise on *Terminalias*. Indian Publishers Distributors, Delhi.

Srivastsav, P.K., Sinha, U.S.P. and Thangavelu, K. (1998b). Foliar characters and constituents in hybrid genotypes of *Terminalia* (*Pentaptera*). Ind. Jour. Seri. 37(1) : 76-78.

Srivatsav, P.K. and Singh, N.I. (1999a). Genetic resources of *Quercus*: their maintenance, evaluation and utilization. Seminar on Breeders-Scientists interaction: Issues related to germplasm maintenance, protection and utilization, Feb. 10, 1999. SMGS, Hosur, pp 45.

Srivastav, P.K. and Singh, N.I. (1999b). Biofertilizer in sericulture. In: Use and Development of microbial inoculants. Eds. Singh, T., Yadav, A.K. and Roy Choudhury, S. Reg. Biofer. Deve. Cent., Imphal and Deptt. of Agri. Guwahati, pp: 99-109.

Srivatsav, P.K. and Thangavelu, K. (2000). Genetic resources and utilization in *Terminalia arjuna* and *T. tomentosa*. In: Sericulture in India. Eds. Agarwal, H.O. and Seth, M.K. M/s. Bishen Singh Mahendra Pal Singh, Dehradun, pp : 793-811.

Srivastav, P.K., Singh, N.I., Singh, T.S. and Singh, T.N. (2000a). Improvement of *Quercus serrata* auct. non Thunb. Syn. *Q. acutissima* Carr. through bio-fertilizer for tasar cultural. Proc.: N.E.R. Conf. on Biofert. Guwahati. 22-23, Jan. 1999. In: Recent advances in biofertilizer technology for promotion and utilization of Resources & Technology, New Delhi.

Srivastav, P.K., Singh, T.S. and Singh, N.I. (2000b) Clonal propagation of *Quercus serrata* auct. non Thunb. Syn. *Q. acutissima* Carr. through air layering. Ind. For. 126(8): 879-884.

Srivatsav, P.K., Singh, N.I. and Thangavelu, K. (2000c). Genetic resources of *Quercus* : Their maintenance, evaluation and utilization. Proc. Ind. Aca. Seri. 6(1): 34-41.

Srivatsav, P.K., Thangavelu, K. and Singh, N.I. (2000d). Non-mulberry food plants: Present status and prospects for improvement in new millenium. Lead paper in: Natl. Seminar on Seri. Res. & Devel : Strategies for improvement in new millenium, CSR&TI, Mysore. Nov. 16-18, 2000, pp: 21-33.

Srivastav, P.K. Ibohal Singh, N. and Sinha, B.R.R.P. (2002a). Effect of different morphotypes of *Quercus serrata* on the rearing of *Antheraea proylei* J. XIX Cong. Intl. Seri. Comm. Bangkok (in press).

Srivastav, P.K., Singh, T.S. and Singh, N.I. (2002c). Macropropagation of *Quercus serrata* and *Q. gifithii* through air layerings and cuttings. Recent Adv. in Life Sci. Ed. Yadav, P.S., Manipur Univ., pp: 186-194.

Srivastav, P.K., Sinha, U.S.P., Beck, S. and Thangavelu, K. (2002b). Foliar characters and constituents in Pentaptera genotypes of *Terminalia*. Ind. Jour. Seri. (in press).

Srivastav, P.K. and Singh, N. Ibohal (2002). Management of genetic resources of oak tasar host plants in India. Base paper in: Works. on Seri. Germp. Manage. and Utiliz., Feb. 6-7, 2002, C.S.G.R.C., Hosur, pp: 1-18.

Srivastava, R.K. (2000). Approach grafting – a new approach for the formation of clonal bank of *Terminalia chebula*. Ind. For. 126(3): 300-304.

Staedler, E. and Hanson, F.E. (1978). Food descrimination and induction of preference for artificial diets in the tobacco hornworm. *Mandusa sexta*. Physiol. Ent. 3: 121-133.

Street, R.J. (1962). Exotic forest trees in British Commonwealth (Ed. Champion, H.), Oxford, pp 662.

Sugun, R. and Kumar, V. (1996). Ind. Silk 34: 19.

Surendranath, B., Samson, M.V. and Jayaprakash, N. (2000). New mulberry silkworm race authorization system in India. Indian Silk 39 (7) : 13-16.

Surendran, T. and Seethalakshmi (1987). Vegetative propagation of some important tree species by rooting cutting, K.F.R.S. Res. Rep. 50, Peechi, Kerala.

Suresh, K.K., Jambulingam, R. and Sekar, I. (2001). Effect of tree age on seed and seedling quantity in *Acacia nilotica*. Ind. For. 127(6): 685-689.

Suri, I.K. (1999-2000). Neem – A nitrogen regulator in fertilizers and production of Neem coated urea. J. Eco. Bot. & Phytochem. 10-11: 125-133.

Taide, Y.B., Babu, L.C. and Achamma, O. (1999). Comparative anatomy of haustorial formation of Sandal on some hosts. Jour. Non-Timber For. Prod. 6(3/4): 179-182.

Tewari, S.K., Shergill, M.S., Sastry, C.R. and Narasimhanna, M.N. (1978). Nursery technique of *Terminalia arjuna* Bedd. plantation. In: Second For. Conf., FRI&C., Dehradun.

Tewari, S.K., Shergill, M.S., Sastry, C.R. and Narasimhanna, M.N. (1985). Studies on effect of seed treatment and gradation on germination behaviour of *Terminalia tomentosa* W. & A. Sericologia 25(1): 117.

Thakur, R.S., Singh, S.B., Goswami, A. (1981). *Azadirachta indica* A. Juss. A. Review. Curr. Res. Med. Arom. Plants 3: 135-140.

Thangavelu, K. Chakraborty, A.K., Bhagowati, A.K. & Isa, M. (1988). Handbook of muga culture. Central Silk Board, Bangalore.

Thangavelu, K. (1997). Silkworm breeding in India - An overview. Indian Silk 36 (2) : 5-13.

Thangavelu, K., Srivastav, P.K. and Srivastava, A.K. (2000). Management of tropical tasar silkworm and host plant germplasm. Base paper in: Natl. Works. on Manage. of Seri. Germp. for Post. CSGRC, Hosur.

Thangavelu, K. and Rai, S. (2000). Prospects of tasar culture in new millennium. In Souvenir: Natl. Conf. on Strat. for Seri. Res. and Develop., C.S.R. & T.I., Mysore, pp 65-71.

Thangavelu, K. (2000). Present status and future research strategy in non-mulberry silkworm. Lead paper in: National Conference on Strategies for Sericultural Research and Development, Central Sericultural Research & Training Institute, Mysore, pp: 49-59.

Tikader, A. (1991). Turmeric as an intercrop with mulberry. Indian Silk 30 (5) : 29-30.

Tikader, A., Rao, A.A. and Mukherjee, P. (1999). *Ex situ* conservation of oldest mulberry tree. Ind. Silk 38 (2) : 17-18.

Tikader, A., Saraswat, R.P., Amarnath, A. and Thangavelu, K. (2002). Management and utilization of mulberry germplasm. Base paper in Works. on Seri. Germp. Manage. and Utiliz. C.S.G.R.C., Hosur, pp 1-23.

Tomar, U.K. (1995). Tissue culture studies on Neem (*Azadirachta indica* A. Juss.), National Seminar on Neem, New Delhi, Jan. 5, 1995.

Troup, R.S. (1913). Indian Forest Utilization (2nd Ed.), Calcutta.

Troup, R.S. (1921). The silviculture of Indian trees, I-III. Clarendon Press, Oxford, U.K.

Tsushida, T., Murai, T., Omori, M. and Okamoto, J. (1987). Production of a new type tea containing a high level of gamma-aminobutyric acid. Nippon Nogeikagaku Kaishi 61: 817-822.

Umadevi, P. *et al.* (1992). *In vitro* growth inhibitory effect of *Withania somnifera* (ashwagandha) as a transplantable mouse tumor, Sarcoma, 180. Ind. J. Expt. Biol. 30: 169-172 (original not consulted).

Uma, H.P., Gouramma, V., Bajpai, A.K. and Sinha, A.K. (1993). Resham ke hastashilpa: Labhkari aya ke shroth. Ind. Silk 32: 53-56.

Ullal, S.R. and Narasimhanna, M.N. (1994). Handbook of Practical Sericulture, Central Silk Board, Bangalore.

Van. der nat, J.M. *et al.* (1991). Ethnopharmacognostical survey of *Azadirachta indica* A. Juss. (Meliaceae) – Review. J. Ethnopharmacol 35: 1-24 (original not consulted).

Vasistha, A. and Kaushal, A.N. (2002). Effect of Khair tree (*Acacia catechu* Willd.) age on Katha and cutch yield and biomass production. Jour. Non-Timber For. Prod. 9(3/4): 106-108.

Vathsala, T.V. (1997). Creativity in cocoon crafts. Ind. Silk 36: 17-22.

Venkataramanam, K.S., Warrier, K.C.S., Ajith Kumar, K.G. and Gurumurthi, K. (1998). Difference in rooting response between the juvenile and adult tissues of *Casuarina equisetifolia* Forst. Ann. For. 6(2): 156-158.

Venkatesh, C.S., Arya, R.S. and Emmanuel, C.J.S.K. (1978). A note on budding and air layering in Semul. Ind. For. 104(2): 142-144.

Venkatesh, M., Chinnaswamy, K.P., Raju, M. and Devaiah, M.C. (1993). Mulberry for environmental food security. Indian Silk 31(12) : 34.

Wallia, S. and Mukherjee, S.K. (1984). Ferraxanthone – A 1,3,5,6 tetraoxygenated xanthone from *Mesua ferrea*. Phytochemistry 23: 1816.

Yadav, G.S. Isa, M. and Thangavelu, K. (1985). Studies on the taxonomy and floral biology of Som, *Machilus bombycina* King Laurales: Lauraceae. Sericologia 25 (1) : 63-70.

Yadav, G.S. and Goswami, B.C. (1992). Effect of ringing and auxins on the vegetative propagation of foliated cuttings of Som (*Machilus bombycina*, Laurales: Lauraceae). Sericologia 32(3): 437-442.

Yadav, G.S., Reddy, K.J., Singh, B.M.K., Roy G.C. and Sinha, S.S. (1997). Frequency distribution of tasar flora. Ind. Silk 35(12): 22-24.

Yadav, G.S. (1999). Seed research in *Persea bombycina* – An analytical review. Bull. Ind. Acad. Seri. 3(1): 16-26.

Yamaguchi, A. (2000). Future direction of bivoltine silkworm breeding in India. Souvenir in: National Conference on Strategies for Sericulture Research and Development. 16-18, Nov. 2000, Central Sericultural Research & Training Institute, Mysore, pp: 17-19.

Yokoyama, T. (1962). Synthesized science of sericulture. Central Silk Board, Bangalore, India, pp. 242-253.

Youzhe, Chen (1996). Unify the understanding for restoring prestige of China Silk. Indian Silk 34 (12) : 23-26.